Solid Electrolytes and Their Applications

Solid Electrolytes and Their Applications

Edited by
E. C. Subbarao
Indian Institute of Technology
Kanpur, India

PLENUM PRESS · NEW YORK AND LONDON

6373-4998

CHEMISTRY

Library of Congress Cataloging in Publication Data

Main entry under title:

Solid electrolytes and their applications.

  Includes bibliographies and index.
  1. Electrolytes. 2. Solids. I. Subbarao, Eleswarapu Chinna, 1928-
QD565.S5665   541.3'72   80-14879
ISBN 0-306-40389-7

© 1980 Plenum Press, New York
A Division of Plenum Publishing Corporation
227 West 17th Street, New York, N.Y. 10011

Printed in the United States of America

Dedicated
to
Kamala

Contributors

K. P. Abraham, Department of Metallurgy, Indian Institute of Science, Bangalore 560012, India

M. S. Chandrasekharaiah, Chemistry Division, Bhabha Atomic Research Centre, Trombay, Bombay 400085, India

G. Chattopadhyay, Chemistry Division, Bhabha Atomic Research Centre, Trombay, Bombay 400085, India

C. B. Choudhary, Department of Metallurgical Engineering, Indian Institute of Technology, Kanpur 208016, India. Present address: Research and Development, Hindustan Steel Ltd., Ranchi 834002, India

A. Ghosh, Department of Metallurgical Engineering, Indian Institute of Technology, Kanpur 208016, India

K. P. Jagannathan, Department of Metallurgical Engineering, Indian Institute of Technology, Kanpur 208016, India. Present address: Research and Development, Hindustan Steel Ltd., Ranchi 834002, India

H. S. Maiti, Department of Metallurgical Engineering, Indian Institute of Technology, Kanpur 208016, India. Present address: Materials Science Centre, Indian Institute of Technology, Kharagpur 721302, India

A. V. Ramana Rao, Department of Metallurgical Engineering, Regional Engineering College, Warangal 506004, India

T. A. Ramanarayanan, Department of Metallurgical Engineering, Indian Institute of Technology, Kanpur 208016, India. Present address: Corporate Research Laboratories, Exxon Research and Engineering Company, Linden, New Jersey 07036

A. K. Ray, Department of Metallurgical Engineering, Indian Institute of Technology, Kanpur 208016, India. Present address: IBM Thomas J. Watson Research Center, Yorktown Heights, New York 10598

H. S. Ray, Department of Metallurgical Engineering, Indian Institute of Technology, Kanpur 208016, India

S. Seetharaman, Metallurgisk Kemi, Institution for Metallurgy, Kungl Teckniska Hogskolan, Stockholm 70, Sweden

O. M. Sreedharan, Chemistry Division, Bhabha Atomic Research Centre, Trombay, Bombay 400085, India. Present address: Reactor Research Centre, Kalpakkarn 603102, India

E. C. Subbarao, Department of Metallurgical Engineering, Indian Institute of Technology, Kanpur 208016, India

V. B. Tare, Department of Metallurgical Engineering, Institute of Technology, Banaras Hindu University, Varanasi 221005, India

S. K. Tiku, Interdisciplinary Programme in Materials Science, Indian Institute of Technology, Kanpur 208016, India. Present address: Department of Materials Science, University of Southern California, Los Angeles, California 90007

Preface

Defect solid state has been an area of major scientific and technological interest for the last few decades, the resulting important applications sustaining this interest. Solid electrolytes represent one area of defect solid state.

The early work on defect ionic crystals and, in particular, the classic results of Kiukkola and Wagner in 1957 on stabilized zirconia and doped thoria laid the foundation for a systematic study of solid electrolytes. In the same year, Ure reported on the ionic conductivity of calcium fluoride. Since then, intense worldwide research has advanced our understanding of the defect structure and electrical conductivity of oxygen ion conductors such as doped zirconia and thoria and of the fluorides. This paved the way for thermodynamic and kinetic studies using these materials and for technological applications based on the oxygen ion conductors. In the last few years we have seen the emergence of two new classes of solid electrolytes of great significance: the β-aluminas and the silver ion conductors. The significance of these discoveries is that now (i) solid electrolytes are available which at room temperature exhibit electrical conductivity comparable to that of liquid electrolytes, (ii) useful electrical conductivity values can be achieved over a wide range of temperature and ambient conditions, and (iii) a wide variety of ions are available as conducting species in solids. The stage is therefore set for a massive effort at developing applications. The present activity in this direction includes the development of oxygen gauges for liquid and gas streams, oxygen pumps, fuel cells, high-energy storage batteries, hybrid nuclear–solid-electrolyte systems for power generation, electrochromic displays, coulometers, etc. Like solid state electronics, the device field of solid state ionics is quickly developing.

Solid electrolytes and their applications is an interdisciplinary area encompassing electrochemistry, solid state physics and chemistry, metallurgy, and ceramics. Chemical engineers and electrical power and device engineers are also examining applications of these materials. The interest in this field is rapidly growing, as evidenced by the international conferences being held at frequent intervals (1972, 1976, 1979).

The subject of solid electrolytes has matured sufficiently to find its way into the curricula of high-temperature chemistry, electrochemistry, solid state chemistry and physics, metallurgy, ceramics, and materials science. For pedagogic purposes, a systematic up-to-date treatment of the subject is needed.

The purpose of this book is to present a critical, concise, systematic treatment of solid electrolytes and their applications. The emphasis is on massively defective solids, sometimes called superionic conductors; pure and slightly doped ionic solids such as alkali halides are not generally included. The principles governing defects in ionic solids and the tools used for their characterization are covered, and this is followed by a fairly detailed discussion of the structure and properties of the more important solid electrolyte materials (Chapter 1).

Chapter 2 identifies the experimental problems encountered in the use of solid electrolytes for a variety of measurements. Electronic conduction, ambient conditions, electrodes, porosity, etc., are specifically discussed.

The thermodynamic studies of binary and ternary alloys and intermetallic compounds using oxide and fluoride solid electrolytes are presented in Chapter 3. Comprehensive tables of thermodynamic data are included.

Corresponding results on oxides, silicates, and spinels form the subject matter of Chapter 4. Again, detailed thermodynamic information is presented in tabular form.

Chapter 5 deals with kinetic studies, including polarization, diffusion, and diffusion-controlled reactions. Potentiostatic, galvanostatic, and potentiometric techniques are illustrated for kinetic studies.

The technological applications based on solid electrolytes are discussed in Chapter 6. Oxygen meters for determining and monitoring oxygen in flowing gaseous and liquid streams are becoming important in a variety of fields. One of the more prominent uses is for a quick determination of oxygen in molten steel. Coulometric addition and removal of oxygen to or from gases and liquid metals represents another significant exploitation of solid electrolytes. Silver ion conductors, which exhibit high conductivity at room temperature, have made possible many new solid state ionic devices. Electrochemical energy conversion by fuel cells and energy storage devices such as the sodium–sulfur batteries are technologically the most demanding application. Commercial feasibility, coupled with environmental advantages, is attracting intense effort in these energy related systems.

The final chapter presents information about the fabrication of solid electrolytes, electrodes, and related topics.

Throughout the book a proper balance has been maintained between the fundamental studies and practical aspects and between the much-studied zirconia- and thoria-type solid electrolytes and the newer electrolytes such as fluorides, β-aluminas, and silver ion conductors.

The book is meant for scientists and engineers entering or engaged in research, development, or production of solid electrolytes and devices based on them. The coverage is such as to introduce newcomers to the subject, but at the same time it is designed to be of interest to workers in the field in broadening and updating their background. The systematic approach employed may allow the book to be used as curricular material for students of this subject. A large number of references are provided for those readers interested in further study. The book is believed to be timely since the field is poised for a major effort in the development of applications by multidisciplinary groups. The fact that the available information in this field is spread over a variety of journals poses problems for the new as well as active workers in the field. A serious attempt has been made in this book to collate, synthesize, and analyze the widely dispersed data on solid electrolytes and their applications.

Efforts were made to make the references up to date. The authors are grateful to a number of active researchers who assisted us by providing reprints of their work.

A preliminary version of this book was discussed in a two-day conference in October 1973 with about 40 persons interested or engaged in this field. This exercise was useful in sharpening the format. I am grateful to Professors S. L. Malhotra and V. B. Tare for the excellent arrangements made for the conference.

The cooperative spirit with which the multidisciplinary team of authors have participated in the preparation of this book has made my editorial task light and pleasant, and I am grateful to them for this. The credit for the excellent typing goes to Mr. R. S. Misra and Mr. R. N. Srivastava, ably assisted by Mr. Vishwa Nath Singh, Mr. A. S. Nayyar, and Mr. Samar Das. The essential, but laborious, editorial and proofreading tasks were made easier thanks to the willing and competent help of my graduate students, in particular, C. B. Choudhary, H. S. Maiti, A. K. Ray, S. K. Tiku, and V. Veluri, and of my children, Veni, Ram, and Kanta. My own research efforts in this field have been supported by the U.S. National Bureau of Standards and the U.S. Aerospace Laboratory. The encouragement provided by Dr. P. K. Kelkar, Dr. M. S. Muthana, and Dr. Jagdish Lal, the successive Directors at this Institute, has been a great help. The main unfailing support for all my efforts has been the central figure in my life—my wife Kamala.

E. C. Subbarao

Contents

3. Thermodynamic Studies of Alloys and Intermetallic Compounds
M. S. Chandrasekharaiah, O. M. Sreedharan, and G. Chattopadhyay

4. Thermodynamic Properties of Oxide Systems
S. Seetharaman and K. P. Abraham

5. Kinetic Studies
V. B. Tare, A. V. Ramana Rao, and T. A. Ramanarayanan

6. Technological Applications of Solid Electrolytes
K. P. Jagannathan, S. K. Tiku, H. S. Ray, A. Ghosh, and E. C. Subbarao

7. Fabrication
A. K. Ray and E. C. Subbarao

Defect Structure and Transport Properties

C. B. Choudhary, H. S. Maiti, and E. C. Subbarao

1. Introduction

Point defects are primarily responsible for electrical conduction in solid electrolytes. Ionic solids contain these defects at all temperatures above 0°K. Aliovalent impurities also introduce excess defects whose concentration is fixed mainly by composition and is often independent of temperature. Presence of ionic defects gives rise to ionic conductivity, while that of electronic defects results in electronic conductivity, which is undesirable in a solid electrolyte. In practice, for a solid electrolyte to be useful, the ratio of ionic to electronic conductivity should be 100 or greater.

The total electrical conductivity (σ_T) in an electrolyte is given by the equation

$$\sigma_T = \sum_i n_i(z_i e)\mu_i \qquad (1.1)$$

where n_i, z_i, and μ_i are the concentration, valency, and mobility, respectively, of the ith charge carrying species and e is the electronic charge.

The relationship between the structure—including defects—of solid electrolytes and their transport properties is examined here. The emphasis is on the massively defective ionic solids (with defect concentration up to several percent) rather than those with defects at the ppm level. The principles of defect formation and the experimental techniques to elucidate the defect structure and transport behavior of solid electrolytes are discussed, followed

C. B. Choudhary, H. S. Maiti, and E. C. Subbarao · Department of Metallurgical Engineering, Indian Institute of Technology, Kanpur 208016, India. Dr. Choudhary's present address: Research and Development, Hindustan Steel Ltd., Ranchi 834002, India. Dr. Maiti's present address: Materials Science Centre, Indian Institute of Technology, Kharagpur 721302, India.

by a comprehensive review of the properties of important classes of solid electrolyte materials. Finally, guidelines for the occurrence of fast ion transport in solids are indicated. Some aspects of this chapter are also covered in References 4, 5, 8, 9, 92, 107, and 289.

2. Defect Structure

The different kinds of ionic and electronic defects which may be present in an ionic solid are conveniently presented using Kröger–Vink notation[1-3], which specifies the nature, location, and effective charge of a defect. The nature of point defects and their concentration in any solid are determined by the consideration of chemical equilibrium between the various species. Clustering of defects and lattice disorder take place at relatively higher defect concentrations.

2.1. Types of Point Defects

The various kinds of point imperfections possible in an ionic crystal MX (M and X are monovalent), taking into account the requirement of charge neutrality, are shown in Fig. 1.1. These are as follows.

(*a*) *Vacancies.* A missing M^+ ion in a pure binary compound MX from its normal site is depicted as V'_M. Similarly a vacant anion site is represented as $V_X^{\cdot}$. Here, V^0 stands for a vacancy, the subscript for the missing species, and the superscript for the charge, prime for effective negative charge and dot for effective positive charge.

(*b*) *Interstitials.* When an ion M^+ (or X^-) occur in an interstitial site, then the defect is $M_i^{\cdot}$ (or X_i').

(*c*) *Misplaced Atoms.* When the atom M occupies a normal X site or vice versa, one has M_X or X_M.

(*d*) *Schottky Defects.* A cation vacancy + an anion vacancy ($V'_M V_X^{\cdot}$). These are the predominant defects in alkali halides.

(*e*) *Frenkel Defects.* A cation vacancy + an interstitial cation ($V'_M M_i^{\cdot}$). Silver halides generally have Frenkel defects. Anti-Frenkel defects ($V_X^{\cdot} X_i'$) occur in ThO_2 and CaF_2.

(*f*) *Impurities.* Impurity atoms or ions may replace an atom or an ion in a normal lattice site (substitutional impurities) or enter interstitial sites. If the impurity has the same valency as the original ion, there is no effective charge associated with the defect. On the other hand, aliovalent impurities give rise to the following pairs of defects. If the valency of the impurity cation (L) is higher (e.g., 2) than that of the host cation, one may have either (i) impurity cation + interstitial anion ($L_M^{\cdot} X_i'$) or (ii) impurity cation + cation vacancy ($L_M^{\cdot} V'_M$). However, if M is a higher valent metal it is possible that the

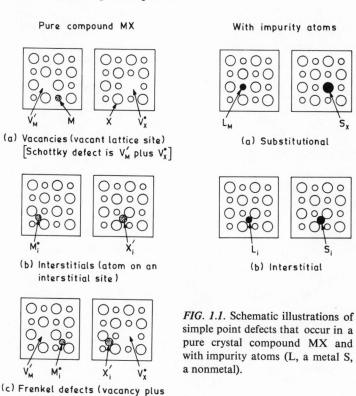

Pure compound MX

(a) Vacancies (vacant lattice site)
[Schottky defect is V'_M plus $V^{\bullet}_X$]

V'_M M   X $V^{\bullet}_X$

(b) Interstitials (atom on an interstitial site)

$M^{\bullet}_i$ X'_i

(c) Frenkel defects (vacancy plus interstitials)

V'_M $M^{\bullet}_i$ X'_i $V^{\bullet}_X$

With impurity atoms

(a) Substitutional

L_M S_X

(b) Interstitial

L_i S_i

FIG. 1.1. Schematic illustrations of simple point defects that occur in a pure crystal compound MX and with impurity atoms (L, a metal S, a nonmetal).

valency of the impurity cation is less than that of the host cation, resulting in defect pairs of either (iii) impurity cation + anion vacancy or (iv) impurity cation + cation interstitial. Aliovalent impurities involving replacement of anions of one valency by those of another, as well as simultaneous substitution of both cations and anions by appropriate ions, are in principle possible but have not been investigated much so far. The resulting defects, however, are covered by the ones listed above.

The size of the impurity ions relative to that of an available site (substitutional or interstitial) is important in determining the kind of defects which are favored. Since most ionic solids can be considered as composed of a close packing of the anions with the cations occupying some of the interstitial sites, defect types based on interstitial anions [such as b, c, f(i), and f(iv) above] are generally less probable. However, if the structure is a relatively open one, such as CaF_2 (Fig. 1.2a), interstitial ions can indeed be accommodated. Further, increased temperature tends to loosen a structure, making a defect model based on interstitial ions sometimes possible at elevated temperatures. While

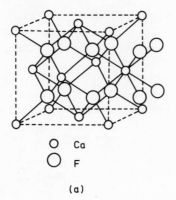

O Ca

O F

(a)

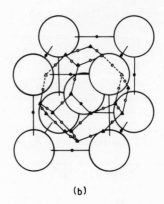

(b)

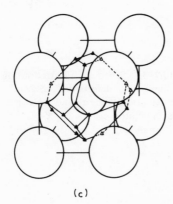

(c)

FIG. 1.2. Crystal structure of (a) CaF$_2$, (b) α-AgI, and (c) Ag$_3$SI.

many types of defects may coexist, energy considerations generally favor only one type of defect under a given set of conditions of temperature, pressure, type of impurity, and crystal structure.

In addition to the above ionic defects (types a–f) a crystalline solid may contain electronic defects such as

(g) *free electron* (e′) and
(h) *electron hole* (h·).

Pure ionic solids contain very few electronic defects and have wide forbidden energy gaps generally >3 eV.[7] However, at higher temperatures it is possible for electrons at the valence band to be thermally excited to the higher energy conduction band, producing a free electron in the conduction band and an electron hole in the valence band. These are known as "intrinsic" electronic defects, while "extrinsic" defects are formed mainly during the process of ionization of various atomic defects, e.g., a free electron may be produced if a neutral metal atom (M_i) is ionized to charged interstitial defect $(M_i^{\cdot})$. Further, in the presence of impurities or in a nonstoichiometric solid, free electrons or electron holes are produced as a result of "charge compensation," to neutralize the effect of charged ionic defects. As an example, the effective negative charge of a metal vacancy (V_x') may be neutralized by the presence of an electron hole in the crystal. However, it may again be emphasized that at a given temperature, atmosphere, and composition, the concentration of either an ionic or an electronic defect predominates.

2.2. Lattice Disorder and Association of Defects

Certain ionic solids exhibit some degree of lattice disorder even in their pure state. An outstanding example is α-AgI (Fig. 1.2b), in which the anion sublattice is perfectly ordered with total disorder (approaching melting) in the cation sublattice.[6] The disorder in the cation sublattice is enhanced in $HgAg_2I_4$ in which the three cations and one cation vacancy are randomly distributed over the four cation sites in a unit cell. The lattice disorder, instead of being isotropic as in α-AgI, doped ZrO_2, and ThO_2, may be in one direction as in crystals with a "tunnel"-type structure[10] (e.g., tungsten bronze and hollandite type), or in two directions as in crystals with a plane containing a loose, random packing of atoms sandwiched between close-packed planes (e.g., β-alumina type[11]). These structure types are discussed in detail in Section 4.

The defects resulting from aliovalent impurities are in general randomly distributed over the appropriate sublattice sites and often carry a net charge. In the case of $ZrO_2 + CaO$, Ca^{2+} ion occupying a Zr^{4+} ion site (Ca_{Zr}'') carries a net effective charge of -2 and the corresponding oxygen ion vacancy $(V_O^{\cdot\cdot})$ has a net effective charge of $+2$. Similarly, in the case of $ZrO_2 + YO_{1.5}$, the Y_{Zr}' has a net effective charge of -1, while the $V_O^{\cdot\cdot}$ has an effective charge of $+2$. Electrostatic attraction between such charged defects may promote the formation of defect pairs or larger clusters which may be electrically neutral or carry a net charge. Due to such association of defects, the concentration of free or quasifree defects does not increase linearly with increasing defect concentration and may decrease with decreasing temperature. Lidiard[12] has shown that, if the concentration of RX_2-type impurity in MX-type compound is x, the degree of association β which is defined in such a way that $x\beta$ represents the mole fraction of complexes, is given by

$$\beta/(1 - \beta)^2 = \Omega \exp(G/kT) \tag{1.2}$$

where Ω is the number of distinct orientations of the complex, G is the Gibbs free energy of association, i.e., the work gained under conditions of constant

pressure and constant temperature in bringing a vacancy from a particular
distant position to a particular nearest-neighbor position of the impurity ion,
k is Boltzmann's constant, and T is absolute temperature. Further,

$$\sigma/\sigma_0 = (A + B/A)/(1 + B) \tag{1.3}$$

where σ and σ_0 are the ionic conductivities of the impure and pure crystals,
$A = x_1/x_0 = x_0/x_2$, with x_1 and x_2 being mole fractions of the two intrinsic
defects in the dissociated state, such that $x_1 x_2 = x_0^2$ and B = ratio of the
mobilities of defects 2 and 1. It follows that, at a given temperature, the conductivity varies with the impurity addition at first linearly (up to about 1%
defect concentration) and then at a decreasing rate until it becomes nearly
asymptotic. This may be compared with the behavior of CaO- and $YO_{1.5}$-
doped ThO_2 ceramics drawn in Fig. 1.3. The above analysis appears to break
down when the defect concentration is beyond about 3%–4%, since the conductivity then decreases with increasing defect concentration. At these high
defect contents, clustering or ordering of defects starts. The number of anions
taking part in cluster formation as a function of anion vacancy concentration
in fluorite-type phases was discussed by Barker.[14] Nearest-neighbor interaction in a simple cubic lattice was considered by O'Keefe.[15] Recent calculations based on near-neighbor interactions show that the effective charge
carrier concentration goes through a maximum at a certain dopant level in
massively defective solids.[26] When very large defect contents are involved,
new compounds can occur.

Dipoles (positive and negative charges separated by a distance) generated

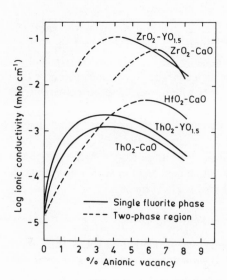

FIG. 1.3. Log ionic conductivity vs. percent anionic vacancy for various fluorite
solid solutions (Reference 13).

by charged defects respond to an alternating electrical field or mechanical stress and give rise to dielectric and mechanical loss, which is a function of frequency, temperature, and concentration (see Section 3.7). The behavior of larger complexes (e.g., in ZrO_2 and ThO_2 doped with $YO_{1.5}$) is expected to be different from that of dipoles.

2.3. Defect Equilibria

The concentration of a thermally generated point defect (intrinsic defect) is a function of the energy of defect formation. Using standard statistical thermodynamic treatment, it can be shown[12] that, in the case of Schottky defects in an ionic crystal M^+X^-,

$$n_S/N = \exp(\Delta S_{th}/2k)\exp(-W_S/2kT) \tag{1.4}$$

and in the case of Frenkel defects,

$$n_F/(NN_i)^{1/2} = \exp(\Delta S_{th}/2k)\exp(-W_F/kT) \tag{1.5}$$

where n_S and n_F are the numbers of Schottky and Frenkel pairs, respectively, N is the total number of ions in the crystal, N_i is the number of interstitial positions in the crystal, W_S is the energy required to form positive and negative ion vacancies (Schottky pair), W_F is the energy required to form a Frenkel pair, and ΔS_{th} is the increase in thermal entropy per vacancy.

Thermodynamically, each type of point defect is considered as an individual chemical species, and thus a defect equilibrium is represented by a form of chemical equation. Application of the law of mass action and the concept of equilibrium constant together with the electrical neutrality condition enable one to calculate the equilibrium concentration of each of the defects as functions of the partial pressures of the components (either P_M or P_X). For simplicity, infinitely dilute solutions of defects are considered so that activities can be equated to concentrations. Kröger and Vink[2] and recently Brook[16] have presented the results of such calculations for various combinations of defects. As an example we shall consider here the case of "pure" and Y_2O_3-doped ThO_2.[17]

At high oxygen pressures, pure ThO_2 exhibits positive hole (electronic) conduction due to equilibrium between oxygen in the surrounding atmosphere and interstitial oxygen ions, O_i'', in the lattice,

$$\tfrac{1}{2}O_2(g) \rightleftharpoons O_i'' + 2h^\bullet \tag{1.6}$$

and

$$K_1 = [O_i'']p^2p_{O_2}^{-1/2} \tag{1.7}$$

where p represents the number of positive holes per unit volume. Electrical neutrality in pure ThO_2 requires that $[O_i''] = \tfrac{1}{2}p$, so that $p \propto p_{O_2}^{1/6}$. On the other hand, at low oxygen pressures and high temperatures, oxygen vacancies are formed, which are electrically compensated by electrons dissociated from vacant oxygen sites,

$$O_O = \tfrac{1}{2}O_2(g) + V_O^{\cdot\cdot} + 2e' \tag{1.8}$$

and

$$K_2 = [V_O^{\cdot\cdot}]n^2 p_{O_2}^{1/2} \tag{1.9}$$

where O_O is an oxygen ion on an oxygen site and n is the number of excess electrons per unit volume. Again, if $[V_O^{\cdot\cdot}] = \tfrac{1}{2}n$, $n \propto p_{O_2}^{-1/6}$ at low oxygen pressures. Finally, at any oxygen pressure, an equilibrium between interstitial oxygen ions and oxygen vacancies must also be satisfied, such that

$$O_O \rightleftharpoons V_O^{\cdot\cdot} + O_i'' \tag{1.10}$$

with

$$K_3 = [V_O^{\cdot\cdot}][O_i''] \tag{1.11}$$

K_1, K_2, and K_3 above are equilibrium constants. When aliovalent impurities are added as in ThO_2–Y_2O_3, the electrical neutrality requires that

$$p + 2[V_O^{\cdot\cdot}] = n + 2[O_i''] + [Y_{Th}'] \tag{1.12}$$

If the added Y_2O_3 dopant is more than a trace, $[Y_{Th}']$ predominates on the right side of the above equation, so that electrical neutrality is established either by electronic defects, as in $p = [Y_{Th}']$ or by ionic defects as in $2[V_O^{\cdot\cdot}] = [Y_{Th}']$. Of these two possible modes of compensation, the latter is confirmed by experiment.[17-23] Analysis similar to that used for pure ThO_2 is applied to the case of Y_2O_3-doped ThO_2 to show that, at high oxygen pressures, $p \propto p_{O_2}^{+1/4}$ and at low oxygen pressures, $n \propto \bar{p}_{O_2}^{1/4}$. These two terms cancel each other at intermediate oxygen pressures when the material becomes predominantly an ionic conductor. The foregoing results can conveniently be summarized in a Kröger–Vink-type diagram, as was done by Rapp and Shores[24] for the case of Y_2O_3-doped ThO_2 (Fig. 1.4a).

Since point defects are in thermal equilibrium with the crystal, their concentrations are greatly influenced by temperature. In a pure material the concentrations of intrinsic Schottky and Frenkel pairs are given by Eqs. (1.4) and (1.5), respectively. Similar expressions may be obtained for a highly doped material like ThO_2–Y_2O_3. In the ionic conduction region of Fig. 1.4a the concentration of oxygen vacancies is given by

$$[V_O^{\cdot\cdot}] = \tfrac{1}{2}[Y_{Th}'] = [Y]_T = \text{const} \tag{1.13}$$

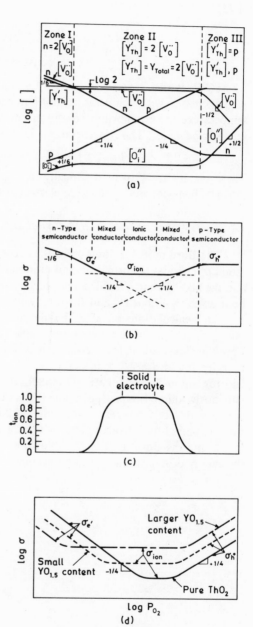

FIG. 1.4. Schematic representation of (a) equilibrium defect concentrations, (b) partial electrical conductivities, and (c) ionic transference number in a $YO_{1.5}$-doped ThO_2 crystal. (d) Schematic illustration of the effect of dopant concentration on partial conductivities in ThO_2–$YO_{1.5}$ solid solutions (Reference 24).

and therefore is independent of temperature. However, according to Eq. (1.11)

$$[O_i''] = \frac{K_3}{[V_O^{\cdot\cdot}]} = \frac{K_3^\circ \exp(-\Delta G_3/kT)}{\tfrac{1}{2}[Y_{Th}']}$$

$$= \text{const} \exp(-\Delta G_3/kT) \qquad (1.14)$$

where ΔG_3 is the free energy change for the reaction (1.10). It follows that $[O_i'']$ increases with temperature until it becomes equal to $[V_O^{\cdot\cdot}]$. At that temperature the equilibrium condition changes and the material shows an intrinsic behavior. The concentrations in this range are given by

$$[V_O^{\cdot\cdot}] = [O_i''] = (K_3^\circ)^{1/2} \exp(-\Delta G_3/2kT) \qquad (1.15)$$

These results are shown graphically in Fig. 1.5. In practice, however, the intrinsic behavior in doped oxide electrolytes is rarely observed.

2.4. Energy of Formation and Motion of Defects

The Born model of ionic solids has been used successfully for the calculation of energies of formation and motion of simple defects in alkali halides and the alkaline earth halides. An excellent account of these calculations has been given by Lidiard[12] and Boswara and Franklin.[25]

Such calculations for alkali halides and alkaline earth halides have been fairly successful for a number of simple defects. Following this method, Franklin[33] has calculated the energies of formation for defects in CaF_2. The energies found are 2.7 ± 0.4 eV/pair, 7.1 ± 1.0 eV/pair, and 5.8 ± 0.8 eV/trio for anion Frenkel, cation Frenkel, and Schottky defects, respectively. Obviously anion Frenkel pairs must be the dominant defects. This is consistent with Ure's[34] experimental value of 2.8 eV for the formation of anion Frenkel pairs found from conductivity measurements.

Using the Born model and following the same method of calculations, the energy of migration of anion vacancy and interstitial have been recently

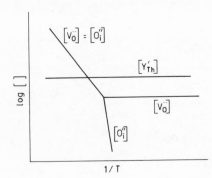

FIG. 1.5. Schematic diagram showing the effect of temperature on defect concentrations in ThO_2–Y_2O_3 at a constant P_{O_2}.

calculated for alkaline earth fluorides.[35,36] The method consists mainly of calculating the energy of the defect at a saddle point position in an assumed path of the defect. The path which gives the smallest value of the saddle point energy is the most probable path for the migration of the defect. Using this method, Chakravorty[35] has found the values of activation energies for migration of anion vacancy and interstitials in CaF_2. The values are 2.08 and 1.56 eV for migration by interstitial and interstitialcy mechanism, respectively. The latter mechanism seems to be operating since the experimentally found value is 1.55 eV.[37] Not much progress is achieved in such calculations for oxides with fluorite structure or other solid electrolyte types. There seem to be some basic difficulties in attempting such a calculation on oxides.[25] There are some attempts at such calculations in the case of one-dimensional tunnel and AgI-type compounds. Kikuchi and co-workers[40-43] applied the path probability method, instead of the more common random walk method, to the estimation of diffusion and ionic conductivity in lattices with sparcely populated tunnels and planes, the latter with particular reference to β and β'' Al_2O_3.

3. Experimental Methods

The defect structure of an ionic solid can be established by direct methods, such as density and diffraction methods. Since transport of mass and charge is a consequence of the existence of defects, measurements of transport properties—like electrical conductivity (as a function of temperature, frequency, and nonmetal activity), diffusion, polarization, thermoelectric power, dielectric loss, and internal friction—can be used to deduce the defect structure. In nonstoichiometric compounds, having once determined the predominant defect, its concentration can be simply related to the deviation from stoichiometry. Electrochemical methods may be used to determine the exact stoichiometric ranges. Resonance and spectroscopic techniques are also used to study the transport properties.

3.1. Density

The various kinds of defects which result from the addition of aliovalent impurities are listed in Section 2.1. A choice among these alternatives can be made on the basis of density differences, provided the impurity is 5% or more, even though very careful measurements were employed in the case of alkali halides doped with much lower impurity contents.[12] The density, ρ, in gm/cm^3 is calculated for the various defect models using the formula

$$\rho = \sum A/N_0 V \tag{1.16}$$

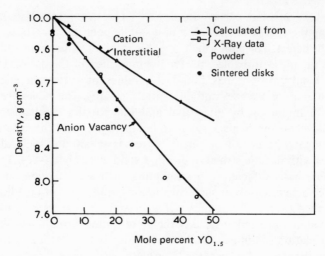

FIG. 1.6. Calculated and measured densities of ThO_2–$YO_{1.5}$ solid solutions (Reference 22).

where $\sum A$ is the sum of the atomic weights of the atoms in the unit cell, N_0 is Avogadro's number, and V is the volume of unit cell in cm^3 (a^3 in the case of a cubic cell with a lattice parameter of a). Comparison of these values with density, measured pycnometrically, establishes the defect model. The defect structure of massively defective materials based on zirconia and thoria was determined by density studies. Typical results for the anion-deficient solid solutions in the system ThO_2–$YO_{1.5}$ are shown in Fig. 1.6.[22] Entrapment of air, particularly when powdered samples are used, reaction between the material and the displacement fluid, and temperature fluctuations require attention in the accurate determination of density. It is evident that lattice disorder, as encountered in β-alumina and silver-halide-type compounds, cannot be detected from density data. On the other hand, weight or density changes accompanying ion exchange in β-alumina indicates their defect structure.[44,45]

3.2. X-Ray and Neutron Diffraction

The application of x-ray and neutron diffraction techniques to the study of defect solids has been reviewed by Robertson[46] and Willis.[47] The most important use of these techniques is in the determination of the crystal structure of the defect solids, e.g., β-alumina and $RbAg_4I_5$ discussed in Sections 4.2 and 4.3. Choice of an appropriate defect model can also be made on the basis of x-ray diffraction intensities. For example, the structure factors,

F_{obs}, are obtained from the observed x-ray diffraction intensities, I_{obs}, using the relationship

$$|F_{obs(hkl)}| = K[I_{obs(hkl)}PM]^{1/2} \qquad (1.17)$$

where P is the Lorentz-polarization factor, M is the multiplicity factor, and K a scale factor. In the case of CaO–ZrO$_2$ solid solutions,[48] the structure factors were calculated ($F_{calc(hkl)}$) for the anion vacancy and cation interstitial models using the formula

$$F_{calc(hkl)} = \sum_{i}^{N} f_n \exp[2\pi i(hu_n + kv_n + lw_n)] \qquad (1.18)$$

and

$$f_n = f_0 \exp[-B(\sin^2 \theta/\lambda^2)] \qquad (1.19)$$

where f_0 is the ionic structure factor, B is the Debye–Waller temperature factor, u, v, and w are the coordinates of the ions, θ is Bragg angle, λ is the wavelength of radiation employed, and N is the total number of atoms in the crystal. Comparison of the F_{calc} with F_{obs} favored the anion vacancy model for these solid solutions.[48] Later studies by Carter and Roth[49] indicated that in samples quenched from 1900°C and in those annealed at 1400°C, 20% of the unit cells contained interstitial Ca ions and the rest anion vacancies. However, according to Diness and Roy[50] very rapid quenching (1000°C/sec) led to a totally cation interstitial structure. The order–disorder transformation[48,51] exhibited by calcia-stabilized zirconia was studied by x-ray and neutron diffraction methods and was attributed to distortions of the cation–oxygen polyhedra. "Forbidden" reflections appear in the case of the ordered phase. The defect structure in the system CaF$_2$–YF$_3$ was confirmed by x-ray methods.[52]

The introduction of lattice defects weakens the x-ray intensities due to an increase of the Debye–Waller temperature factor, B, which is equal to $8\pi^2\langle\bar{u}\rangle^2$, where $\bar{u}$ is the average value of the amplitude of the vibrating structural unit. Hund and Mezger[20] have shown that with increasing defect concentration, B increases and activation energy for conduction decreases in the system ThO$_2$–Y$_2$O$_3$. The large amplitude of thermal vibrations reported in the channel direction in NaTi$_2$Al$_5$O$_{12}$[53] and Na$_x$Fe$_x$Ti$_{2-x}$O$_4$[54] may be suggestive of high ionic conductivity.

The statistical disorder of oxygen ions in ZrO$_2$–CaO produces marked diffuse scattering in the neutron diffraction patterns, which disappears on high-temperature annealing.[49] Electron diffuse scattering has been used to study the order–disorder transformation in ZrO$_2$–CaO electrolyte.[55,56] Similar studies were also made on the CaF–YF$_3$ system.[57,58]

3.3. Electrical Conductivity and Diffusion

Discussions on the transport properties of solid electrolytes in general were earlier presented by Rickert,[59] Haven,[60] and LeClaire.[61] Here we shall consider some of the specific results on electrical conductivity. Equation (1.1) gives the expression of conductivity irrespective of any specific defect model. However, in a NaCl structure, if one considers the presence of Schottky defects and if both the cations and the anions are assumed to be mobile, the conductivity in the presence of some aliovalent impurity of mole fraction x is given by[62]

$$\sigma = Ne(n_c\mu_c + n_a\mu_a)$$

$$= \frac{4N\gamma_c b^2 e^2}{kT} \exp\left(-\frac{\Delta G_c}{kT}\right) \frac{x}{2}\left[\left(1 + \frac{4n_0^2}{x^2}\right)^{1/2} + 1\right]$$

$$+ \frac{4N\gamma_a b^2 e^2}{kT} \exp\left(-\frac{\Delta G_a}{kT}\right) \frac{x}{2}\left[\left(1 + \frac{4n_0^2}{x^2}\right)^{1/2} - 1\right] \qquad (1.20)$$

where the subscripts c and a correspond to the cation and anion, respectively. N is the number of ions per unit volume, b is the cation–anion separation, γ is the frequency of jump, ΔG is the free energy of migration, and n_0 is defined as

$$n_c \cdot n_a = \exp(-W_S/kT) = n_0^2 \qquad (1.21)$$

where W_S is the free energy of formation of a Schottky pair. Although Eq. (1.20) has been derived for a particular crystal structure and a specific type of defect, it represents the general behavior and qualitatively depicts the variation of ionic conductivity with temperature and impurity concentration. The experimental results on conductivity can yield the values of defect parameters like enthalpy and entropy of formation and migration of defect, provided the type of the defect and the charge-carrying species for the particular system are known. This information can be obtained from x-ray diffraction, density measurement, or diffusion studies. In most of the ionic conductors one of the ions is relatively more mobile than the others. For example, in alkali and silver halides and β-aluminas, cations are the current-carrying species, while in doped ZrO_2, ThO_2, and CeO_2 electrolytes, oxygen ions carry practically all of the current.

If we consider cation as the only conducting species then at higher temperatures or with very low impurity concentration (intrinsic region) where $n_0 \gg x$, Eq. (1.20) is converted to an Arrhenius type and can be written as

$$\sigma T = \frac{4N\gamma_c b^2 e^2}{k} \exp\left(\frac{\Delta S_c}{k} + \frac{S_s}{2k}\right) \exp\left(-\frac{\Delta h_c}{kT} - \frac{h_s}{2kT}\right) \qquad (1.22)$$

where ΔS_c and Δh_c correspond to the entropy and enthalpy of migration and S_S and h_S represent the same quantities for the formation of defects. The plot of log σT vs. $1/T$ gives a straight line of negative slope $(\Delta h_c + h_s/2)/k$.

However, at lower temperatures or higher impurity concentration (extrinsic range) where $x \gg n_0$ the equation is

$$\sigma T = \frac{4N\gamma_c b^2 e^2}{k} \exp\left(\frac{\Delta S_c}{k}\right) \exp\left(-\frac{\Delta h_c}{kT}\right) \tag{1.23}$$

In this case the slope is only $\Delta h_c/k$. Experimentally the intrinsic conductivity can only be obtained in systems having very low defect concentration, e.g., alkali halides. It is not possible to observe such a phenomenon in massively doped electrolytes having a very high concentration of defects, e.g., oxides. At high concentrations the situation becomes also complicated due to the association of defects and the Debye–Hückel effect[62] (see also Section 2.2).

Ionic conductivity arises due to the diffusion of ions under an electric potential gradient. The conductivity can be related to the diffusion coefficient (D) by the Nernst–Einstein equation

$$\frac{\sigma}{D_i} = \frac{N_i q^2}{kT} \tag{1.24}$$

where the subscript i corresponds to the interstitial mechanism, N_i is the number of interstitial ions per unit volume, and q is the charge of the diffusing ion. The diffusion coefficient which is measured experimentally is usually the tracer diffusion coefficient (D_t) and is different from either D_i or D_v, the diffusion coefficient due to vacancy mechanism. With D_t, Eq. (1.24) changes to

$$\frac{\sigma}{D_t} = \frac{Nq^2}{kT} \tag{1.25}$$

where N is the total number of diffusing ions per unit volume. However, for a vacancy mechanism each successive jump of a tracer is correlated and the relation becomes

$$\frac{\sigma}{D_t} = \frac{Nq^2}{fkT} \tag{1.26}$$

where f is a correlation factor and to a first approximation has a value of $(1 - 2/Z_{cn})$ with Z_{cn} as the number of nearest neighbors for the diffusing ion. Equations (1.24) to (1.26) are valid for Schottky disorder. Similar equations can also be derived for Frenkel disorder.[63]

Figure 1.7 gives the measured diffusion coefficients of oxygen, calcium, and zirconium ions in CaO-doped zirconia.[64–66] The diffusion coefficient calculated from the measured electrical conductivity by using the Nernst–Einstein relation is also shown and a good comparison may be observed. Defect complexes which may form at lower temperatures help in diffusion process but do not contribute to the conductivity due to their neutral character, resulting in a mismatch between the two values. Sometimes, the ratio D_t/D_σ is called the Haven ratio, H_R. The observed value of H_R (0.61) in Na and Ag β-alumina may be compared with the value of 0.6, calculated for the

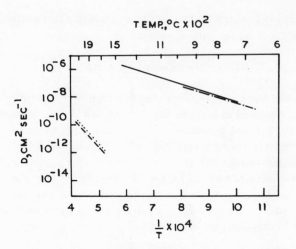

FIG. 1.7. Log D vs. $1/T$ in calcia-doped zirconia. ——, O^{2-} and D calculated from electrical conductivity (Reference 66); —·—, O^{2-} (Reference 64); ···, Ca^{2+} (Reference 65); ---, Zr^{4+} (Reference 65).

interstitialcy diffusion in a two-dimensional triangular lattice.[67,68] On the other hand, for vacancy diffusion of Na ions on a hexagonal lattice, $H_R = 0.33$. β''-alumina and mixtures of β- and β''-alumina show H_R values greater than 0.33 at temperatures above 400°C and smaller values below.[61] There is some uncertainty about the rigorous correlation factors for structures such as β-alumina which are grossly defective.

As discussed earlier, nonmetal activity in the ambient atmosphere has a pronounced effect on the defect concentration and therefore on the electrical conductivity. The mobilities of electronic defects are generally 100–1000 times greater than those of ionic defects and it is often assumed that they are independent of defect concentrations. This assumption is valid at least at low defect concentrations and at high temperatures. For Y_2O_3-doped ThO_2, the relation (1.1) can be written as

$$\sigma_T = \sigma_{Y'_{Th}} + \sigma_{V_{\ddot{O}}} + \sigma_{O''_i} + \sigma_{h^{\cdot}} + \sigma_{e'} \tag{1.27}$$

However, as a consequence of the variation of defect concentration as a function of P_{O_2} (Fig. 1.4a) and due to the fact that $\mu_{Y'_{Th}} \ll \mu_{V_{\ddot{O}}}$ only one type of conductivity is predominant at a particular P_{O_2} range (Fig. 1.4b). At lower P_{O_2}, $\sigma_{e'}$ predominates with $\sigma_{e'} \propto P_{O_2}^{-1/4}$, at high P_{O_2}, $\sigma_{h^{\cdot}}$ predominates with $\sigma_{h^{\cdot}} \propto P_{O_2}^{+1/4}$, while at the intermediate range the conductivity is independent of P_{O_2} and is due only to oxygen vacancy. The ionic transference number in this electrolyte is expressed as

$$t_{\text{ion}} = \frac{\sigma_{V_{\ddot{O}}}}{\sigma_T} \tag{1.28}$$

and can be deduced from Fig. 1.4b. The variation of t_{ion} as a function of P_{O_2} is plotted in Fig. 1.4c. Such a plot enables one to find out the range of P_{O_2} in which the value of t_{ion} is close to unity and therefore to determine the useful range for the solid electrolyte. Since the slope of the log σ vs. P_{O_2} plot in the mixed conduction region is a direct consequence of the particular defect equilibrium within the solid, such measurements are often useful in establishing the defect model of unknown electrolytes.

Addition of certain aliovalent impurities not only increases the conductivity, it also extends the useful P_{O_2} range for the electrolyte. Figure 1.4d shows schematically the effect of $YO_{1.5}$ on ThO_2. Patterson[69] has determined the electrolytic domain for a number of useful solid electrolytes with respect to temperature and nonmetal activity. These are presented in Fig. 1.8.

Analogous to the free-electron theory of metals, a unified theoretical model for the ionic transport in "superionic" conductors like silver halides, β-aluminas, and defect-stabilized oxides has been put forward by Rice and Roth.[70,71] The model is based on the hypothesis that in an ionic conductor, the ions of mass M belonging to the conducting species can be thermally excited from a localized energy state to an excited state and there exists an energy gap between them. In the excited state, the ion acts like a "free-ion" and propagates throughout the solid with a velocity v_m and energy $e_m = \frac{1}{2}Mv_m^2$. However, due to the interaction with the rest of the solid the free ion has only a finite lifetime and mean free path. Following this hypothesis a simple expression for electrical conductivity which is analogous to that

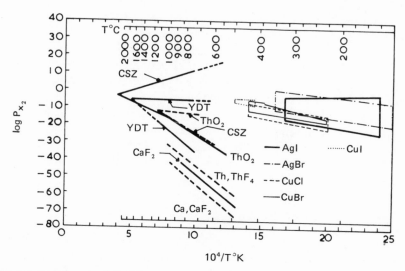

FIG. 1.8. Log P_{x_2} vs. $1/T$ to indicate the electrolytic domains for selected halide and oxide solid electrolytes (Reference 69).

obtained from conventional hopping theory has been derived. The expressions for thermal conductivity and thermoelectric power have also been derived from the same model and attempts have been made to compare the experimental results with the theory. However, the model has been later criticized by other investigators.[72]

Experimental details for the measurement of electrical conductivity and other transport properties of solid electrolytes have been reviewed by Rapp and Shores,[24] Kvist,[73] and more recently by Blumenthal and Seitz.[74] For ionic solids, dc measurement becomes difficult due to the polarization problem, as has been discussed in Section 3.4; ac measurement at a suitable frequency is often carried out as a function of temperature and nonmetal activity. For maintaining a definite gaseous ambient it is necessary to keep the sample inside a closed sample holder. Different types of such holders have been described by Steele[75,76] and Rapp and Shores.[24] For oxides, the P_{O_2} in the sample can be maintained either by placing different metal–metal-oxide pellets, e.g., $Cu–Cu_2O$, $Ni–NiO$, $Co–CoO$, $Cr–Cr_2O_3$, etc., having different equilibrium oxygen partial pressures, on both sides of the sample (the preparation of the sample and the electrodes is described in detail in Chapter 7). The oxygen activity of the gaseous ambient can be maintained either by a controlled oxygen leak into a vacuum chamber or by introducing a known gas mixture of $H_2–H_2O$ or $CO–CO_2$ into the sample holder or by an electrochemical pumping of oxygen through an oxide solid electrolyte from a flowing inert gas.[77-79]

Two-probe measurement is the simplest method to determine the bulk conductivity of a sample. However, difficulties may arise for highly conducting specimens when the lead as well as electrode resistances become significant. Surface conduction may be important at low temperatures when there is a temperature gradient between the surface and the bulk of the specimen. Thermionic emission becomes significant at very high temperatures (above 1000°C) and at a very low total pressure, but in general its contribution is not significant.[24] Surface conduction can be eliminated in a three-probe measurement technique as has been described by Tallan et al.[80-82] A four-probe technique is used to eliminate the lead and electrode resistances. A long cylindrical specimen is generally used for this technique.[23,83,84] Mehrotra et al.[85] have used an ac, four-terminal technique with a disk specimen to eliminate the lead resistances.

3.4. Polarization Measurements

It has already been pointed out that the solid electrolytes are not necessarily fully ionic conductors but very often exhibit partial electronic conduction of either n type or p type. Polarization measurements, developed by Wagner[86-88] and Hebb,[89] offer one means of separating electronic and ionic

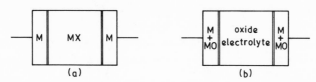

FIG. 1.9. Solid electrolytes between identical reversible electrodes. (a) Cationic conductor, (b) anionic conductor.

contributions to the conductivity. The measurement techniques involve an intelligent selection of electrode conditions to suppress either the electronic or the ionic part of the conduction.

In the general case of metallic electrodes on a solid electrolyte, the contact can be anything between an "ohmic" and a "rectifying" one. The latter allows no significant current to pass in one direction. An additional factor which limits the current is the material transport that is always associated with ionic current. An excellent account of the experimental aspects of conductivity in oxides has been given by Heyne,[90,91] Mitoff,[84] and Kleitz *et al.*[93]

The uncertainties in electrode properties do not interfere seriously when the total conductivity is determined by measurements with high-frequency alternating current. But the ac conductivity must be measured as a function of frequency to separate the ionic and electronic conductivities from relaxation processes at the electrodes. The frequency of measurement must be high enough to fall in the range where the total conductance of the sample is independent of frequency. For stabilized zirconia with porous platinum electrodes satisfactory conductivity values can be obtained at 1 K Hz,[94] while for β-alumina solid electrolytes a frequency as high as 10^6–10^7 Hz (Reference 11) is needed to eliminate the polarization at the electrodes.

In the discussion that follows, we shall use the following symbols: j_k the rate of transport through the solid electrolyte, μ_k the chemical potential, η_k the electrochemical potential, σ_k the partial conductivity, z_k the valence, and t_k the transport number. The subscript k denotes the species and can be either electron (e′), hole (h·), cation (c), or anion (a). E is the voltage across the solid electrolyte, e the electronic charge, and σ_T and σ_e are the total conductivity and electronic conductivity, respectively. With dc measurements the following limiting conditions can be approached.

(1) Solid Electrolyte between Identical and Completely Reversible Electrodes. The examples of such a cell are shown in Fig. 1.9. In this case the sample is in complete equilibrium with the electrodes and no concentration gradients of either the metallic or the nonmetallic components are formed in the sample. The equations for current density j are

$$j_{\text{ion}} = -\sigma_{\text{ion}} \,\text{grad}\, E \quad \text{and} \quad j_e = -\sigma_e \,\text{grad}\, E \qquad (1.29)$$

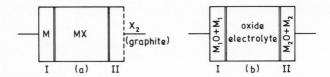

FIG. 1.10. Solid electrolytes between unlike reversible electrodes. (a) Cationic conductor, (b) anionic conductor.

and hence only the total conductivity can be determined even with a dc field:

$$j_T = -(\sigma_{ion} + \sigma_e)\, \text{grad}\, E = -\sigma_T\, \text{grad}\, E \qquad (1.30)$$

This method has the serious disadvantage that the reversibility of electrodes can seldom be achieved with a direct current flowing through the sample, and has not been widely used. This problem has been discussed by Heyne.[90]

(2) *Solid Electrolyte between Unlike, but Reversible, Electrodes.* The examples of such cells are shown in Fig. 1.10. The open circuit voltage for such a cell is given by

$$E = -1/e \int_I^{II} d\eta_{e'} = 1/e \int_{\mu_c^I}^{\mu_c^{II}} \frac{\sigma_{ion}}{\sigma_T z_c}\, d\mu_c$$

$$= 1/e \int_I^{II} d\eta_{h\cdot} = 1/e \int_{\mu_a^I}^{\mu_a^{II}} \frac{\sigma_{ion}}{\sigma_T z_a}\, d\mu_a \qquad (1.31)$$

The ionic transport number can be directly determined from a measurement of E and a knowledge of the chemical potentials of the components on the two sides. If the difference in the chemical potentials on the two sides are not large,

$$E = \frac{1}{z_c e}\, t_{ion}(\mu_c^{II} - \mu_c^I) = \frac{1}{z_c e}\, t_{ion}(\mu_a^{II} - \mu_a^I) \qquad (1.32)$$

To determine t_{ion} as a function of the chemical potential (μ) of a component, μ^I and μ^{II} have to be changed simultaneously. This is possible for the cell in Fig. 1.10b. For the cell in Fig. 1.10a, t_{ion} as a function of μ_a can be obtained from measurements of the emf as a function of μ_a. For the cells discussed above the leads are electronic conductors. However, ionic conductors can also be used as leads, for example, the cell

$$\text{Ag} \mid \text{AgCl} \underset{\text{Ag}}{\mid} \text{Ag}_2\text{S} \underset{\text{s}}{\mid} \text{AgCl} \mid \text{Ag} \qquad [1.1]$$

where the left-hand side of Ag_2S is in contact with metallic silver and the right-hand side with liquid sulfur. The open circuit potential E of such a cell is

$$E = -1/e \int_{\mu_c^I}^{\mu_c^{II}} t_e \frac{d\mu_c}{z_c} = -1/e \int_{\mu_a^I}^{\mu_a^{II}} t_e \frac{d\mu_a}{z_a} \qquad (1.33)$$

and t_e can be obtained as

$$t_{e(\mu_a = \mu_a^{II})} = -z_a e \frac{dE}{d\mu_a} \quad \text{at} \quad \mu_a = \mu_a^{II} \tag{1.34}$$

To date, this is the most widely used technique to determine the transport numbers in solid electrolytes. For specific results of transport numbers of doped ZrO_2 and ThO_2, Sections 4.1.1, 4.1.2, and reference 90 may be referred to.

(3) *Solid Electrolyte between a Reversible Electrode and an Electronic Conductor.* The examples of such cells are shown in Fig. 1.11. The reversible electrode and the electronic conductor which block the flow of ionic charge carriers are M and Pt, respectively, for (a) and (M + MO) and the Pt/Ag layer, respectively, for (b).

The application of an external voltage below the decomposition voltage of the electrolyte in these cells in such a direction that the ionic charge carriers move from the nonreversible blocking electrodes into the electrolyte causes only a temporary ion flow. After some time a concentration gradient is built up and no ion current flows any more; the cell is "polarized." Only an electronic current now flows and a concentration gradient of the components of the solid electrolyte is established automatically as a consequence of the applied voltage. In the stationary state the current is given by

$$j_T = j_e = \frac{\sigma_e}{e} \text{ grad } \eta_{e'} \tag{1.35}$$

and

$$\text{grad } \eta_c = \text{grad } \eta_a = 0 \tag{1.36}$$

If there are electronic probes located at different points of the electrolyte (e.g., one of them being at $x = 0$) the potential difference between two probes is

$$E(x) = -\frac{\eta_{e'}(x) - \eta_{e'}(x = 0)}{e} \tag{1.37}$$

where $x = 0$ denotes the interface between the reversible electrode and the

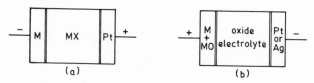

FIG. 1.11. Solid electrolytes between a reversible electrode and an electronic conductor. (a) Cationic conductor, (b) anionic conductor.

electrolyte. Consideration of local thermodynamic equilibrium at different points of the electrolyte gives rise to

$$\mu_e(x) = \mu_c^I - z_c e E(x) \tag{1.38}$$

and

$$\sigma_e[\mu_c(x)] = \frac{-j_T}{\text{grad } E(x)} \tag{1.39}$$

With the help of Eqs. (1.38) and (1.39) the electronic conductivity of a solid electrolyte can be calculated as a function of μ_c or μ_a from probe potentials at different locations for a given current density.

The use of many electronic probes at different points over the length of the electrolyte is somewhat inconvenient, although it was used by Hebb[89] for Ag_2S.

Wagner[87] suggested a modification of this method to obtain the O_2 dependence without the use of probes. For a sample of constant cross section the total current density j_e is independent of x. It can be shown[90] that, on integrating with respect to x, we obtain

$$lj_T = \frac{1}{z_c e} \int_{\mu_c^I}^{\mu_c^{II}} \sigma_e(\mu_c) \, d\mu_c \tag{1.40}$$

where l is the length of the sample, $\mu_c^I \equiv \mu_c \, (x = 0)$ and $\mu_c^{II} \equiv \mu_c \, (x = l)$.

Differentiating Eq. (1.40) with respect to μ_c^{II} and using Eq. (1.38), gives

$$\sigma_e(\mu_c) = l \frac{dj_T}{dE} \tag{1.41}$$

So the electronic conductivity component is obtained from the slope of the j vs. E plot which is measured in the final steady state situation.

If it can be assumed for a particular solid electrolyte under consideration that the electron or hole mobility is independent of the composition of the substance, then the σ_e dependence of the chemical potential of any component results only from the concentrations of the electrons and holes.

Using the relation $\mu_k = \mu_k^\circ + kT \ln n_k$ and Eq. (1.38), it can be shown that the concentration of electrons $n(x)$ is

$$n(x) = n^I \exp \frac{-eE(x)}{kT} \tag{1.42}$$

Similar expressions can be obtained for hole concentration and the total electronic conductivity is given by

$$\sigma_e = \sigma_{e'} + \sigma_{h\cdot} = \sigma_{e'}^I \exp \frac{-eE(x)}{kT} + \sigma_{h\cdot}^I \exp \frac{eE(x)}{kT} \tag{1.43}$$

Substituting this equation in Eq. (1.40) and using Eq. (1.38), we obtain

$$j_T = j_e = \frac{-kT}{eL} \sigma_{e'}^I \left(1 - \exp\frac{-eE}{kT}\right) + \sigma_h^I \left(\exp\frac{eE}{kT} - 1\right) \quad (1.44)$$

The above equation represents a general relationship between current and voltage when conduction due to both electron and electron hole is significant. However, in some of the cationic conductors like $AgBr$,[95,96] $CuCl$,[97,98] Ag_3SI,[99] etc. the partial electronic and electron hole conductions are so dissimilar that one of the terms in Eq. (1.44) may be neglected. In those cases evaluation of the predominant term is possible from the intercept of a log j_T vs. E plot. Several investigators[28-30] have used this technique for various other materials. However, for the materials like ZrO_2–CaO and ThO_2–Y_2O_3,[100] silver β-alumina,[68] $AgCl$,[101] etc. in which both these terms are significant, it is necessary to rearrange the terms as suggested by Patterson *et al.*[100] Accordingly, we define a dimensionless quantity

$$u = \frac{eE}{kT}$$

and dividing both sides of Eq. (1.44) by $[1 - \exp(-u)]$ gives

$$\frac{j_e}{[(1 - \exp(-u)]} = \frac{-kT}{el} [\sigma_{e'}^I + \sigma_h^I \cdot \exp(u)] \quad (1.45)$$

A plot of the quantity $j_e/[1 - \exp(-u)]$ vs. $\exp(u)$ will give a straight line with the slope $(-kT/el)\sigma_h^I$ and an intercept $(-kT/el)\sigma_{e'}^I$.

(4) Solid Electrolyte between a Reversible Electrode and an Ionic Conductor. The examples for such cells are shown in Fig. 1.12. The ionic conductors in a cell of this type must have an ion common with the solid electrolyte to be investigated. The blocking of electronic current can be achieved by the use of a purely ionic conductor exhibiting negligible electronic conductivity. The situation is most simple if the ionic conductance of this electrode is much higher than that of the sample under test. Some related phenomena like nonstoichiometry of samples have also been studied by this method.[102-104] The difficulties with this method have been discussed by Heyne.[90]

Besides the measurement of partial electronic or ionic conductivity, later

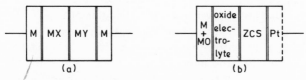

FIG. 1.12. Solid electrolytes between a reversible electrode and an ionic conductor. (a) Cationic conductor, (b) anionic conductor.

developments on Wagner's theory of polarization have made it possible to determine (1) the concentration and mobility of the electronic charge carriers (2) double-layer capacitance between the electrode and the electrolyte, and in some cases (3) the diffusion coefficient of the mobile ion into the electrode material. This is done by measuring the transient response of an asymmetrical polarization cell when the voltage or the current through the cell is suddenly changed from one steady state to another. Current developments of this theory have been recently reviewed by Wagner.[105]

3.5. Impedance Measurement

In recent years, impedance measurement has been increasingly used to investigate the transport phenomenon in solid electrolyte galvanic cells. It has been found that the simultaneous measurement of resistance and capacitance of a cell over a wide range of frequency and the construction of an impedance (or admittance) plot from such data reveals much more information than a dc or single-frequency ac measurement. Bauerle[106] was the first to apply this technique to a solid electrolyte system.

The impedance Z of a RC circuit is given by

$$Z(\omega) = R_s - i(\omega C_s)^{-1} \qquad (1.46)$$

where R_s and C_s are the equivalent series resistance and capacitance, respectively, ω is the frequency, and i represents the imaginary quantity. When the same impedance is represented by an equivalent parallel circuit, the corresponding values are

$$R_p = \frac{1 + \omega^2 R_s^2 C_s^2}{\omega^2 R_s C_s^2} \quad \text{and} \quad C_p = \frac{C_s}{1 + \omega R_s^2 C_s^2} \qquad (1.47)$$

Admittance Y is the inverse of impedance and is expressed as

$$Y(\omega) = \frac{1}{Z} = G_p + i\omega C_p \qquad (1.48)$$

where $G_p = 1/R_p$.

The plots of the real and imaginary parts of the above equations (R_s vs. $1/\omega C_s$ for impedance and G_p vs. ωC_p for admittance) as parametric functions of frequency show distinctive features characterizing particular combinations of the circuit elements. A few simple examples are presented in Fig. 1.13. Arrows in the diagram represent the direction of increasing frequency. Shapes of the plots vary greatly with circuit elements. In general it may be said that each semicircular arc corresponds to a lumped RC combination, a vertical line represents a lone capacitance, and a quarter circle with a 45° line represents a Warburg impedance which is the electrical analog of a diffusion process (Fig. 1.14). It is possible to calculate the resistance and capacitance

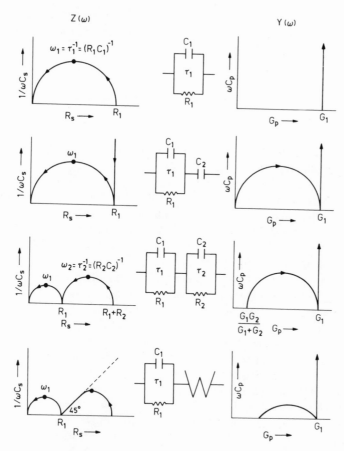

FIG. 1.13. Schematic impedance and admittance diagrams for some simple *RC* circuits (Reference 110).

values from the intercepts on the real axis and the highest point on the semi-circles, respectively. More details regarding the theory of this technique may be obtained in References 108–111. When similar diagrams are constructed with the real and imaginary value of dielectric constant they are known as Cole–Cole plots.

The physical processes which take place in an electrode–electrolyte system by the application of an electric field are very often represented by analogous electrical circuits. Impedance measurement has been most successful in qualitative as well as quantitative determination of such equivalent circuits and therefore provides much better insight into the transport mechanism. In principle any such equivalent circuit may be represented either by a

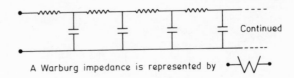

A Warburg impedance is represented by

FIG. 1.14. Circuit diagram for Warburg impedance.

Maxwell model (Fig. 1.15a) or a Voigt model (Fig. 1.15b). Mathematically they are interchangeable so that with the proper choice of element values they can have the same impedance values at all frequencies, although the transformation becomes difficult with $N > 2$. However, their powers of interpretation of the physical processes are not always the same and the final selection between them depends on the nature of particular system under investigation. It may be mentioned that an admittance plot is preferable for a Maxwell model, while an impedance diagram will give more information for a Voigt model. Although a two-layer Voigt model is very often used to describe a solid electrolyte system, the actual circuit may be quite complicated depending on the exact nature of the transport mechanism. Recently Mcdonald[110] has proposed a five-layer Voigt model (Fig. 1.16a) in which each of the RC sections correspond to one of the following five physical processes: (1) charge separation in the bulk (B), (2) charge transfer reaction at the electrode (R), (3) absorption reaction (A/R), (4) generation and recombination (G/R), and (5) diffusion (D). In the last two cases the circuit elements are functions of frequency. An impedance diagram for such a system will result in a series of connected arcs (Fig. 1.16b) each associated with a single process provided the relaxation time ($\tau = RC$) of these processes differ at least by two orders of magnitude. Since each of these relaxation processes follows the Arrhenius principle, impedance measurement as a function of frequency and tempera-

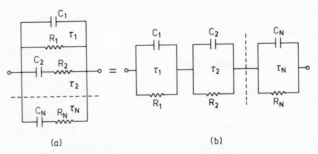

FIG. 1.15. Two types of representation for equivalent circuits. (a) Maxwell model, (b) Voigt model.

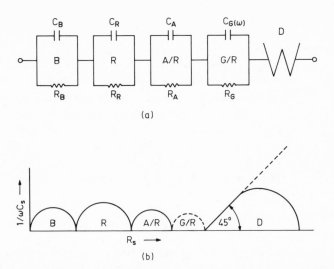

FIG. 1.16. Equivalent circuit (a) and impedance diagram (b) for a solid placed between two plane parallel electrodes (Reference 110).

ture enables determination of activation energy for each of these processes. In practice, however, it is quite likely that all the five arcs may not be observable at the same time. The relative sizes will also vary greatly depending on the experimental conditions (e.g., specimen geometry, microstructure, etc.). The G/R arc may be either semicircular or similar to that of diffusion and may rarely be observed in a solid electrolyte. Although the B arc will always be at the higher frequency end, the order of some of these arcs may change depending on conditions. In actual experiments the arcs are often less than half-circles. This is believed to be due to a distribution of relaxation times instead of a discrete one.

A maximum of three arcs have been observed with CaO- and Y_2O_3-stabilized ZrO_2 electrolytes.[112–115] β-alumina[116–119] and Ag_4RbI_5[119] electrolytes have also been studied by this technique. It has been possible to make accurate estimations of electrode and electrolyte resistances as well as to differentiate between the bulk and grain boundary resistances in these electrolytes. It is also possible to measure the activation energy and the mobility of the charge carriers by this technique. Since the shape and size of the arcs are quite sensitive to the length and surface area of the specimens a careful selection of specimen geometry is essential for a successful experiment. Temperature of the specimen has also a pronounced effect on the frequency range over which the arcs are obtained. An incorrect selection of temperature may give rise to an incomplete arc within the normally available frequency range.

3.6. Thermoelectric Power

Charge carriers in a solid electrolyte tend to redistribute themselves under a temperature gradient resulting in a thermal emf. Total potential difference across the solid is the result of three contributing factors: (i) the homogeneous potential difference ($\Delta\theta$) due to the temperature gradient (ΔT) within the solid, (ii) the algebraic sum of the contact potential differences ($\Delta\psi$) at the electrode–electrolyte interfaces and, (iii) the algebraic sum of the homogeneous potential differences within the lead wires. The magnitude of the third term is, however, insignificant compared to the other two and is generally neglected. For a small value of ΔT the total thermoelectric power α_T is given by

$$\alpha_T = \frac{\Delta\theta}{\Delta T} + \frac{\Delta\psi}{\Delta T} = \alpha_{\text{hom}} + \alpha_{\text{het}} \qquad (1.49)$$

The quantities α_{hom} and α_{het} are termed as homogeneous and heterogeneous thermoelectric powers respectively. DeGroot[120] and Holtan *et al.*[121,122] pioneered in developing the theory of thermoelectric power in ionic solids based on the principles of thermodynamics of irreversible processes. Expressions for thermoelectric power were later derived invoking the concept of a particular defect structure such as Schottky or Frenkel pairs.[123]

Measurement of thermoelectric power provides useful information about the heat of formation of defects, heat and entropy of transport, and the type (positive or negative) of the predominant charge carrier. The technique was used with a number of cationic conductors such as NaBr, AgBr, AgCl, AgI, and various other compounds[124–128] and the results obtained have been recently reviewed by Shahi.[129] In most of the cationic conductors θ and $1/T$ are linearly related and the heat of transport is very close to the activation energy for ionic conduction. Oxide electrolytes have also been studied by this technique[130–136] even though the interpretation of the results in this case becomes difficult due to a very high concentration of defects. Wagner[137] has presented an excellent account of the subject covering different kinds of thermocells including mixed conductors. We shall briefly discuss here the results obtained with some of the oxygen ion conductors.

Tallan and Bransky[134] have derived general expressions of thermoelectric power for the following thermocell containing a MO_2-type mixed conductor, in accordance with the principles put forward by Holtan *et al.*[121,122] and Howard and Lidiard[123]:

$$(P_{O_2})\, Pt \underset{T}{\bigg|}\, \text{``pure'' } ThO_2 \underset{T+\Delta T}{\bigg|}\, Pt\, (P_{O_2}) \qquad [1.2]$$

It is assumed that the oxygen interstitial and electron holes are the predominant defects in the electrolyte and the impurity atoms (L_{Th}) are partially associated with the oxygen interstitials. Under these conditions the homogeneous thermoelectric power is derived to be[134]

$$\alpha_{\text{hom}} = \frac{1}{eT} \left\{ t_{\text{h}\cdot} \frac{\Delta H_{\text{ex}}}{2} + Q(\text{h}^\cdot) - Q \frac{1}{2} \frac{\beta}{1+\beta} \Delta H_{\text{p}} \right.$$

$$\left. - t_{\text{i}}\frac{1}{2}\left[Q(O_i'') + \frac{1}{2}\frac{\beta}{1+\beta} \Delta H_{\text{p}} \right] \right\} \qquad (1.50)$$

where ΔH_{ex} and ΔH_{p} are the enthalpy change of the reactions $\frac{1}{2}O_2 \rightleftharpoons O_i'' + 2\text{h}^\cdot$ and $L_{\text{Th}}^\cdot + O_i'' = (L_{\text{Th}}O_i)'$, respectively. $Q(\text{h}^\cdot)$ and $Q(O_i'')$ are the heats of transport and β is the degree of association given by

$$\beta = \frac{[(L_{\text{Th}}O_i)']}{[L_{\text{Th}}^\cdot]}$$

Similarly the heterogeneous thermoelectric power is given by

$$\alpha_{\text{het}} = -\frac{1}{e}\left[\frac{1}{2} S^*(O_i'')_{\text{ThO}_2} - S(e')_{\text{Pt}} - \frac{1}{4}\Delta S(O_2)\text{g} \right] \qquad (1.51)$$

where $S^*(O_i'')_{\text{ThO}_2}$ is the partial molar entropy of O_i'' in ThO_2. $S(e')_{\text{Pt}}$ is the transported entropy of e' in Pt and $\Delta S(O_2)_{\text{Pt}}$ is the change in entropy of O_2 gas due to the heterogeneous electrode reaction at the electrode–electrolyte interface.[138] Equations (1.50) and (1.51) may be combined to obtain the expression for α_{T} and from that the ionic thermoelectric power α_{ion} is derived by substituting $t_{\text{ion}} = 1$ and $t_{\text{h}\cdot} = 0$. Similarly the electron hole thermoelectric power $(\alpha_{\text{h}\cdot})$ is obtained by putting $t_{\text{ion}} = 0$ and $t_{\text{h}\cdot} = 1$. Accordingly,

$$\alpha_{\text{ion}} = \frac{1}{e}\left[\frac{1}{2} S(O_i'')_{\text{ThO}_2} - S(e')_{\text{Pt}} - \frac{1}{4}\Delta S(O_2)_{\text{g}} + \frac{1}{2}\frac{\beta}{1+\beta}\frac{\Delta H_{\text{p}}}{T} \right]$$

$$(1.52)$$

because $Q(O_i'')/T = S(O_i'') - S^*(O_i'')$. In the absence of any association phenomenon the last term in Eq. (1.52) becomes zero. Such an expression for calcia-stabilized zirconia was earlier derived by Ruka *et al.*[131]

In terms of partial thermoelectric powers α_{T} may be expressed as

$$\alpha_{\text{T}} = t_{\text{ion}}\alpha_{\text{i}} + t_{\text{h}\cdot}\alpha_{\text{h}\cdot}. \qquad (1.53)$$

Since there is no way to know independently the individual quantities on the right-hand side of Eq. (1.53) it is not possible to calculate the value of α_{T} from this equation. However, to illustrate these results Ruka *et al.*[131] suggested the measurement of thermoelectric power under two different oxygen partial pressures in which case all the terms except $\Delta S(O_2)_g$ in Eq. (1.52) remain unaltered. Considering one of these partial pressures as reference, Eq. (1.52) simplifies to

$$(\alpha_{\text{ion}})_g - (\alpha_{\text{ion}})_{\text{ref}} = \frac{1}{4F}[\Delta S(O_2)_g - \Delta S(O_2)_{\text{ref}}] \qquad (1.54)$$

where F is the Faraday constant. This equation suggests that the difference between thermoelectric powers varies linearly with the difference of entropies. It has been verified in the case of calcia-stabilized zirconia[131] as well as in "pure" ThO_2. The results obtained with pure ThO_2 are shown in Fig. 1.17. It is generally convenient to choose the reference gas in the partial pressure range where the electrolyte is completely ionic. Measurement of thermoelectric power as a function of oxygen partial pressure together with the reported values of $S(O_2)_g$ and $S(e')_{\text{Pt}}$ gives the following values for the entropy of transport: $S(O_i'') = 14.5 \pm 0.5$ deg^{-1} mol^{-1} for ThO_2 at 1000°C over P_{O_2} range of 10^{-13}–10^{-23} atm,[134] and $S(V_{\ddot{o}}) = 10.3$–10.9 deg^{-1} mol^{-1} for calcia-stabilized ZrO_2 at 1000°C.[131]

3.7. Relaxation Methods

In ionic crystals, defect complexes are formed due to the existence of coulombic attraction between the oppositely charged ionic defects. Introduction of aliovalent impurities produces compensating defects which are equal but oppositely charged to those of the impurity ions, and the two entities will form a defect complex having relatively strong binding energy between them. Such defect complexes are electrically neutral and do not take part in the conduction process but act as dipoles which change their orientation as a result of defect jumps aided by thermal activation under an alternating electrical or mechanical stress field, giving rise to dielectric or anelastic relaxation, respectively. The theory of relaxation processes in crystalline

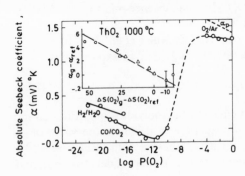

FIG. 1.17. Seebeck coefficient of ThO_2 vs. oxygen partial pressure in different gas mixtures at 1000°C. The inset is a plot of change in Seebeck coefficient with $\Delta S(O_2)_g$–$\Delta S(O_2)_{\text{ref}}$. The dashed line, $\alpha_{h\cdot}$, is the calculated electron hole Seebeck coefficient (Reference 134).

solids has been extensively discussed[139] and the phenomenon is always represented by the well-known Debye equations

$$\varepsilon'(\omega) = \varepsilon_\infty + \frac{(\varepsilon_S - \varepsilon_\infty)}{1 + \omega^2\tau^2} \tag{1.55}$$

$$\varepsilon''(\omega) = \frac{(\varepsilon_S - \varepsilon_\infty)\omega\tau}{1 + \omega^2\tau^2} \tag{1.56}$$

and

$$\tan \delta = \frac{\varepsilon''}{\varepsilon'} = \frac{(\varepsilon_S - \varepsilon_\infty)\omega\tau}{\varepsilon_S + \varepsilon_\infty\omega^2\tau^2} \tag{1.57}$$

where ε' and ε'' are, respectively, the real and imaginary parts of the dielectric constant. The subscripts ∞ and S correspond to the high and low frequency values, respectively. ω is the frequency and τ is called the relaxation time. In the case of dielectric relaxation δ is known as the "loss angle," while in the anelastic case it is called the "internal friction." The function ε'' when plotted against $\log(\omega\tau)$ gives a symmetric peak centered about $\log \omega\tau = 0$, i.e., $\omega\tau = 1$ and is called the "Debye peak." In practice it is often convenient to study a Debye peak not by changing ω but by varying τ by change of temperature. This becomes possible because τ^{-1} usually obeys an Arrhenius equation given by

$$\tau^{-1} = \tau_0^{-1} \exp(-\Delta E/kT) \tag{1.58}$$

where ΔE is an activation energy.

Following Breckenridge[140] a number of studies have been carried out on both the dielectric loss and the internal friction peaks in a number of alkali halides containing different divalent cation impurities. A completely analogous situation exists in fluorite-type lattices and the Ca^{2+} impurity ion-O^{2-} vacancy complex in ThO_2-CaO solid solutions was studied by Wachtman[141] in terms of both the dielectric and anelastic relaxations. During the reorientation process, several defect jumps are generally possible and there may be several modes of relaxation.[142] Wachtman[141] has considered an 8-position model to represent the relaxation process in the ThO_2-CaO system. These studies gave an activation energy $E = 0.95$ eV and $\tau_0 = 10^{-13}$ sec. Figures 1.18 and 1.19 show the internal friction and dielectric loss peaks of ThO_2-CaO samples as a function of frequency and temperature. The height of the peak is proportional to the concentration of CaO. The ratio of τ_{diel}/τ_{anel} was observed to be close to 2.0 as predicted by the 8-position model. It has also been verified in systems like CeO_2-CaO[143] and CaF_2-NaF[144] having the same crystal structure. Wachtman and Corwin[145] have observed analogous internal friction peaks in ZrO_2 with 10%–20% CaO but also found an additional unsymmetric peak.

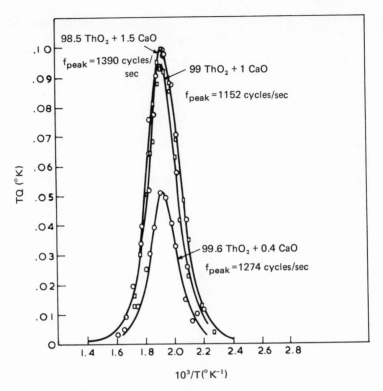

FIG. 1.18. Internal friction (Q) in ThO_2 ceramics containing CaO (Reference 141).

3.8. Miscellaneous Techniques

An increasing number of techniques and tools are being applied to screen and study ionic conductors. For example, ion transport by exchanging the cation in a material with another in a molten salt and determining the weight change gravimetrically is used as a quick screening test for ion mobility in β-aluminas.[68,146]

When the nuclei of the mobile ions in an ionic conductor have suitable magnetic moments (such as H, Li, Na) their movements can be studied by the nuclear magnetic resonance technique.[146] Measurement of absorption linewidth as a function of temperature at a constant frequency enables determination of the jump frequency and the activation energy of the mobile ion. The variation of the linewidth with temperature is schematically represented in Fig. 1.20. The highest-temperature plateau corresponds to the free translation of the ion. Its width is determined by the inhomogeneity of the magnet and is typically of the order of 100 mG. At very low temperatures, where no ionic

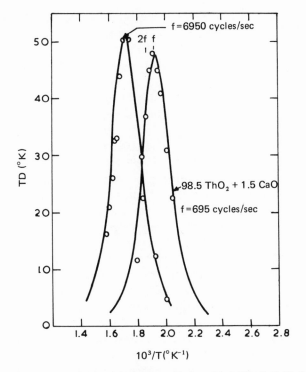

FIG. 1.19. Dissipation factor (D) in ThO_2 ceramics with 1.5% CaO (Reference 141).

motion occurs, the linewidth is solely due to the rigid atomic positions in the crystal. Other plateaus are found where there is some restriction to translation or to rotation of the polynuclear groups like ammonium.

If the linewidth is solely due to dipolar effects the dipolar linewidth (ΔH_d) is given by

$$(\Delta H_d)^2 = (\Delta H_1)^2 (2/\pi) \tan^{-1}\left[\alpha\left(\frac{\gamma \Delta H_d}{2\pi \nu}\right)\right] \qquad (1.59)$$

where ΔH_1 is the low-temperature linewidth, ν the jump frequency, γ the gyromagnetic ratio, and α a constant whose value varies with line shape and

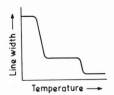

FIG. 1.20. NMR linewidth vs. temperature (Reference 146).

falls between 1 and 10. This expression suggests that the linewidth can fall to zero, which it does not in practice. To avoid this the equation is empirically modified with the substitution of ΔH_d by $(\Delta H - \Delta H_r)$ and ΔH_1 by $(\Delta H_1 - \Delta H_r)$ in the above equation, where ΔH is the measured linewidth and ΔH_r is the high-temperature residual linewidth. By knowing the values of ΔH, ΔH_1, and ΔH_r from the experiment the value of jump frequency as well as the activation energy may be calculated. This technique has been successfully used in case of sodium β-alumina,[147,148] PbF_2,[149] and some inorganic bronzes.[150,151] A typical line-narrowing curve for β-alumina is given in Fig. 1.21. In this case the situation is complicated by the nuclear quadruple moment of the sodium ion, resulting in excessive broadening, particularly at low temperatures. The activation energy calculated from the high temperature end of the curve is only slightly lower than that obtained by other techniques.

Light- as well as neutron-scattering methods are also used to study the transport properties of the ionic conductors. In the light-scattering process only the excitations close to the center of the Brillouin zone are observed, while by neutron scattering the entire Brillouin zone can be investigated. Study of the optical mode of lattice vibrations by light-scattering technique is termed Raman scattering, while that of the acoustic mode is referred to as Brillouin scattering.

It was seen that the dielectric measurements over a wide range of frequencies produces useful information regarding the transport properties of the solid electrolytes. The frequency range of this measurement can be extended by the infrared reflectivity technique in which one determines the real and imaginary parts of the complex refractive index

$$\eta = n + ik \tag{1.60}$$

and therefore one can find out the real and imaginary parts of the dielectric constant using the following equations:

$$\varepsilon' = n^2 - k^2 \quad \text{and} \quad \varepsilon'' = 2nk \tag{1.61}$$

X-ray and electron diffuse scattering studies are also very important tools

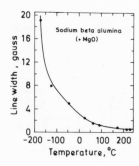

FIG. 1.21. NMR line-narrowing curve for sodium β-alumina (Reference 146).

to study the order–disorder phenomenon in the ionic conductors. All the above techniques have been extensively used to study the transport properties of the β-alumina, AgI, and PbF$_2$ groups of superionic conductors,[152–159] and the results obtained have been recently reviewed by Hayes.[39]

4. Materials

There have been several review articles and books published covering various aspects of solid electrolyte materials. Some of the recent ones are those by Hooper and co-workers,[31,32,38] Hayes,[39] and Geller.[225] Rapp and Shores[24] earlier presented a listing of solid electrolytes, with an indication of the conducting ions, known up to about 1970. A more recent compilation of the properties of various alkali metal, copper, and silver ion conductors has been published by McGeehin and Hooper.[31] Here, an updated listing is presented in Table 1.1. The properties of major types of ionic conductors are discussed in the following sections.

4.1. Fluorite-Type Oxides

Most of the electrolytes exhibiting predominant oxygen ion conduction have crystal structures of either the fluorite type (ThO$_2$, CeO$_2$) or the distorted fluorite type (ZrO$_2$, HfO$_2$, δ-Bi$_2$O$_3$). Amongst them, ZrO$_2$- and ThO$_2$-based electrolytes have been studied most extensively and found suitable for a wide range of applications like high-temperature fuel cells, oxygen monitors and pumps, and for various thermodynamic and kinetic measurements. ZrO$_2$-base electrolytes have the advantage of higher conductivity and are used in the high oxygen partial pressure range (near atmospheric), while ThO$_2$-base electrolytes possess greater thermodynamic stability and are useful at very low partial pressures. In recent years CeO$_2$ electrolytes have been found very promising due to their higher conductivity than most of the ZrO$_2$-base electrolytes even though there is greater possibility of electronic conduction due to variable valency state of the cerium ion. In the following sections the properties of each of these electrolytes are discussed separately. Earlier Etsell and Flengas[180] published an exhaustive review covering these materials.

4.1.1. Zirconia-Base Materials

At room temperature pure ZrO$_2$ has a monoclinic crystal structure which changes to a tetragonal form above 1200°C,[181] and finally to a cubic one above 2300°C. However, addition of some of the aliovalent oxides stabilizes the high-temperature cubic fluorite phase in zirconia. It is believed[182] that for better stabilization, the impurity cation should have a stable valency which is less than 4, producing vacancies in the oxygen sublattice. The stability of the

Table 1.1. Solid Electrolytes

Conducting species	Compound	Reference[a]
Fluorine (F⁻)	CaF_2, MgF_2, PbF_2, NaF, BaF_2, SrF_2	
	Cryolite, Na_3AlF_6	
	LaOF, KPb_3F_7, LaF_3, CeF_3, YF_3, ErF_3	160, 161
Chlorine (Cl⁻)	$BaCl_2$, $PbCl_2$, $SrCl_2$	
Bromine (Br⁻)	$BaBr_2$, $PbBr_2$, NaBr, KBr	
Iodine (I⁻)	PbI_2, KI	
Oxygen (O^{2-})	$Zr_{1-x}M_x^{2+}O_{2-x}$, $Zr_{1-x}M_x^{3+}O_{2-x/2}$	
	$Th_{1-x}M_x^{2+}O_{2-x}$, $Th_{1-x}M_x^{3+}O_{2-x/2}$	
	$Hf_{1-x}M_x^{2+}O_{2-x}$, $Hf_{1-x}M_x^{3+}O_{2-x/2}$	177
	$Ce_{1-x}M_x^{2+}O_{2-x}$, $Ce_{1-x}M_x^{3+}O_{2-x/2}$	
	$Bi_{2-x}Sr_xO_{3-x/2}$, $Bi_{2-x}W_xO_3$ $_{3x/2}$	
	$Bi_{2-x}Y_xO_3$, $Bi_{2-x}Gd_xO_3$,	
	LaOF	
Sulfur (S^{2-})	$CaS-Y_2S_3$	172
Carbon	BaF_2-BaC_2	173
Nitrogen	AlN	173
Hydrogen (H⁺)	KHF_2, KH_2PO_4, $(NH_4)_2H_3IO_6$, H_xWO_3	
Silver (Ag⁺)	α- and β-AgI, AgCl, AgBr, Ag_2S, Ag_2Se,	
	Ag_2Te, Ag_3SBr, Ag_3SI, Ag_2HgI_4,	
	$Ag_4HgSe_2I_2$, $Ag_8HgS_2I_6$, Ag_4HgTeI_2	190–193
	KAg_4I_5, $RbAg_4I_5$, $NH_4Ag_4I_5$	
	$β-Ag_2O \cdot 11Al_2O_3$	11, 45
	$Q\ Ag_6I_7$	162
	Q = tetramethyl, ethyltrimethyl,	
	diethyldimethyl, trimethyl isopropyl,	
	trimethylpropyl, tetraethyl,	
	diethylmethyl isopropyl, triethylpropyl	
	$C_aH_bNI \cdot AgI$	163
	Pyrrolodinium, piperidium,	
	1,1- dimethyl pyrrolidinium, quinuclidinium,	
	1,1- dimethyl piperidinium,	
	N-methyl quinuclidinium,	
	5-azomiaspiro[4,4]nonane,	
	5-azoniaspiro[5,5]undecane	
	$C_aH_bNI \cdot nAgI$	163
	3-methyl pyridinium,	
	1-methyl pyridinium,	
	1,3-dimethyl pyridinium,	
	1,2,6-trimethyl pyridinium,	
	quinolinium,	
	1,2,4,6-tetramethyl pyridinium,	
	1,2,3,6-tetramethyl pyridinium,	

————continued

[a] References are given only for those materials which have not been included in the list presented by Rapp and Shores[24] and McGeehin and Hooper.[31]

Table 1.1. (*continued*)

Conducting species	Compound	Reference[a]
	1-methyl quinolinium,	
	1,2,3,4-tetrahydro-1-methyl quinolinium,	
	1-ethyl quinolinium,	
	1,2-dimethyl quinolinium,	
	1,3,4-trihydro-1,2-dimethyl quinolinium,	
	2,3,4-trihydro-1, 1-dimethyl quinolinium,	
	1-ethyl-2,6-dimethyl quinolinium,	
	polymethonium diiodide $\cdot$ 12AgI	164
	$Ag_7I_4PO_4$, $Ag_7I_4AsO_4$, $Ag_7I_4VO_4$, $Ag_{19}I_{15}P_2O_7$, $\left.\begin{array}{l}\\\\\end{array}\right\}$	165, 166
	$Ag_6I_4WO_4$	
	MCN $\cdot$ 4AgI, RbCN $\cdot$ 4AgI, CsCN $\cdot$ 4AgI, $Ag_2SeO_4 \cdot$ 2AgI	176, 168
Copper (Cu$^+$)	β-CuI, CuCl, β-, and γ-CuBr	
	CuS, Cu_2Se	174–176
	$HgCu_2I_4$, KCu_4I_5(257–332°C), 7CuBr $\cdot C_6H_{12}N_4$XBr	
	(X = CH_3, H, C_2H_5), 7CuCl $\cdot C_6H_{12}N_4$HCl	
	17CuI $\cdot 3C_6H_{12}N_4CH_3$I	
Lithium (Li$^+$)	LiH, Li_4SiO_4, Li_2SiO_5, Li_2SiO_3, $LiAlSi_2O_6$	
	$LiAlSiO_4$, Li_2SO_4, (Li, Ag)$_2SO_4$, (Li, Na)$_2SO_4$	
Sodium (Na$^+$)	NaF, NaCl, NaBr,	
	β-$Na_2O \cdot 11Al_2O_3$, β''-$Na_2O \cdot 5Al_2O_3$	178
	β''-$Na_2O \cdot MgO \cdot 5Al_2O_3$	178
	$NaSbO_3$, $NaSbO_3 \cdot I_6NaF$,	
	$NaTa_2O_5F$, $Na_3Zr_2PSi_2O_{12}$	
Potassium (K$^+$)	KCl, KBr, KI	
	β-1.3$K_2O \cdot O$, $2Li_2O \cdot 10Al_2O_3$	
	$K_{1-x}Mg_{1-x}Al_{1+x}F_6$, $K_2Al_2Ti_6O_{16}$	
	$K_{2x}Mg_xTi_{8-x}O_{16}$	
	$K_{2x}Al_{2x}Ti_{8-x}O_{16}$	
Alkali metal ions	SiO_2 (quartz), Pyrex glass	179
(Na$^+$ and K$^+$)	$3Al_2O_3 \cdot 2SiO_2$ (mullite)	
Rubidium (Rb$^+$)	β-$Rb_2O \cdot 11Al_2O_3$	44
Ammonium		
(NH$_4^+$)	β-$(NH_4)_2O \cdot 11Al_2O_3$	44
Thallium (Tl$^+$)	β-$Tl_2O \cdot 11Al_2O_3$	44
Magnesium		
(Mg^{2+})	MgO	
Aluminum		
(Al^{3+})	Al_2O_3	

solid solution increases with lowering of the eutectoid temperature and also with higher electropositive character of the cation. Theoretically CaO should be the best stabilizing oxide for ZrO_2.

ZrO_2–CaO System. Kiukkola and Wagner[183] first demonstrated the possible use of this electrolyte in thermodynamic study of various oxides.

Since then, this has been the most thoroughly studied and widely used electrolyte system and is commonly known as calcia-stabilized zirconia (CSZ). The preparation techniques for this electrolyte are discussed in Chapter 7. Numerous phase diagram studies have been carried out on this system to determine the composition range of the cubic phase stability[48,49,184-186] but there exists a considerable disagreement between the observed values. However, a stability range of 12–20 mol % CaO may be considered representative. A phase diagram for the system is shown in Fig. 1.22. At lower CaO concentrations the cubic phase is in equilibrium with the tetragonal ZrO_2 solid solution, while at higher concentrations it coexists with a compound $CaZrO_3$. There have been several conflicting reports about the instability of the cubic phase at lower temperatures. However, it is believed that the cubic phase is thermodynamically stable only above 900°C and it decomposes to ZrO_2 and $CaZrO_3$ below this temperature, even though prolonged heating at elevated temperatures (between 700 and 1200°C up to 3000 hr) does not destroy the cubic phase significantly.[189] This is because the cubic phase is

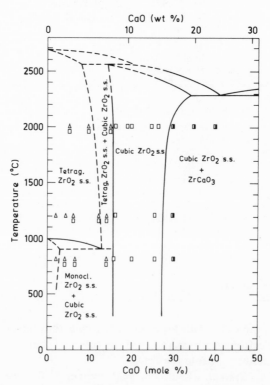

FIG. 1.22. Phase diagram for the system ZrO_2–CaO (Reference 184).

kinetically metastable at these temperatures due to a very low cation diffusion rate.[65] Existence of a second compound suggested earlier by some of the investigators[186-190] has now been confirmed by electron diffuse scattering study.[55,56] This is believed to be an ordered phase within a microdomain formed as a result of prolonged high temperature aging (at 1000°C for 35–40 days). The lattice parameter of the solid solutions has been measured by various investigators.[48,49,64,185] It increases with increasing CaO content due to the larger size of Ca^{2+} ion compared to that of Zr^{4+}. The lattice parameter for the most widely used composition $Zr_{0.85}Ca_{0.15}O_{1.85}$ is 5.131 Å.

The predominant defects present in the solid solutions are anion vacancies.[191] This has been confirmed by density[64] and x-ray intensity[48,192] measurements. The impurity cations and the anion vacancies are considered to be randomly distributed over the available lattice sites. However, at higher temperatures (> 1800°C) presence of cation interstials has been reported.[50] Electrical conductivity of ZrO_2–CaO electrolytes has been measured by numerous investigators. A representative plot for a 15 mol % CaO solid solution is shown in Fig. 1.23 together with those of other important oxygen conductors. As expected it follows an Arrhenius-type equation [Eq. (1.23)]. The activation energy for these electrolytes is in the range of 25–30 kcal/mol. Within the cubic phase field there is a systematic increase of conductivity with decreasing CaO content and it reaches a maximum (5.5×10^{-2} ohm^{-1} cm^{-1} at 1000°C) at 12–13 mol % CaO, which coincides with the minimum quantity of CaO necessary to stabilize the fluorite phase.[48,185,187] However, some investigators have observed a conductivity maximum within the fluorite phase field.[49,193] The ionic conductivity of the most widely used 15 mol % CaO electrolyte is about 2.4×10^{-2} ohm^{-1} cm^{-1} at 1000°C. The lowering of conductivity at higher CaO content is believed to be due to decreasing mobility of the oxygen ion as a result of lattice distortion[48,194] and also to the formation of (Ca''_{Zr}–$V_O^{\cdot\cdot}$) complexes[145,195] or vacancy clustering.[48,49,193] Measurement of conductivity on single and polycrystalline specimens does not show any significant effect of grain boundaries on conductivity.[49] Grain growth[196] and sintering[65,194] appear to be controlled by cation diffusion in these solid solutions.

Various investigators[48,49,197-199] have reported the formation of an ordered superstructure in ZrO_2–CaO electrolytes on prolonged annealing around 1000°C. The lattice parameter of the superstructure is four times that of the original fluorite phase.[48] It develops more easily in a single crystal and at high CaO content. Ionic conductivity of the electrolyte is lowered due to ordering without any significant change in activation energy.[48,99,51,200] However, the ordered phase disappears and the conductivity is restored by heating to higher temperatures (around 1400°C). Carter and Roth[49] studied this phenomenon by measuring dielectric relaxation. On aging, the area of the loss

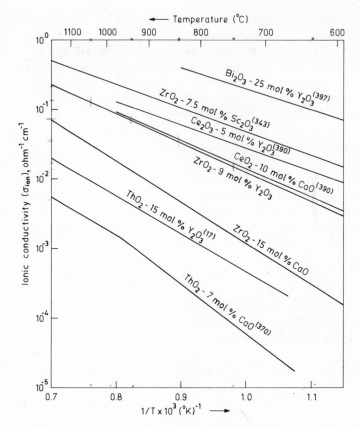

FIG. 1.23. Log σ vs. $1/T$ plots for various oxide electrolytes.

peak increases and it moves to a lower frequency, while on deaging it regains its original size and position. They have given a probable explanation for the anomalously large conductivity of the disordered phase and its aging behavior. The structure is visualized as constructed of deformed cubes of oxygen ions each containing a cation sharing edges with similar adjacent cubes. Anion vacancies, although preferentially bound to cubes which contain Ca^{2+} ions, are mobile within the $(Ca''_{Zr}\text{-}V\overset{..}{_O})$ complex. The development of a three-dimensional periodic structure on annealing results in trapping vacancies to particular Ca^{2+} ions and therefore a decrease in conductivity. Recently Hudson and Moseley[55,56] have suggested the formation of microdomains of $CaZr_4O_9$ and growth of monoclinic ZrO_2 within the disordered fluorite phase as a result of aging.

The anion diffusion coefficient in this material is at least five to six orders of magnitude higher than that of cations (Fig. 1.7). The transference number for oxygen ion is very close to unity (>0.99) down to an oxygen partial pressure of 10^{-20} atm at 1000°C. Compared to fully stabilized zirconia, partially stabilized zirconia (PSZ) electrolytes (7–8 mol % CaO) have greater resistance to thermal fluctuations and higher mechanical strength with a loss in electrical conductivity. Few investigations have been carried out to measure their ionic transference number and oxygen permeability,[201,202] and they have been found suitable as solid electrolytes.

ZrO_2–Y_2O_3 System. Historically, these electrolytes are the first to be investigated that exhibit abnormally high ionic conductivity. This was observed by Nernst[203] as early as 1899. The electrolyte was used in the first solid oxide electrolyte fuel cell constructed by Bauer and Preis.[204] Except for the fact that they have much higher conductivity (Fig. 1.23), their properties are very similar to those of calcia stabilized zirconia.

Several authors have investigated the phase diagram of the system,[185, 205–210] a recent version of which is shown in Fig. 1.24. The most recent investigation[207] gives the lower and upper limits of the cubic fluorite phase at 13 and 68 mol % $YO_{1.5}$, respectively. The similarity between the fluorite structure and the cubic *C*-type rare earth structure of Y_2O_3 partly accounts for the high solubility of Y_2O_3 in cubic ZrO_2. The possible existence of the pyrochlore-type compound $Y_2Zr_2O_7$ has been suggested by several investigators[207,211,212] but could not be proved by neutron diffraction. However, an ordered phase of $Y_4Zr_3O_{12}$ has recently been reported.[213] The lattice parameter of the solid solutions has also been measured by many workers.[185,205,206,208,214,215] Their values vary from 1.530 to 1.545 Å for a 10 mol % Y_2O_3 solid solution.

Similar to ZrO_2–CaO, the highest conductivity in this system also occurs near the lower limit of the cubic phase region. In practice ZrO_2–9-mol %-Y_2O_3 has been regarded as having the highest conductivity in the system. Although addition of 9 mol % Y_2O_3 produces only 4.1% anion vacancies

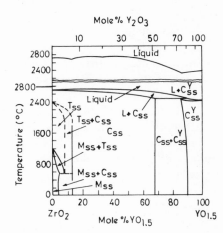

FIG. 1.24. Revised phase diagram for the system ZrO_2–$YO_{1.5}$.[207] Liquidus curve due to Noguchi *et al.*[210] is shown above the broken line.

compared to 6% in 12 mol % CaO, the conductivity of the former is almost twice that of the latter at 1000°C. Weaker defect interaction is believed to be the main reason for high anion mobility in these solid solutions. Compared to CSZ a lower activation energy of 18–25 kcal/mol was measured in these electrolytes. A linear Arrhenius plot has always been observed by various workers[185,187,216,217] throughout the temperature range investigated. Measurements of transport properties show very similar behavior ($t_{ion} > 0.99$) to that of CSZ[185,208,274,218,219] although few of the investigators observed electronic conductivity at lower oxygen partial pressures.[220,221]

One of the main drawbacks of these electrolytes is their aging behavior on prolonged heating around 1000°C with a consequent decrease in conductivity. According to Markin et al.[222], the resistivity of an 8 mol % Y_2O_3 specimen rose from 18 ohm cm to 32 ohm cm on annealing in hydrogen at 1000°C for 3 days and thereafter remained constant up to 11 days. This is again believed to be due to defect ordering and possible formation of a pyrochlore phase. This phenomenon was earlier studied by Baukal.[223]

One of the recent applications of ZrO_2–Y_2O_3 electrolytes is in the high-temperature electrolytic dissociation of water vapor into hydrogen and oxygen in which the electrolyte is required to conduct both oxygen ions and electrons. Attempts have been made to introduce electronic conductivity by doping ZrO_2–Y_2O_3 electrolytes by electronic conductors such as CeO_2, Mn_2O_3, ZnO, Cr_2O_3, Fe_3O_4, etc.[224] It is found that CeO_2–Y_2O_3–ZrO_2 and Cr_2O_3–Y_2O_3–ZrO_2 are the most promising systems for this purpose.

Miscellaneous ZrO_2-Base Electrolytes.   Several authors have studied the phase diagram for the ZrO_2–MgO system.[184,226,227] Cubic solid solutions are stable only at high temperatures and decompose to a tetragonal ZrO_2 solid solution and MgO at temperatures lower than 1300–1400°C.[184,226–229]

These solid solutions are ionic conductors. At 1000°C conductivities of 2.0×10^{-2}, 4.0×10^{-2}, and 3.4×10^{-2} (15 mol % MgO)[193,230,231] and 3.8×10^{-2} ohm^{-1} cm^{-1} (20 mol % MgO)[193] have been measured, while activation energies of 19.6 and 33.7 kcal/mol (References 193 and 230, respectively) for ZrO_2 + 15 mol % MgO have been reported. Determination of the ionic and electronic transport numbers indicates that the p-type conductivity is absent at 1500 and 1600°C, but n-type conductivity readily occurs under reducing conditions.

The existence of the cubic fluorite phase in the system ZrO_2–La_2O_3 had been reported by many authors,[232–234] but Roth[235] showed that the La^{3+} ion cannot form fluorite solid solution due to its large ionic radius (1.21 Å). Later authors[235–237] verified the existence of the compound $La_2Zr_2O_7$ with cubic pyrochlore structure.

Studies of transport properties[238–240] show that the values of conductivities for this system are much lower than the earlier described ones. Some typical values at 1000°C are 4.4×10^{-3} ohm^{-1} cm^{-1} (5 mol % La_2O_3),

2.5×10^{-3} and 1.1×10^{-3} ohm^{-1} cm^{-1} (10 mol % La$_2$O$_3$),[208,232] and 2.5×10^{-3}, 1.1×10^{-4}, and 3.0×10^{-3} ohm^{-1} cm^{-1} (15 mol % La$_2$O$_3$).[208] Such low values of conductivities make these compositions unsuitable as solid electrolytes. The reason is that the anion vacancies are ordered in the pyrochlore structure. The activation energies for a ZrO$_2$ + 13 mol % La$_2$O$_3$ composition were 15.9, 34.6, and 54.9 kcal/mol at 1200–1400, 1400–1525, and 1525–1650°C, respectively.

Several studies have been made on the ZrO$_2$–Sc$_2$O$_3$ system,[205,208,241–244] the most recent one being that by Ruh *et al.*[245] Due to the small radius of the Sc^{3+} ion (0.81 Å) a distorted fluorite phase appears in this system. This phase has been placed in different composition ranges by different authors. More work is needed to define the range and stability of phases in this system.

Measurements of electrical conductivity[187,208,242] indicate that these materials have much higher conductivity compared to most of the common oxide electrolytes (Fig. 1.23). But they could not be used in practice due to the very high cost of Sc$_2$O$_3$ and extreme thermal instability of solid solutions forming more ordered phases like Sc$_2$Zr$_7$O$_{17}$ on high-temperature aging. Table 1.2 presents the conductivity data available on ZrO$_2$-rare-earth-oxide electrolytes.

Figure 1.25 shows the variation of isothermal conductivity vs. ionic radius for different ZrO$_2$–M$_2$O$_3$ systems. It may be seen that the conductivity generally decreases with an increase in the radius of the dopant cation.

Table 1.2. Ionic Conductivities (ohm^{-1} cm^{-1} $\times$ 10^2) of ZrO$_2$–M$_2$O$_3$ Electrolytes at 1000°C (Activation Energies, kcal/mol, in Parentheses) (Reference 180)

	Composition, mol % M$_2$O$_3$			
M$_2$O$_3$	8	9	10	15
Nd$_2$O$_3$			0.60	
				1.7
				3.8
				1.4 (24.9)
Sm$_2$O$_3$			5.8 (22.0)	2.3 (26.1)
Gd$_2$O$_3$	11		11	3.1
Yb$_2$O$_3$		15		3.2
	8.8 (17.3)		11 (19.6)	3.9 (26.1)
				4.9
Lu$_2$O$_3$	1.5			1.2
Sc$_2$O$_3$			24	13
	25 (15.2)		25 (14.9)	15 (15.8)
	1.1			0.84

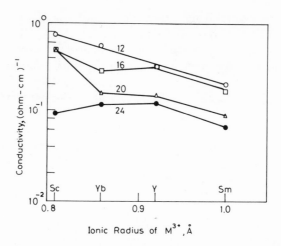

FIG. 1.25. Conductivity at 1300°C vs. ionic radius for the system M_2O_3–ZrO_2. Numbers on curves denote mol % M_2O_3 (Reference 208).

Few attempts have been made to determine the transport number in these systems. However, more systematic investigation is needed in this direction.

4.1.2. *Thoria and Thoria-Base Materials*

Unlike ZrO_2, pure thorium oxide has the cubic CaF_2-type structure (Fig. 1.2a) up to its melting point, therefore does not need any stabilization. "Pure" ThO_2 exhibits mixed conduction, as was observed by Danforth and Bodine[246,247] from their polarization measurements. Electrical conductivity of "pure" thoria as a function of temperature and oxygen partial pressure has been measured and activation energy calculated by a number of investigators.[13,18,19,22,23,27,83,85,248] It exhibits positive hole conductivity at higher oxygen partial pressures, while a predominantly ionic conduction takes place below about 10^{-10} atm of oxygen pressure. The range of ionic conduction decreases with increasing temperature, and the *n*-type conduction has been detected only above 1600°C.[83] Determination of the defect structure of pure nonstoichiometric thoria has been attempted by Lasker and Rapp[17] and later by Bransky and Tallan[83] on the basis of their electrical conductivity measurements. In the higher partial pressure range, the presence of interstitial oxygen ion (O_i'') indicates a 1/6 pressure dependence of total conductivity. However, experimentally, either 1/4 or 1/5 dependence has been observed (Fig. 1.26). The discrepancy is due to the presence of some unexpected impurities resulting in an extrinsic behavior. Bransky and Tallan[83] suggested that a 1/6 dependence could only be obtained if $\sigma_T/\sigma_{ion} \gg 10^3$, a condition which may be satisfied

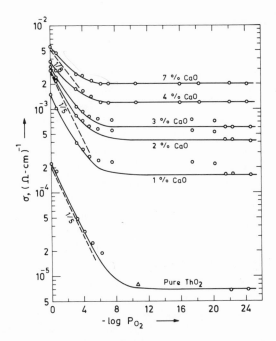

FIG. 1.26. Electrical conductivity of ThO_2–CaO solid solutions as a function of oxygen partial pressure at 1000°C (Reference 254).

in a much purer sample. Thus with the purest ThO_2 samples currently available there is no possibility of observing the nonstoichiometric behavior of pure thoria.

The ionic transference number of nominally pure thorium dioxide has been measured by cell emf techniques.[17,22] Thorium dioxide exhibits predominantly ionic conductivity in the pressure range of 10^{-10}–10^{-20} atm at 1000°C. Measurement of the Seebeck coefficient of pure thoria has been done by Tallan and Bransky[134] (see Section 3.6).

Addition of impurities greatly increases the defect concentration in pure ThO_2. The most important thoria-base electrolyte, which has found wide application, is ThO_2–Y_2O_3 solid solution. Detailed investigation on the electrolytic properties of this material has been done and the solubility limits in the ThO_2–Y_2O_3 have been determined by the x-ray diffraction technique.[20,22] On the ThO_2 rich side, the cubic fluorite-type solutions were found up to 40–50 mol % $YO_{1.5}$ at 2200°C. The limit of solubility decreases to 20–25 mol % $YO_{1.5}$ at 1400°C. The defect structure of the ThO_2 solid solution has been determined by density measurement[19,20,22,249] (Section 3.1). Measurement of electrical conductivity indicates that the oxygen ion vacancies are the mobile defects[22,249] and the conductivity is exclusively anionic over a wide range of oxygen activity and temperature.[13,17] The p-type conductivity is observed only above 10^{-6} atm of oxygen partial pressure. As expected the conductivity and the P_{O_2} range of ionic conduction increase with increasing $YO_{1.5}$ content of the solid solution. The highest conductivity is obtained at 15 mol % $YO_{1.5}$ or 3.75% anion vacancies. At higher P_{O_2} the p-type semiconduction in ThO_2–Y_2O_3 electrolytes is proportional to $P_{O_2}^{1/4}$. As has been mentioned earlier, the conductivity values of ThO_2-base materials are relatively lower than those of ZrO_2-base electrolytes (Figs. 1.3 and 1.23). At higher dopant concentration the lowering of the conductivity value is attributed to vacancy ordering[48,208] or dopant–vacancy interaction.[26] The activation energy for ionic conduction varies between 23 and 28 kcal/mol. These electrolytes do not show any significant electronic conduction at oxygen potentials as low as those corresponding to Cr–Cr_2O_3 equilibrium. Ionic transference numbers of the solid solutions have been calculated from the conductivity data[17] and also from the galvanic cell emf measurement[22] at different oxygen partial pressures. The results are shown in Fig. 1.27 along with that for pure thoria. Partial electronic and hole conductivities of ThO_2–15-mol %-$YO_{1.5}$ electrolyte have been measured by using Wagner's polarization technique.[100] The ionic transference number of this particular electrolyte remains more than 0.99 down to a P_{O_2} value of $10^{-34.3}$ atm at 1000°C, $10^{-39.5}$ atm at 900°C, and $10^{-44.7}$ atm at 800°C. The electrolytic domain of ThO_2–15-mol %-$YO_{1.5}$ (YDT) has been determined[69,250] and the results are shown in Fig. 1.8 (see also Fig. 2.1).

Addition of divalent metal oxides such as CaO also produces oxygen ion vacancies in ThO_2. The limited conductivity measurement[13,85,183,251–253] indicates that these electrolytes are in general relatively less conductive than ThO_2–Y_2O_3 solid solutions (Figs. 1.3 and 1.23). Recent measurements[85,254] show that the ionic conductivity of 7 mol % CaO at 1000°C is 2.0×10^{-3} ohm^{-1} cm^{-1} compared to 6.3×10^{-3} ohm^{-1} cm^{-1} for ThO_2–15-mol %-$YO_{1.5}$ observed by Lasker and Rapp.[17] These values are much higher than

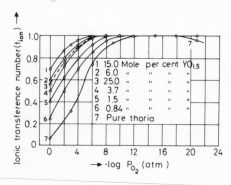

FIG. 1.27. Ionic transference number as a function of oxygen partial pressure at 1000°C (Reference 17).

those measured earlier by Steele and Alcock.[13] In all other respects, ThO_2–CaO electrolytes behave very similar to ThO_2–Y_2O_3 electrolytes. The activation energy for conduction varies between 1 and 1.4 eV (23–32 kcal/mol) above 1000°C. For all the compositions studied, higher activation energies were observed at lower temperatures (below 1000°C). Similar observations were made earlier in pure ThO_2[18,83] and ThO_2–Y_2O_3.[249] Thermoelectric power of ThO_2–CaO electrolytes has been measured in the mixed conduction region and the values were found to vary with transference numbers.[255]

Conductivity data on La_2O_3-doped ThO_2 electrolytes are available to a limited extent and it is found that the behavior is quite similar to that of ThO_2–Y_2O_3.[13,19,21,183,256] However, the conductivity is about 10% greater than that of ThO_2–Y_2O_3. The presence of oxygen ion vacancies has also been detected in ThO_2–Gd_2O_3 and ThO_2–Yb_2O_3 systems.[257] However, these materials could not be promising solid electrolytes due to their high hole conductivity in the useful partial pressure range. Hund[258] and Roy and Roy[259] earlier reviewed various other types of solid solutions formed between fluorite-type oxides and aliovalent impurities.

4.1.3. Ceria and Bismuth-Oxide-Base Materials

Pure CeO_2 also crystallizes with a fluorite structure. However, unlike ZrO_2 and ThO_2 it exhibits large deviations from stoichiometry. It is an oxygen deficit semiconductor with chemical formula CeO_{2-x}, where x may be as large as 0.3 depending on temperature and oxygen partial pressure. Accordingly the fluorite phase is stable over a wide range of temperature and oxygen partial pressure.[260–262] It is believed to be an n-type semiconductor in which oxygen vacancies ($V_O^{\cdot\cdot}$) and electrons trapped at the tetravalent cerium sites (Ce'_{Ce}) are the predominant defects.[263–271] Therefore pure CeO_2 is not useful as a solid electrolyte. However, addition of either CaO or Y_2O_3 to CeO_2 greatly increases the concentration of oxygen vacancies with a consequent decrease in the electronic defects resulting in good solid electrolytes. The electrical properties of CeO_2–CaO electrolytes have been investigated by Blumenthal and co-workers,[272,273] while Tuller and Nowick[274] have measured that of the CeO_2–Y_2O_3 system. The solid solubility in the CeO_2–Y_2O_3 system was earlier studied by Bevan et al.[275] It is interesting to note that both these electrolytes exhibit much higher conductivity than either ZrO_2–CaO or ZrO_2–Y_2O_3 electrolytes, CeO_2–Y_2O_3 having the highest (Fig. 1.23). The solubility of CaO in CeO_2 is about 15 mol % at 1600°C. However, the maximum conductivity is obtained at 10 mol % CaO. In the case of Y_2O_3 although the solubility is more than 25 mol % $YO_{1.5}$, electrical properties of only 5 mol % Y_2O_3 composition have been investigated so far. The activation energy of 0.76 eV (17.5 kcal/mol) for 5 mol % Y_2O_3 solid solution is the lowest among the oxide electrolytes. The existence of oxygen vacancy as the predominant defect in CeO_2–CaO electrolytes has been confirmed by x-ray and density measurements.[273] The transference number for oxygen ion is very close to unity near 1 atm of P_{O_2}; however, it rapidly changes to an *n*-type conductor with decreasing P_{O_2}. As expected the effect is more significant at higher temperatures. Figure 1.28 shows the plots of ionic transference number against $\log P_{O_2}$ for CeO_2–5-mol %-Y_2O_3 electrolytes. The results indicate that these electrolytes would be more useful at low temperatures, especially below 800°C. Measurement of high-temperature dielectric proper-ties[276] and a thermodynamic study[277] of CaO-doped CeO_2 electrolytes have also been carried out as a part of the investigation of their electrolytic properties.

The electrical conductivity of CeO_2 electrolytes doped with various rare earth elements such as La, Nb, Sm, Eu, Gd, Dy, Ho, Er, and Yb has also been studied recently and predominant oxygen ion conduction has been observed in all cases.[278] CeO_2–Nd_2O_3 solid solution is found to have the highest conductivity among these electrolytes. A systematic variation of conductivity with lattice parameter is observed. The conductivity is maximum when the lattice constant is around 5.46 Å. Sc_2O_3 does not form a fluorite lattice with CeO_2 due to the large difference in their ionic radii.

Besides the above oxygen ion conductors, some recent investigations made by Takahashi and co-workers[279–286] show that δ-Bi_2O_3, which has a distorted fluorite structure and is stable only above 730°C, is an efficient oxygen ion conductor with conductivity few orders of magnitude higher than that of ZrO_2 electrolytes. Figure 1.23 shows that it has the highest con-ductivity among all the oxygen ion conductors. However, they are not prac-tically useful due to their thermodynamic instability at room temperature. Several attempts have been made by Takahashi and co-workers to stabilize the fluorite phase by adding various oxides such as CaO, SrO, Y_2O_3, WO_3, MoO_3, Cr_2O_3, V_2O_5, Nb_2O_5, etc. Amongst them, Y_2O_3 and WO_3 have been

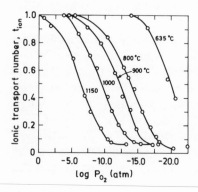

FIG. 1.28. Ionic transference number of CeO_2–5-mol %-Y_2O_3 as a function of P_{O_2} (Reference 274).

found promising since the transformation temperatures has been brought down to below 400°C.

4.2. β-Alumina-Type Oxides

β-alumina was originally thought to be a polymorphic form of alumina.[287] Since Na^+ (or other mono- or divalent) ions are always present in β-alumina structure, that name is now known to be a misnomer. Comprehensive reviews of these materials are presented by Kummer,[11] Kennedy,[288] and Collongues *et al.*[290]

The composition of β-alumina corresponds to $Na_2O \cdot 11Al_2O_3$ but varies up to $Na_2O \cdot 9Al_2O_3$ due to the presence of 15%–30% excess Na_2O relative to the ideal formula. The related β″-alumina is $Na_2O \cdot 5Al_2O_3$. In β″-Al_2O_3, Al^{3+} ions can be partially replaced by Mg^{2+} ions and charge balance achieved by the addition of extra Na^+ ions. β′, β‴, and β⁗[291] varieties are also mentioned in the literature but are not firmly established. The crystal structure of β-Al_2O_3 originally studied some 30 years ago[292-294] was refined recently by Felsche,[295] Peters *et al.*,[296] and Roth,[45] among others.[297-299] The crystal structure of β″-Al_2O_3 ($Na_2O \cdot 5Al_2O_3$–$Na_2O \cdot 7Al_2O_3$) and its stabilized version ($Na_2O \cdot MgO \cdot 5Al_2O_3$) was also established.[291,300-303] Some of the properties of these materials are listed in Table 1.3 and the crystal structures are shown in Figs. 1.29, 1.30.

The crystal structure of hexagonal β-alumina consists of four cubic

Table 1.3. Properties of β- and Stabilized β″-Alumina

Property	β-alumina $Na_2O \cdot 11Al_2O_3$[a]	Stabilized β″-Al_2O_3 $Na_2O \cdot MgO \cdot 5Al_2O_3$[b]
Melting point (°C)	2000°C	
Density (g/cm³)	3.25	
Index of refraction, ε	1.635–1.650	
Index of refraction, ω	1.676	
Coefficient of expansion (10^{-6}/°C):		
a axis	7.7	7.8
c axis	5.7	7.3
% Na_2O, theoretical	5.24	
% Na_2O, actual	6.2–6.8	8.10
% MgO, actual	—	5.00
Hexagonal lattice constants (Å):		
a axis	5.594	5.622
c axis	22.530	33.510

[a] Reference 11.
[b] Reference 304.

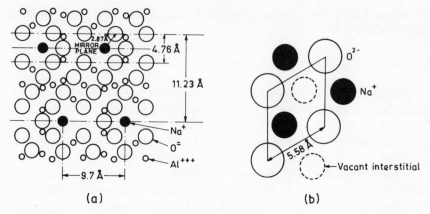

FIG. 1.29. (a) Arrangement of ions on (110) plane of β-alumina crystal (Reference 303). (b) Arrangement of ions in the [NaO] layer (Reference 303).

close-packed layers of oxygens with $3Al^{3+}$ ions occupying some of the resulting octahedral and tetrahedral positions between each pair of oxygen layers, giving a typical composition of Al_3O_4. This arrangement is the same as in the spinel, $MgAl_2O_4$. These spinel-like blocks are separated by basal planes containing a loose packing of Na and O ions, the spacing between these mirror

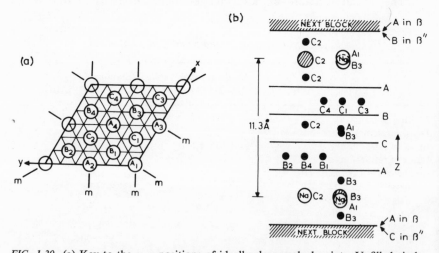

FIG. 1.30. (a) Key to the x, y positions of ideally close-packed points. Unfilled circles indicate aBR, C indicates MO, and B indicates BR. Vertical mirror planes are indicated by short lines marked m (Reference 303). (b) Schematic projection of β- or β″-alumina on to the b_0–c_0 plane. Solid lines indicate the oxygen layers. Aluminums are small solid circles. Oxygens are large hatched circles. The placement of sodiums holds for the β″-alumina structure only (Reference 303).

planes being 11.2 Å. The close-packed oxygen layers above and below the Na–O planes are mirror images of each other, 4.76 Å apart, and are bound together not only by the Na ions but also by Al–O–Al columns, one of which occurs for each Na^+ ion. The linkage of the spinel-like blocks can alternatively be described as two AlO_4 tetrahedra with a common oxygen vertex in the mirror plane, the two sets of three basal oxygen atoms being part of each spinel block. The nearest Na–O distance, 2.87 Å, is much larger than the sum of their ionic radii, 2.35 Å. This fact, coupled with the large number of the available but unoccupied sites for Na^+ ions (Fig. 1.29b),[305] is responsible for the large Na^+ ion diffusion and ionic conductivity in the plane perpendicular to the *c* axis.[44,306–311] It also explains the variability of the Na content of β-alumina. The actual number and position [Beevers–Ross(BR), mid-oxygen (MO), and anti-Beevers–Ross(aBR)] of Na^+ ions varies from unit cell to cell (sometimes due to differences in crystal growth[298,312]) and the electron density due to Na^+ ions appears smeared out. The Debye–Waller temperature factors for Na^+ ions are anisotropic and anomalously large.[296] The Na^+ ions are located in the Beevers–Ross positions, whereas Ag^+ ions are in the anti-Beevers–Ross positions.[45] The excess Na^+ ions have to be compensated by counterion defects, which may be Al^{3+} ion vacancies[296] or oxygen interstitials.[45] The possibility of Al^{3+} ion interstitials, tetrahedrally coordinated by oxygen ions located in mid-oxygen positions, among others, is also suggested.[298,299] The location of Na^+ ions in the conducting plane is elucidated by studying other β-aluminas containing heavier ions, in particular, Ag^+ ions.[45] The distribution of Na^+ ions in the various sites determined at room temperature becomes increasingly disordered with increasing temperature until it simulates a two-dimensional liquid.[158,290,320,346] Harata[313] reported on the lattice parameters of $(1.16 + x)$ $Na_2O \cdot 11Al_2O_3$ with $0.19 < x < 0.59$. Yao and Kummer,[44] among others, have shown that Na^+ in β-Al_2O_3 can be completely exchanged by Li^+, Ag^+, Tl^+, K^+, Rb^+, NH_4^+, In^+, NO^+, Ga^+, Cu^+, and H_3O^+, and partially by Cs^+, by treatment with appropriate nitrate melts at 300–350°C or in sulfuric acid. Such ion exchange leads to small changes in lattice parameters, particularly *c*. Lattice expansion due to ion exchange (e.g., K^+ for Na^+) can cause fracture of polycrystalline samples. MgO is soluble in β-Al_2O_3 up to about 2.5% at 1700°C.[11] Replacement of Al^{3+} by Mg^{2+} (or Ni^{2+}, Zn^{2+}, or Cu^{2+}) ion in the spinel block is accompanied by increased Na^+ in the Na–O layer (e.g., for 2% of MgO, 7.6% Na_2O instead of 5.24%) and decreased resistivity (125 ohm cm, which is approximately an order of magnitude less than for "pure" β-Al_2O_3).[305,314,315] Water enters the β-Al_2O_3 lattice easily and can be detected by infrared and NMR techniques.[11,316] The desorption of water on heating fine powder of β-Al_2O_3 in dry nitrogen was followed by NMR and consists of three steps with 75% loss at 100–150°C, 7% at 150–200°C, and balance between 200 and 500°C.[316]

Other techniques employed for the detailed structural study of β- and

β''-Al$_2$O$_3$ are neutron diffraction,[317] neutron diffuse scattering,[318,319] x-ray diffuse scattering,[158,320] infrared[321] and Raman spectroscopy,[152] electron microscopy,[322,323] besides NMR[148,324–326] and light scattering.[330]

β''-alumina decomposes to β-Al$_2$O$_3$ and sodium aluminate (NaAlO$_2$) at temperatures beyond 1500°C (Figure 1.31).[287,327] However, introduction of divalent cations such as Mg^{2+} in place of Al^{3+} ions in the spinel lattice was found to stabilize the β'' phase up to at least 1700°C so that sintered material and single crystals can be prepared. The true symmetry of β''-Al$_2$O$_3$ is rhombohedral, but is described in terms of a hexagonal cell to facilitate comparison with β-Al$_2$O$_3$. The crystal structure of β'' is similar to that of β-Al$_2$O$_3$ (Fig. 1.30) except for an increase of the c axis by 50% (Table 1.3). β'' has three 11-Å spinel blocks related by a threefold screw axis, while β has two 11-Å spinel blocks related by a twofold screw axis parallel to the c axis. Therefore the oxygen planes enclosing the Na$^+$ ions in β''-Al$_2$O$_3$ are staggered, while they are mirror planes in β-Al$_2$O$_3$. In β''-Al$_2$O$_3$ there are two Na$^+$ sites per unit cell, located 2.57 Å from the central oxygen (apex of the tetrahedron) and 2.69 Å from the three oxygens in the top or bottom layer. On the other hand, in β-Al$_2$O$_3$, the Na$^+$ ion site is located at the center of a triangular prism of six oxygen ions, so that a traveling Na$^+$ ion moves through a position (called the anti-Beevers and Ross position) which has oxygens above and below 2.38 Å away. Thus the interionic separation is comfortable for Na$^+$ as well as K$^+$ in β''-Al$_2$O$_3$, while it is just adequate for Na$^+$ in β-Al$_2$O$_3$. Partial replacement of Al^{3+} by Mg^{2+} in the spinel block (as in Na$_2$O·MgO·5Al$_2$O$_3$) requires extra Na$^+$ ions in the Na–O for maintaining charge neutrality. Due to the larger concentration of charge carriers (Na$^+$ ions) and the greater space available for its motion, the ionic conductivity of β'' is larger than

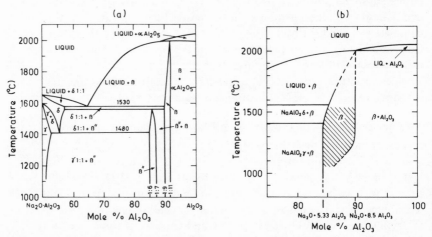

FIG. 1.31. Phase diagrams of Na$_2$O·Al$_2$O$_3$–Al$_2$O$_3$. (a) From Reference 287 and (b) from Reference 327.

that of β-Al$_2$O$_3$.[178] The absence of interstitial oxygen in the conductive plane of β''-Al$_2$O$_3$ aids its increased conductivity.

Radioactive tracer diffusion studies (e.g., ^{24}Na) in β-Al$_2$O$_3$ clearly demonstrated that Na$^+$ ions diffuse two dimensionally in the direction perpendicular to the c axis with very little, if at all, diffusion parallel to the c axis, i.e., through the close-packed oxygen layers in the spinel blocks. The self-diffusion coefficients for the various monovalent ions in β-Al$_2$O$_3$ are shown in Fig. 1.32. From the expression relating the diffusion coefficient D with temperature T ($^\circ$K)

$$D = D_0 \exp(-E_a/kT) \tag{1.61a}$$

where k is the Boltzmann constant; and the preexponential factor D_0 and the activation energy E_a are derived from Fig. 1.32 and listed in Table 1.4. Since the distance between oxygen layers through which the cation has to diffuse remains nearly the same in all these compounds (~ 4.76 Å), the repulsion between the diffusing cation and the outer electrons of the oxygen ions increases as the cation size increases. Consequently the activation energy E_a increases. The small Li$^+$ ion has an exceptional behavior in that it occupies

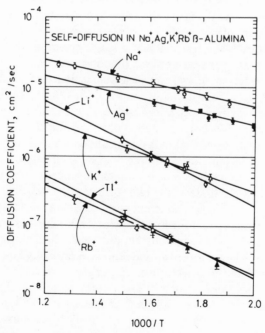

FIG. 1.32. Self-diffusion coefficients of various cations in substituted β-aluminas (Reference 44).

an asymmetric position (close to three oxygens and away from the other three) which necessitates additional energy to get it free from the three closest oxygen ions. The changes in electrical resistivity of these materials with hydrostatic pressure (increase in K^+, decrease in Li^+, and no change in Na^+ β-Al_2O_3) qualitatively support the above picture.[328] Detailed theoretical treatment of diffusion in the β-alumina is complicated by the presence of a large concentration of defects.[41,329] Ion exchange and diffusion data are not available on β''-Al_2O_3.

Measurement of ionic conductivity of superionic conductors such as β- and β''-Al_2O_3 require either reversible electrodes[29] (e.g., molten sodium,[11] molten $NaNO_3/NaNO_2$ mixture,[331] and tungsten bronzes[67]) or use of high enough frequencies where the conductivity is independent of frequency (e.g., ~ 1 MHz[44,332]). Whittingham and Huggins[178] have reported the temperature dependence of conductivity of Li, Na, K, Rb, and Ag β-Al_2O_3 over a wide temperature range (Fig. 1.33 and Table 1.4). It may be noted that Na β-alumina exhibits the highest conductivity and lowest activation energy among all the β-aluminas. The nearly same activation energy measured for electrical conductivity and diffusion (Table 1.4) strongly suggests that the mechanism is the same for both of these transport properties.

The ionic conductivity of single-crystal and polycrystalline form of MgO-stabilized β''-Al_2O_3 is higher than that of β-Al_2O_3, as expected from the structural differences cited above (Fig. 1.34). The data below 315°C appear to be dependent on frequency when reversible electrodes were used with hot-pressed β''-alumina.[67] The resistivity of a single crystal in the c direction is 100–1000 times that in the basal plane (about 6 ohm cm at 300°C). Values intermediate between these two are observed in the case of sintered poly-crystalline β-Al_2O_3 (about 25 ohm cm at 300°C).[331] Mixtures of β- and β''-Al_2O_3 have resistivity values intermediate between those of the pure compounds (16 ohm cm for β-Al_2O_3 and 3 ohm cm for β''-Al_2O_3 at 350°C). Whittingham has suggested that in fast ion conductors such as β-alumina, the grain boundary diffusion may be slower than bulk diffusion, unlike in other materials.[333] The conductivity was not much dependent upon the grain size of β''-Al_2O_3.[331] A 4-electrode ac method for measuring impedance as a function of frequency at various fixed temperatures was employed by Powers and Mitoff,[118] who observed that the low-frequency value corresponds to the sum of the bulk and grain boundary contributions, while the high-frequency value is that of the bulk. They obtained an activation energy of 4.4 kcal/mol for bulk conduction (compared to 3.8 kcal/mol for single crystals) and 6.6 kcal/mol for grain boundary conduction.[344] The curvature in the $\log(\sigma T)$ vs. $1/T$ plots for β- and β''-Al_2O_3 single-crystal and polycrystalline form are noteworthy (Fig. 1.34). Incorporation of divalent cations for Al^{3+} in β-Al_2O_3 generally increases the conductivity, the most effective species being Mg^{2+}.

Table 1.4. Transport Properties of β-Alumina

Ion	D_0 (cm²/sec)	$\sigma_0 T$ (ohm⁻¹ cm⁻¹) °K	σ, 25°C (ohm⁻¹ cm⁻¹)	Activation energy (kcal/mol)			Me⁺–O distance (Å)		r_{Me^+} + $r_{O^{2-}}$ (Å)
				Diffusion	Conductivity	Dielectric loss	β-Al₂O₃	Me₂O	
Na	2.4 × 10⁻⁴	2.4 × 10³	1.4 × 10⁻²	3.81 (25–400°C)	3.78 (−150–820°C)	3.70 (85–126°K)	2.87	2.40	2.35
K	0.78 × 10⁻⁴	1.5 × 10³	6.5 × 10⁻⁵	5.36 (200–400°C)	6.78 (−70–820°C)	8.48 (170–230°K) 12.10 (247–303°K)	2.91	2.78	2.73
Rb	0.34 × 10⁻⁴	—	—	7.18 (200–400°C)	—		2.94	2.91	2.88
Tl	0.65 × 10⁻⁴	6.8 × 10²	2.2 × 10⁻⁶	8.22 (200–400°C)	8.18 (−20–800°C)	7.38 (170–220°K)	—	—	2.84
Ag	1.65 × 10⁻⁴	1.64 × 10³	6.7 × 10⁻³	4.05 (25–400°C)	3.94 (25–800°C)	3.78 (85–126°K)	2.86	2.05	2.66
Li	14.5 × 10⁻⁴	9.7 × 10³	—	8.71 (200–400°C)	8.55 (180–800°C)	—			
Li	—	5.4 × 10¹ (−100–180°C)	1.34 × 10⁻⁴	—	4.3 (−100–180°C)	8.65 (280–340°K)	2.88	2.0	2.0

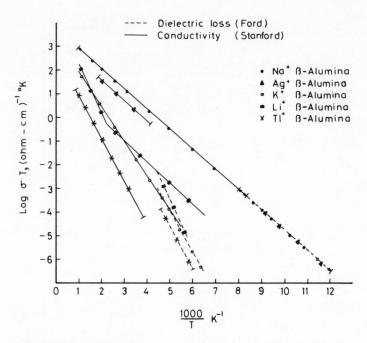

FIG. 1.33. Temperature dependence of conductivity and dielectric loss of various β-aluminas (adopted from Reference 178).

The β''-Al$_2$O$_3$ content and therefore the conductivity of a hot-pressed or sintered β-Al$_2$O$_3$ ceramic can be increased by annealing the sample at 1350–1500°C. The presence of Mg^{2+} ions in β''-Al$_2$O$_3$ enables incorporation of additional Na$^+$ ions for charge compensation, may remove blocking interstitial O^{2-} ions, and remove Al^{3+} vacancies which attract Na$^+$ ions.[288] It may be mentioned that the gallium and iron analogs of β-Al$_2$O$_3$ are also investigated.[288] The role of various ionic additions to β-alumina on the relative proportion of β- and β''-alumina and on the electrical conductivity has also been studied.[342,345] The location of a particular foreign ion in the lattice depends on its size, charge, method of incorporation, and the heat treatment given.[341–343]

Dielectric relaxation studies[334] on β-aluminas containing various monovalent ions were carried out, which yield an activation energy in good agreement with that for conduction and diffusion (Table 1.4).[335]

The mechanism for ionic conductivity (interstitialcy or vacancy) is not unequivocally established, nor the value of the correlation factor (or Haven ratio) in the Nernst–Einstein equation relating ionic conductivity and diffusion in these materials.[61,68,338–340] Whittingham and Huggins[68] have shown that the ionic conductivity of Ag β-Al$_2$O$_3$ between 600 and 800°C is independent

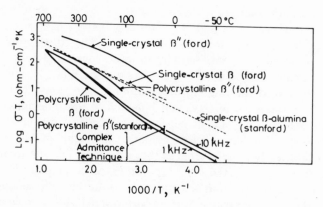

FIG. 1.34. Variation of electrical conductivity with temperature in single and polycrystalline β- and β″-aluminas. (Adapted from References 67 and 11.)

of oxygen partial pressure in the range $0.2-10^{-24}$ atm, and that $t_e \approx 10^{-6}$ at 600°C. The method of Hsueh and Bennion[336] was improved by Galli *et al.*[337] to measure the t_e of β-Al_2O_3 by using a cell Hg-$Na(x_2)/\beta$-Al_2O_3/Hg-Na/x and comparing the measured and calculated emf values. For hot-pressed pure β-Al_2O_3 of composition $Na_2O \cdot 8.3Al_2O_3$, they obtained $(1 - t_e) = 1.0048 \pm 0.0037$, whereas for a MgO-doped sample, it is 0.9989 ± 0.0054. For β-Al_2O_3 doped with CoO, Fe_2O_3, or $CoO + TiO_2$ also the value is nearly the same. Further, the ionic conductivity of β-Al_2O_3 is influenced much by these additives.

4.3. Silver-Iodide-Type Materials

Among the silver halides, AgCl and AgBr crystallize with the NaCl structure and contain cationic Frenkel pairs, i.e., Ag^+ ion interstitial and Ag^+ ion vacancy. Early measurements by Tubandt *et al.*[347] established that conduction in these materials takes place by migration of Ag^+ ions by interstitialcy mechanism. Addition of cation (e.g., Cd^{2+}) or anion (e.g., S^{2-}, Se^{2-}, or Te^{2-}) impurities enhances the conductivity of these materials but not beyond 10^{-5} ohm^{-1} cm^{-1}.

AgI exists in three crystalline forms:[6,348–352] two at room temperature depending upon the method of preparation—face centered cubic γ and hexagonal β, and a third from 147°C to the melting point at 555°C—body centered cubic α. β-AgI transforms to α-AgI at 147°C, accompanied by about 1000-fold increase in ionic conductivity (Fig. 1.35). The α–β transition is lowered to 100°C by adding 5 mol % PbI_2. In α-AgI, the I^- ions occupy the regular positions of the bcc lattice (Fig. 1.2b). According to powder x-ray diffraction studies, the two Ag ions are randomly distributed among the large number of available sites: 12 tetrahedral sites, 24 shared faces of tetrahedra, and 6 octahedral sites. The high Ag^+ ion conductivity is attributed to the quasimolten state of Ag^+ ions in the structure. The equal probability of occupancy of the three types of sites is, however, considered unlikely due to differences in coordination number, size, and energy. While Hoshino[353] and Burley[352] questioned the random distribution of Ag^+ ions, it is only the recent single-crystal study of Cava and Wuensch[354] which demonstrated the nonrandom nature of Ag^+ ion arrangement and the asymmetric smearing of electron densities. The latter fact suggests that the highly mobile ions move about in a sufficiently large volume as they pass from one site to another.

The cation disorder in the α-AgI-type lattice can be increased by reducing the number of cations per unit cell and, as a result, the ionic conductivity is enhanced. The transition temperature is also lowered. An example of such a material[355] is $HgAg_2I_4$.[8,356,357] The ionic conductivity is 3×10^{-2} ohm^{-1} cm^{-1} and electronic conductivity $\approx 10^{-4}$ ohm^{-1} cm^{-1} at 80°C[174,361] (Fig. 1.35). The crystal structure has been studied, including with single crystals, showing that the cations occupy the tetrahedral sites corresponding to zinc blende type of structure and that they migrate from one tetrahedral site to another via an octahedral site.[8,356-359]

KAg_4I_5 and $RbAg_4I_5$ were synthesized and electrical behavior evaluated by Bradley and Greene[175,362] and Owens and Argue.[363] KAg_4I_5 is stable between 38°C and its melting point, 253°C; below 38°C it disproportionates at a sluggish rate into β-AgI and K_2AgI_3. $RbAg_4I_5$ is stable up to its incongruent melting point of 228°C. The β–α transformation at -136°C for KAg_4I_5 and at -155°C for $RbAg_4I_5$ was detected in the conductivity–temperature data (Fig. 1.35). The conductivity cell consists of Ag foil/Ag, $RbAg_4I_5$/electrolyte/ Ag, $RbAg_4I_5$/Ag foil. Room temperature conductivity is 0.26 ohm^{-1} cm^{-1} (which is the highest value observed for any solid electrolyte) with an activation energy of 1.7 kcal/mol. Electronic conductivity is estimated at 10^{-11} ohm^{-1} cm^{-1} (reference 335). Single crystals show an isotropic conductivity of the same magnitude as polycrystalline samples.[364] The self-diffusion coefficient D_c of Ag in $RbAg_4I_5$ is 2.6×10^{-6} at 25°C and 7.6×10^{-6} cm^2/sec at 150°C with an activation energy of 2.0 kcal/mol. An additional reversible transformation was detected at -65°C by x-ray diffraction[365] and coulometry with only a change in slope of the $\sigma(T)$ plot, and transitions under high pressures were also studied.[355,366] The room temperature conductivity values varied from 0.124 to 0.26 ohm^{-1} cm^{-1} depending upon the electrodes used and the preparation of the samples.[175,355,362-365,368-371] Reaction of the electrode with the electrolyte and decomposition of the electrolyte under pressure are identified as responsible for the observed differences. It may be noted that there is no jump in conductivity at the melting point of these compounds showing that the crystal structure of these compounds is in a very disordered state.

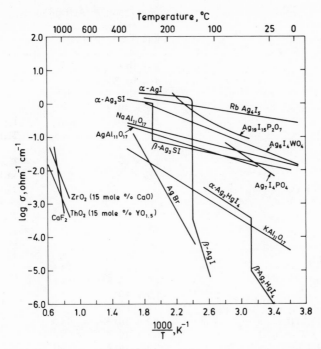

FIG. 1.35. Electrical conductivity vs. $1/T$ for various solid electrolytes.

The crystal structure of $RbAg_4I_5$ is similar to that of β–Mn.[175,362,365] It is cubic with $a = 11.24$ Å and has four $RbAg_4I_5$ per unit cell. The x-ray density of 5.384 g/cm^3 may be compared with the measured value of 5.30 g/cm^3. The 16 Ag^+ ions are located in 56 tetrahedral interstices of three types in a nonrandom fashion and 4 Rb^+ ions in a unit cell in distorted iodine octahedra. Ag^+ ions migrate across the shared faces of the tetrahedra. There are two channels per unit cell, running parallel to each of the three axes; these are the preferred diffusion paths for the Ag^+ ion.

A search for high conductivity materials stable at room temperature (and lower) led to the discovery of $Q\,Ag_6I_7$ compounds, where Q is a tetramethyl or a tetraethyl radical.[162,372] They have $\sigma = 2 \times 10^{-2}$ to 6×10^{-2} ohm^{-1} cm^{-1} at room temperature (Fig. 1.36). The hexagonal crystal structure of these compounds was studied.[373] Solid solutions are also possible (see Table 1.5). Combination of hydrocarbon-substituted ammonium iodides and partially substituted acyclic ammonium iodides with AgI did not lead to any conducting phases.[163] Polymethonium diiodide and silver iodide in the ratio 1:12 give good conductors[164] (Table 1.5). There is thus evidence to suggest that ionic volume alone does not determine which ions form the conductive

Table 1.5. *Electrical Conductivity of Ag Ion Conducting Solid Electrolytes*

Compound	σ at room temperature (ohm^{-1} cm^{-1})	Activation energy (kcal/mol)	Remarks[a]
AgCl	3×10^{-8}	8	
AgBr	4×10^{-9}	7	
AgI	2×10^{-6}	10	T_{Tr}, -147
$HgAg_2I_4$	1.5×10^{-6}	10.2	
$RbAg_4I_5$	2.6×10^{-1}	1.7	T_{Tr}, -155
			M.P., 232
			T_d, 25
KAg_4I_5	2.4×10^{-1}	—	T_{Tr}, -136
			M.P., 253
			T_d, 38
$NH_4Ag_4I_5$			M.P., 232
			T_d, 32
Ag_3SI	10^{-2}	3.3	T_{Tr}, -235
			M.P., 700
			T_d, 60
Ag_3SBr	2×10^{-3}	5.5	
$Ag_6I_4WO_4$	4.7×10^{-2}	3.7	
$Ag_7I_4PO_4$	1.9×10^{-2}	3.8	
$Ag_{19}I_{15}P_2O_7$	9×10^{-2}	3.3	
$6.5AgI \cdot (CH_3)_4NI$	4.2×10^{-2}	4.0	
$7AgI \cdot (CH_3)_2(C_2H_5)_2NI$	6.4×10^{-2}	3.6	
$7AgI \cdot (C_2H_5)_4NI$	2.2×10^{-2}	5.7	
$7AgI \cdot QI$ (Q = 1,1-dimethyl-pyrrolidinium)	6×10^{-2}		
$8AgI \cdot QI$ (Q = pyridinium)	4×10^{-2}		
$6.7AgI \cdot QI$ (Q = 4-methyl-4-azomiacyclohexane)	6×10^{-2}		
$(C_2H_5)_2CH_3N-CH_2-NCH_3(C_2H_5)_2I_2 \cdot 12AgI$	4.5×10^{-2}		
$(CH_3)_3N-(CH_2)_2-N(CH_3)_3I_2 \cdot 12AgI$	2.7×10^{-2}		
$(CH_3)N-(CH_2)_3-N(CH_3)_3I_2 \cdot 12AgI$	1.3×10^{-2}		
$(CH_3)N-(CH_2)_4-N(CH_3)_3I_2 \cdot 12AgI$	2.7×10^{-2}		
$(CH_3)N-(CH_2)_5-N(CH_3)_3I_2 \cdot 12AgI$	2.9×10^{-2}		
$(CH_3)N-(CH_2)_9-N(CH_3)_3I_2 \cdot 12AgI$	1.4×10^{-2}		
$(CH_3)N-(CH_2)_{10}-N(CH_3)_3I_2 \cdot 12AgI$	1.1×10^{-2}		
$KCN \cdot 4AgI$	1.4×10^{-1}	1.94	
$RbCN \cdot 4AgI$	1.8×10^{-1}		
$CsCN \cdot 4AgI$	9×10^{-4}		
$Ag_8HgS_2I_6$	1.47×10^{-1}	2.95	
$Ag_{1.85}Hg_{0.4}Te_{0.65}I_{1.35}$	9.4×10^{-2}	3.24	
$Ag_{1.8}Hg_{0.4}Te_{0.65}I_{1.35}$	1.0×10^{-1}	3.34	
$Ag_2Hg_{0.5}Se_{0.5}I_1$	4.3×10^{-2}	3.58	
$AgAl_{11}O_{17}$	6.4×10^{-3}	3.98	

[a] T_{Tr} = transformation temperature, °C; M.P. = melting point, °C; T_d = decomposition temperature, °C.

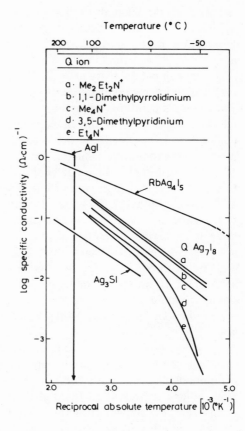

FIG. 1.36. Temperature dependence of specific conductivity of solid electrolytes of QAg_7I_8 type (Reference 162).

electrolyte. The crystal structure of tetramethyl-ammonium-iodide–silver-iodide $[(CH_3)_4N]Ag_{13}I_{15}$ and of $(C_5H_5NH)Ag_5I_6$ was determined by Geller[374,375] and was similar but more complicated than that of $RbAg_4I_5$. A break in the conductivity vs. $1/T$ plot at 50°C is attributed to a change in the degree of disorder. The cation-substituted silver iodides are generally characterized by face-sharing polyhedra formed by iodide ions, with the silver ions moving along the available pathways.[376] Anion substitutions (e.g., S, Se, etc.) are also attempted in some of the organic compounds exhibiting Ag^+ ion conduction.[382]

Ag_3SI[377,378] was found to have a crystal structure similar to that of AgI, with S^{2-} and I^- ions taking the place of I^- ions in AgI and the three Ag^+ ions randomly distributed over all the anion sites. It can also be visualized as an antiperovskite structure (Fig. 1.2c).[360] Equimolar mixtures of AgI and Ag_2S heated in a closed vessel at 550°C for 17 hr under sulfur vapor pressure yields Ag_3SI.[360] While the Wagner technique gave a value of 10^{-8} for t_e at room temperature, the Tubandt method gave $t_{ion} = 0.977$ and there was substantial difference in the thermodynamic and observed emf for the cell $Ag/Ag_3SI/I_2$. Dissolution of I_2 into Ag_3SI can generate holes, causing the above discrepancy.[355] Ag_3SI was found to decompose to AgI and S at about 60°C.[379]

The high conductivity of $Ag_6I_4WO_4$ at temperatures up to about 300°C makes it an interesting material.[166] The β–α transformation occurs at 235°C. The room temperature conductivity is 10^{-2} ohm^{-1} cm^{-1} with an ionic transference number of 0.997 and an electronic conductivity of 10^{-8} ohm^{-1} cm^{-1} (Table 1.5). Silver amalgam anode and a cathode of acetylene black-iodine mixture were found most suitable for cell measurements.

Takahashi *et al.*[165] investigated the AgI–silver-oxyacid systems and found two compounds of high conductivity ($Ag_7I_4PO_4$ and $Ag_{19}I_{15}P_2O_7$) (Table 1.5). Alkali cyanides (KCN and RbCN) in combination with AgI (in the ratio of 1:4) form interesting Ag^+ ion conductors[167] (Table 1.5). Phase diagrams in these systems were investigated using x-ray diffraction, differential thermal analysis, and electrical conductivity. An investigation of ternary systems AgI–HgI with Ag_2S, Ag_2Se, and Ag_2Te led to conducting compounds such as $Ag_8HgS_2I_6$, which is stable at least between 20 and 100°C.[169–171] The thermodynamic properties of the various AgI-type materials are recently reviewed by Owens.[380] The Ag-type solids have recently been reviewed extensively.[355,360,376,380–387]

Analogous to the silver compounds, copper compounds were obtained with high conductivity essentially due to copper ions.[176] Some typical double compounds of copper(I) halide with N-alkyl(or hydro)hexamethylene tetramine halide compounds are listed in Table 1.6 together with the conductivity at 20°C and the activation energy for conduction.

On the other hand, Scrosati[388] partially replaced silver in $RbAg_4I_5$ by copper using a cell in which $RbAg_4I_5$ is sandwiched between a pellet of Ag

Table 1.6. Some Typical Double Compounds of Copper(I) Halide with N-Alkyl(or Hydro)-hexamethylene Tetramine Compounds

Compound	$\sigma_{20°C}$ (ohm^{-1} cm^{-1})	Activation energy (kcal/mol)
$7CuBr \cdot C_6H_{12}N_4CH_3Br$	1.7×10^{-3}	4.1
$7CuBr \cdot C_6H_{12}N_4HBr$	4.6×10^{-3}	4.1
$7CuBr \cdot C_6H_{12}N_4C_2H_5Br$	3.0×10^{-4}	
$7CuCl \cdot C_6H_{12}N_4HCl$	3.2×10^{-3}	6.2
$17CuI \cdot 3C_6H_{12}N_4CH_3I$	1.1×10^{-3}	6.1

and $RbAg_4I_5$ and a pellet of Cu and $RbAg_4I_5$. The resulting material is made use of in a cell

$$Cu \mid Cu_xRbAg_{4-x}I_5 \mid I_2\text{-B} \qquad [1.3]$$

where I_2-B is iodine benzidine. An open-circuit voltage of 0.6 V was observed at room temperature. These results are interpreted as the presence of Cu^+ ions, which are mobile, in substituted $RbAg_4I_5$. The copper ion conductors are surveyed by Takahashi[382] and Matsui and Wagner.[389]

4.4. Fluorides

In the same year (1957) in which Kiukkola and Wagner[183] established the oxygen ion conductivity of doped ZrO_2 and ThO_2, Ure[34] demonstrated the purely fluorine ion conductivity in pure and doped CaF_2. Following this, fluorides of Mg, Ca, Sr, Ba, and Pb, which crystallize with the fluorite structure as well as NaF and cryolite have been employed as electrolytes in galvanic cells for the study of thermodynamic properties, particularly of strongly electropositive metals such as Ca, Th, Mg, etc. It may be pointed out that since F^- ion has the smallest size and lowest charge, fluorides can be expected to be good ionic conductors. These aspects have been reviewed by Markin,[390] Reau and Portier,[391] O'Keefe,[426] and Takahashi,[387] among others.[367] Density and lattice parameter measurements[52,392] established that anti-Frenkel-type disorder of interstitial fluoride ions exists in CaF_2 doped with YF_3. Based on electrical conductivity data and diffusion measurements with radioactive ^{45}Ca, Ure[34] found that the transport number for fluoride ion (t_{F^-}) is almost unity in the temperature range 690–920°C with $t_{Ca^{2+}}$ of the order of 10^{-6}. These results have since been confirmed[393,394] and extended up to 1100°C[393] as also over a wide range of YF_3 content.[414,415] The isothermal conductivity of $Ca_{1-x}Y_xF_{2+x}$ is invariant with x at $x < 0.05$ and $x > 0.25$ and increases sharply with x at intermediate values. Corresponding changes in the activation energy were also noted. This is explained in terms of the kind of defect complexes formed, which were examined by neutron diffraction.[416,417] The self-diffusion coefficient of ^{45}Ca in single-crystal CaF_2 between 987 and 1246°C is given by[395]

$$D = 5.35 \times 10^3 \exp(-4.15/kT) \text{ cm}^2/\text{sec} \qquad (1.62)$$

in agreement with the results of Martze and Lindner[396] and with Sr^{2+} ion diffusion in CaF_2.[397] The calcium diffusion coefficient at 1100°C[398] increases from 8.6×10^{-11} for pure CaF_2 to 1.9×10^{-10} and 1.6×10^{-9} cm^2 sec^{-1} for CaF_2 doped with 20% and 40% YF_3, respectively.[52] Schottky-type disorder in cation sublattice

$$O = V''_{Ca} + 2V_F \qquad (1.63)$$

can account for this behavior. This limits the usefulness of YF_3-doped CaF_2 as an anionic conductor. The role of NaF doping is not clearly established.[34,52] The self-diffusion data in CaF_2 are summarized in Fig. 1.37.[399] The interdiffusion in the system CaF_2–SrF_2 is also studied.[400]

Using the color center migration data of Mollwo,[401] Wagner[402] has estimated that the electronic conductivity of CaF_2 electrolytes is insignificant when used with Th/ThF_4 or even Mg/MgF_2 electrodes at 600°C, but not with Ca/CaF_2 electrodes. Later measurements of dc polarization, ac conductivity, and open-circuit emf by Hinze[403] give an overall indication that the electrolytic domain[69,403] for CaF_2 extends down to the stability limit determined by the equation

$$\log P_{F_2} = 9.0 - \frac{6.34 \times 10^4}{T} \tag{1.64}$$

More recently, Hinze and Patterson[404] carried out ac conductivity measurements from 550 to 800°C on a number of symmetrical cells of the type

$$M, MF_n \mid CaF_2 \mid M, MF_n \tag{1.4}$$

where the coexistence electrodes Co, CoF_2; Ni, NiF_2; Mn, MnF_2; Th, ThF_4; Y, YF_3; and Ca, CaF_2 were used together with optical-quality, high-purity CaF_2 single crystal. Open-circuit emf measurements were performed from 400 to 815°C on nonsymmetric cells of the type

$$Ca \mid CaF_2 \mid Th, ThF_4 \tag{1.5}$$

The variation of conductivity with P_{F_2} indicates a flat portion at 800°C down to Ca, CaF_2 coexistence, which is the stability limit of CaF_2. At lower temperatures, a sloping region is observed. This slope is attributed to the

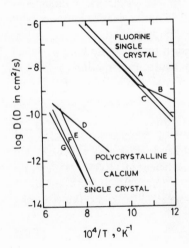

FIG. 1.37. Temperature dependence of diffusion coefficients in single and polycrystalline CaF_2. A, C are from conductivity measurements and B, D, E, F, G are from tracer studies (Reference 399).

movement of impurity oxygen ions and neither to cations (which have a very low diffusivity in CaF_2) nor to electrons (since the slope is $-1/16$ instead of the theoretical value of $-1/2$). Based on these considerations, it may be concluded that solid CaF_2 can be employed as a solid electrolyte within the temperature range 550–800°C with fluorine chemical potentials ranging from Ni, NiF_2 to Ca, CaF_2 coexistence. Thus CaF_2 is characterized by a wider P_{x_2} and temperature range where $t_{ion} \approx 1$ than other anionic conductors such as the ZrO_2 and ThO_2 types. The electrical conductivities of pure CaF_2 single crystals and CaF_2 doped with NaF and YF_3 were measured from 600 to 1000°K.[127]

The high chemical activity at elevated temperatures limits the usefulness of CaF_2 as a solid electrolyte. Oxygen is soluble in CaF_2 up to about 0.2 mol % above 800°C,[405] occupying the interstitial or anion sites and generating holes. However, doping of CaF_2 with 0.2% CaO does not seem to affect its t_{ion}.[406] Further, CaF_2 has been successfully used as a solid electrolyte in pure oxygen below 1000°C,[407–409] though its ionic conduction deteriorates in a moist atmosphere. Also, CaF_2 tends to react with the electrodes forming intermediate compounds.[407,410–413] As a result of these considerations, the electrolytic domain region of CaF_2 is narrowed. A practical approach to this problem consists of separating the electrolyte and electrode in a galvanic cell.

The conductivity of β-PbF_2 is several orders of magnitude larger than that of CaF_2 at comparable temperatures. The activation energy for conduction in single crystals is 0.66 eV and in ceramics 0.45 eV, with that from NMR linewidth studies assuming intermediate values (0.59–0.63 eV).[418–421] The higher polarizability of Pb^{2+} compared to Ca^{2+} might favor the higher mobility of F^- ions in this case. The conductivity behavior of the $Pb_{1-x}Bi_x$ F_{2+x} system is similar to that of $Ca_{1-x}Y_xF_{2+x}$.[422]

The anion conductivity of SrF_2 and SrF_2 doped with LaF_3 in the range 300–1100°C, BaF_2 and BaF_2 doped with GdF_3 and KF in the range 800–1100°C, and β-PbF_2 doped with BiF_3 was also studied.[160,393,423,424] The energy of formation of defects and the variation of defect concentration with temperature has been deduced for these fluorides.

On the other hand, both Na^+ and F^- ions were found to contribute to electrical conductivity in NaF with NaCl-type structure and in cryolite, Na_3AlF_6, with a complex structure. In the case of NaF, t_{Na^+} decreased from 0.995 at 550°C to 0.916 at 600°C and 0.861 at 625°C.[160,161]

Cryolite undergoes a transformation from monoclinic to cubic symmetry at 560°C. Landan and Ubbelohde[161] have pointed out that in the cubic form, the $(AlF_6)^{3-}$ octahedra lie at the corners and face centers of the lattice with Na ions located at the edge centers and body centers with an anion coordination of 6 (Na_1–F distance = 2.21 Å) and near space diagonals with an anion coordination of 12 (Na_2–F distance = 2.68 Å). This large Na_2–F distance (even compared to the Na–F distance of 2.31 Å in NaF) enables formation of Frenkel defects without a distortion of the surrounding F^- ions. Activation energy for electrical conductivity was found to be 16.5 kcal/mol for the monoclinic phase and 28 and 51 kcal/mol for the cubic phase below and above 880°C, respectively. Possible contribution from Na^+ and F^- ions to electrical conductivity is suggested at high temperatures. The conductivity is enhanced 50 times by doping cryolite with CaF_2.[425]

4.5. Miscellaneous

While solid electrolytes with a variety of cations as the conducting species are available, anion conductivity has been limited essentially to the oxygen and fluorine ions so far (Table 1.1). Exploratory research into solid conductors for the sulfur ion in $CaS-Y_2S_3$,[172] for the carbon ion in BaF_2-BaC_2, and for the nitrogen ion in AlN, and for protons in KH_2PO_4, KHF_2, etc., has been started.

The alkali-oxide–tungsten-oxide systems form cubic, tetragonal, and hexagonal tungsten bronze structures. In these, WO_6 octahedra are linked together through corners in such a way as to give rise to channels of different sizes. Any W^{6+} to W^{5+} conversion is accompanied by the presence of an appropriate number of alkali ions in the channels, through which they can travel relatively easily. However, the mixed valency of tungsten ions gives rise to electronic conductivity, justifying the name "tungsten bronze."[151,427–429] The electronic conductivity may be eliminated in "tantalum bronzes" since Ta^{5+} is not known to be reduced in air, vacuum, or inert atmospheres. Roth *et al.*[430] have studied many systems and listed a number of "tunnel" compounds,[431] which may exhibit high alkali ion conductivity. Among these are $Li_2O \cdot 14Nb_2O_5$, $Li_2O \cdot 19Ta_2O_5$, $Li_2O \cdot 5Ta_2O_5 \cdot 2MoO_3$, $Na_2O \cdot 3Ta_2O_5$–$Na_2O \cdot 4Ta_2O_5$, and $K_2O:Ta_2O_5$ of 1:2, 1:5, 1:7, and 1:8.

Following the work of Dryden and Wadsley[432] on the anisotropy of electrical conduction in $Ba_xMg_xTi_{8-x}O_{16}$ of hollandite structure, modified materials[151] with ac conductivities up to 10^{-2} $ohm^{-1} cm^{-1}$ and activation energies of 4 to 6 kcal/mol were explored, for example, $K_xMg_{x/2}Ti_{(8-x/2)}O_{16}$, $K_xAl_xTi_{(8-x)}O_{16}$, and $(K, Li)_{2-x}(Li, Ti)_8O_{16}$.[433–435]

Among the skeletal structures explored, one of the interesting materials is the solid solution series $NaZr_2(PO_4)_3$–$Na_4Zr_2(SiO_4)_3$ with the composition $Na_3Zr_2PSi_3O_{12}$ exhibiting a conductivity of 0.3 $ohm^{-1} cm^{-1}$ at 300°C, which is superior to β''-Al_2O_3. The activation energy is 0.24 eV, slightly higher than the 0.18 eV of β''-Al_2O_3. This material, which is resistant to moisture attack and stable to sodium polysulfides, can be densified at 1250°C in ceramic form.[436] Unlike β''-Al_2O_3, it is a three-dimensional conductor.

Inorganic silicate glasses are made up of a linking of SiO_4 tetrahedra through corners in an irregular manner with the alkaline and alkaline earth ions randomly distributed in the cavities of the silicate network. As a result, silicate glasses are known to be electrolytic conductors with the alkali ions as the migrating species.[437] This is made use of to build a prototype battery[179] using glass capillary tubes as the electrolyte. Rapid quenching of some alkali niobates and tantalates leads to noncrystalline materials exhibiting significant ionic conductivity behavior and a switch to considerably lower conductivity on crystallization.

While high polymers are good insulators, it has been found that doped nonconjugate polymer complexes such as polyvinyl 4-pyridine, polyvinyl chloride, and nylon 6-6 complexes exhibit conductivity due to the movement of electrons or ions. These materials can therefore be used as electrodes or electrolytes in solid state batteries.[438]

5. Concluding Remarks

The foregoing discussion shows that for a solid electrolyte to be useful it should possess the following characteristics:

(a) The value of t_{ion} should be greater than 0.99; the conduction should be due to only one ionic species; transference number for electronic defects should be 10^{-3} or less.

(b) The forbidden energy gap should be large, say, 3 eV or $T/300$ eV, where T is absolute temperature.[7]

(c) The activation energy for ion migration should be less than the activation energy for electronic conductivity.[7]

(d) The electrolyte should be stable with respect to the surroundings and in the temperature range of use.

Taking into account the above characteristics and the list of known solid electrolytes (Table 1.1), one may venture to suggest the following guidelines in a search for new solid electrolytes:

(a) The crystal structure should be an open one, containing vacant lattice points (e.g., CaF_2), tunnels (e.g., tantalum bronzes), or layers (e.g., β-alumina).

(b) The ratio of the number of mobile ions to the number of structurally equivalent sites which are available to the mobile ions must be small (e.g., β-Al_2O_3).[439,440]

(c) Disorder or liquid-like nature in a sublattice promotes fast ion transport.[441] The energy difference between the ordered and disordered distribution of the mobile ions over the available sites should be small.[440]

(d) The compound should predominantly have ionic character, i.e., the electronegativity difference of anions and cations should be greater than 2.

(e) The material should be transparent or white.

(f) The material should not contain atoms with small ionization potentials, i.e., the ions should not easily change their valency.

(g) The activation energy for ion migration must be less than that for electron motion.[7]

ACKNOWLEDGMENTS

The work on solid electrolytes in the authors' laboratory has been supported in part by the U.S. National Bureau of Standards and the U.S. Aerospace Laboratory, for which we are grateful.

References

1. F. A. Kröger, *Chemistry of Imperfect Crystals*, North-Holland, Amsterdam (1964).
2. F. A. Kröger and H. J. Vink, in *Solid State Physics*, Vol. 3, eds. F. Seitz and D. Turnbull, Academic Press, New York (1956), p. 307.
3. W. Van Gool, *Principles of Defect Chemistry of Crystalline Solids*, Academic Press, New York (1966).
4. L. Eyring and M. O'Keefe, eds., *The Chemistry of Extended Defects in Non-Metallic Solids*, North-Holland, Amsterdam (1970).
5. W. Van Gool, ed., *Fast Ion Transport in Solids*, North-Holland, Amsterdam (1973).
6. L. W. Strock, *Z. Phys. Chem.* **B25**, 441 (1934); **B31**, 132 (1936).
7. L. Heyne, in *Fast Ion Transport in Solids*, ed. W. Van Gool, North-Holland, Amsterdam (1973), p. 123.
8. L. Mandelcorn, ed., *Non-stoichiometric Compounds*, Academic Press, New York (1964).
9. J. B. Wachtman and A. D. Franklin, eds., *Mass Transport in Oxides*, Natl. Bur. Stand. (U.S.) Spec. No. 296 (1968).
10. A. D. Wadsley, in *Non-stoichiometric Compounds*, ed. L. Mandelcorn, Academic Press, New York (1964), Chap. 3.
11. J. T. Kummer, in *Progress in Solid State Chemistry*, Vol. 7, eds. H. Reiss and J. O. McCaldin, Pergamon Press, New York (1972), p. 141.
12. A. B. Lidiard, in *Handbuch der Physik*, Vol. 20, ed. S. Flügge, Springer-Verlag, Berlin (1957), p. 246.
13. B. C. H. Steele and C. B. Alcock, *Trans. Metall. Soc. AIME* **233**, 1359 (1965).
14. W. W. Barker, *Mater. Sci. Eng.* **2**, 208 (1967).
15. M. O'Keefe, in *The Chemistry of Extended Defects in Non-metallic Solids*, eds. L. Eyring and M. O'Keefe, North-Holland, Amsterdam (1970), p. 609.
16. R. J. Brook, in *Electrical Conductivity in Ceramics and Glass*, ed. N. M. Tallan, Marcel Dekker, New York (1974), Part A, p. 179.
17. M. F. Lasker and R. A. Rapp, *Z. Phys. Chem. N.F.* **49**, 198 (1966).
18. J. Rudolph, *Z. Naturforsch.* **A14**, 727 (1959).
19. F. Hund and W. Dürrwächter, *Z. Anorg. Allgem. Chem.* **265**, 67 (1951).
20. F. Hund and R. Mezger, *Z. Phys. Chem.* **201**, 268 (1952).
21. F. Hund, *Z. Anorg. Allgem. Chem.* **274**, 105 (1953).
22. E. C. Subbarao, P. H. Sutter, and J. Hrizo, *J. Am. Ceram. Soc.* **48**, 443 (1965).

23. J. E. Bauerle, *J. Chem. Phys.* **45**, 4162 (1966).
24. R. A. Rapp and D. A. Shores, in *Techniques of Metals Research*, Vol. 4, Part 2, ed. R. A. Rapp, Interscience, New York (1970), p. 123.
25. I. M. Boswara and A. D. Franklin, in *Mass Transport in Oxides*, eds. J. B. Wachtman and A. D. Franklin, Natl. Bur. Stand. (U.S.) Spec. Publ. No. 296 (1968), p. 25.
26. E. C. Subbarao and T. V. Ramakrishnan, unpublished research.
27. N. S. Choudhuri and J. S. Patterson, *J. Am. Ceram. Soc.*, **57**, 90 (1974).
28. C. Wagner, *Z. Elektrochem.* **63**, 1027 (1959),
29. D. Raleigh, in *Progress in Solid State Chem.*, Vol. 3, ed. H. Reiss, Pergamon Press, New York (1967), p. 83.
30. D. Raleigh, *J. Phys. Chem. Solids* **26**, 329 (1963).
31. P. McGeehin and A. Hooper, *J. Mater. Sci.* **12**, 1 (1977).
32. R. M. Dell and A. Hooper, in *Solid Electrolytes*, eds. P. Hagenmuller and W. Van Gool, Academic Press, New York (1978).
33. A. D. Franklin, *J. Phys. Chem. Solids* **29**, 823 (1968).
34. R. W. Ure, Jr., *J. Chem. Phys.* **26**, 1363 (1957).
35. D. Chakravorty, *J. Phys. Chem. Solids* **32**, 1091 (1971).
36. M. J. Norgett and C. R. A. Cartlow, *J. Phys. C.: Solid State Phys.* **6**, 1325 (1973).
37. B. Rossing, Ph.D. thesis, Massachusetts Institute of Technology (1966).
38. A. Hooper, *Contemp. Phys.* **19**, 147 (1978).
39. W. Hayes, *Contemp. Phys.* **19**, 469 (1978).
40. H. Sato and R. Kikuchi, *J. Chem. Phys.* **55**, 677 (1971).
41. R. Kikuchi and H. Sato, *J. Chem. Phys.* **55**, 702 (1971).
42. R. Kikuchi, in *Fast Ion Transport in Solids*, ed. W. Van Gool, North-Holland, Amsterdam (1973), p. 555.
43. R. Kikuchi, in *Fast Ion Transport in Solids*, ed. W. Van Gool, North-Holland, Amsterdam (1973), p. 249.
44. Y. Y. Yao and J. T. Kummer, *J. Inorg. Nucl. Chem.*, **29**, 2453 (1967).
45. W. L. Roth, *J. Solid State Chem.* **4**, 60 (1972).
46. J. M. Robertson, in *Non-stoichiometric Compounds*, ed. L. Mandelcorn, Academic Press, New York (1964), p. 1.
47. B. T. M. Willis, in *The Chemistry of Extended Defects in Non-metallic Solids*, eds. L. Eyring and M. O'Keefe, North-Holland, Amsterdam (1970), p. 272.
48. T. Y. Tien and E. C. Subbarao, *J. Chem. Phys.* **39**, 1041 (1963).
49. R. E. Carter and W. L. Roth, in *Electromotive Force Measurements in High-Temperature Systems*, ed. C. B. Alcock, Institute of Mining and Metals, London (1968), p. 125.
50. A. M. Diness and R. Roy, *Solid State Commun.* **3**, 123 (1965).
51. E. C. Subbarao and P. H. Sutter, *J. Phys. Chem. Solids* **25**, 148 (1964).
52. J. M. Short and R. Roy, *J. Phys. Chem.* **67**, 1860 (1963).
53. W. G. Mumme and A. D. Wadsley, *Acta Crystallogr.* **B23**, 754 (1967).
54. W. G. Mumme and A. D. Wadsley, *Acta Crystallogr.* **B24**, 625 (1968).
55. B. Hudson and P. T. Moseley, AERE Harwell Report No. R 8014 (1975).
56. B. Hudson and P. T. Moseley, AERE Harwell Report No. R 8236 (1976).
57. A. K. Cheetham, quoted by B. T. M. Willis, in *Electromotive Force Measurements in High-Temperature Systems*, ed. C. B. Alcock, Institute of Mining and Metals, London (1968).
58. A. K. Cheetham, B. E. F. Fender, D. Steele, R. I. Taylor, and B. T. M. Willis, *Solid State Commun.* **8**, 171 (1970).
59. H. Rickert, in *Fast Ion Transport in Solids*, ed. W. Van Gool, North-Holland, Amsterdam (1973), p. 3.

60. Y. Haven, in *Fast Ion Transport in Solids*, ed. W. Van Gool, North-Holland, Amsterdam (1973), p. 35.
61. A. D. LeClaire, in *Fast Ion Transport in Solids*, ed. W. Van Gool, North-Holland, Amsterdam (1973), p. 51.
62. L. W. Barr and A. B. Lidiard, in *Physical Chemistry: An Advanced Treatise*, Vol. 10, eds. H. Eyring, D. Henderson, and W. Jost, Academic Press, New York (1970), p. 151.
63. P. G. Shewmon, *Diffusion in Solids*, McGraw-Hill, New York (1963).
64. W. D. Kingery, J. Papis, M. E. Doty, and D. C. Hill, *J. Am. Ceram. Soc.* **42**, 393 (1959).
65. W. H. Rhodes and R. E. Carter, *J. Am. Ceram. Soc.* **49**, 244 (1966).
66. L. A. Simpson and R. E. Carter, *J. Am. Ceram. Soc.* **49**, 139 (1966).
67. M. S. Whittingham and R. A. Huggins, *J. Chem. Phys.* **54**, 414 (1971).
68. M. S. Whittingham and R. A. Huggins, *J. Electrochem. Soc.* **118**, 1 (1971).
69. J. W. Patterson, *J. Electrochem. Soc.* **118**, 1033 (1971).
70. M. J. Rice and W. L. Roth, *J. Solid State Chem.* **4**, 294 (1972).
71. M. J. Rice, in *Fast Ion Transport in Solids*, ed. W. Van Gool, North-Holland, Amsterdam (1973), p. 263.
72. C. Warren Haas, *J. Solid State Chem.* **7**, 155 (1973).
73. A. Kvist, in *Physics of Electrolytes*, Vol. I, ed. J. Hladik, Academic Press, New York (1972), Chap. 8.
74. R. N. Blumenthal and M. A. Seitz, in *Electrical Conductivity in Ceramics and Glass*, ed. N. M. Tallan, Marcel Dekker, New York (1974), Part A, p. 35.
75. B. C. H. Steele, in *Electromotive Force Measurement in High Temperature Systems*, ed. C. B. Alcock, Institute of Mining and Metals, London (1968), p. 3.
76. B. C. H. Steele, in *Mass Transport in Oxides*, eds. J. B. Wachtman and A. D. Franklin, Natl. Bur. Stand. (U.S.) Spec. Publ. No. 296 (1968), p. 165.
77. D. Yuan and F. A. Kröger, *J. Electrychem. Soc.* **116**, 594 (1969).
78. Y. K. Agrawal, D. W. Short, R. Gruevalle, and R. A. Rapp, *J. Electrochem. Soc.* **121**, 354 (1974).
79. C. B. Choudhary, H. S. Maiti, and E. C. Subbarao, unpublished research.
80. H. C. Graham, N. M. Tallan, and R. Russell, *J. Am. Ceram. Soc.* **50**, 156 (1967).
81. R. W. Vest, N. M. Tallan, and W. C. Tripp, *J. Am. Ceram. Soc.* **47**, 635 (1964).
82. R. W. Vest and H. C. Graham, *Rev. Sci. Instrum.* **38**, 661 (1967).
83. I. Bransky and N. M. Tallan, *J. Am. Ceram. Soc.* **53**, 625 (1970).
84. S. P. Mitoff, in *Fast Ion Transport in Solids*, ed. W. Van Gool, North-Holland, Amsterdam (1973), p. 415.
85. A. K. Mehrotra, H. S. Maiti, and E. C. Subbarao, *Mater. Res. Bull.*, **8**, 899 (1973).
86. C. Wagner, *Z. Phys. Chem.*, **B21**, 42 (1933); **B23**, 469 (1934).
87. C. Wagner, in *Proceedings of the 7th Meeting of the International Commission on Electrochemistry, Thermodynamics and Kinetics, Lindau (1955)*, Butterworths, London (1956), p. 361.
88. C. Wagner, *Z. Elektrochem.* **60**, 4 (1956).
89. M. H. Hebb, *J. Chem. Phys.* **20**, 185 (1952).
90. L. Heyne, in *Mass Transport in Oxides*, eds. J. B. Wachtman and A. D. Franklin, Natl. Bur. Stand. (U.S.) Spec. Publ. No. 296 (1968), p. 149.
91. L. Heyne and N. M. Beekmans, *Proc. Brit. Ceram. Soc.* No. 19, 229 (1971).
92. M. Kleitz and J. Dupuy, eds. *Electrode Process in Solid State Ionics*, D. Reidel, Dordrecht/Boston (1976).
93. M. Kleitz, P. Fabry, and E. Schouler, in *Fast Ion Transport in Solids*, ed. W. Van Gool, North-Holland, Amsterdam (1973), p. 439.

94. R. W. Vest and N. M. Tallan, *J. Appl. Phys.* **36**, 543 (1965).
95. B. Ilschner, *J. Chem. Phys.* **28**, 1109 (1958).
96. D. Raleigh, *J. Phys. Chem. Solids* **26**, 329 (1965).
97. J. B. Wagner and C. Wagner, *J. Chem. Phys.* **26**, 1597 (1957).
98. A. V. Joshi and J. B. Wagner Jr., *J. Electrochem. Soc.* **122**, 1071 (1975).
99. T. Takahashi and O. Yamamoto, *Electrochem. Acta* **11**, 779 (1966).
100. J. W. Patterson, E. C. Borgen, and R. A. Rapp, *J. Electrochem. Soc.* **114**, 752 (1967).
101. Y. J. Vander Mullen and F. A. Kroger, *J. Electrochem. Soc.* **117**, 69 (1970).
102. C. Wagner, *J. Chem. Phys.* **21**, 1819 (1953).
103. C. B. Alcock and T. N. Belford, *Trans. Faraday Soc.* **60**, 822 (1964).
104. C. B. Alcock, S. Zador, and B. C. H. Steele, *Proc. Brit. Ceram. Soc.* No. 8, 231 (1967).
105. J. B. Wagner Jr., in *Electrode Process in Solid State Ionics*, eds. M. Kleitz and J. Dupuy, D. Reidel, Dordrecht/Boston (1976), p. 185.
106. J. E. Bauerle, *J. Phys. Chem. Solids* **30**, 2657 (1969).
107. G. D. Mahan and W. L. Roth, eds., *Superionic Conductors*, Plenum Press, New York (1976).
108. J. R. McDonald, *J. Chem. Phys.* **61**, 3977 (1974).
109. J. R. McDonald, in *Electrode Process in Solid State Ionics*, eds. M. Kleitz and J. Dupuy, D. Reidel, Dordrecht/Boston (1976), p. 149.
110. J. R. McDonald, in *Superionic Conductors*, eds. G. D. Mahan and W. L. Roth, Plenum Press, New York (1976), p. 81.
111. A. K. Jouscher, *J. Mater. Sci.* **13**, 553 (1978).
112. E. Schouler, M. Kleitz, and C. Deportes, *J. Chim. Phys.* **70**, 923 (1973).
113. E. Schouler, G. Giroud, and M. Kleitz, *J. Chim. Phys.* **70**, 1309 (1973).
114. E. Schouler, A. Hammou, and M. Kleitz, *Mater. Res. Bull.* **11**, 1137 (1976).
115. S. H. Chu and M. A. Seitz, *J. Solid State Chem.* **23**, 297 (1978).
116. A. Hooper, *J. Phys. D: Appl. Phys.* **10**, 1487 (1977).
117. R. D. Armstrong and R. Mason, *J. Electroanal. Chem.* **41**, 231 (1973).
118. R. W. Powers and S. P. Mitoff, *J. Electrochem. Soc.* **122**, 226 (1975).
119. R. D. Armstrong, in *Electrode Process in Solid State Ionics*, eds. M. Kleitz and J. Dupuy, D. Reidel, Dordrecht/Boston (1976), p. 261.
120. S. R. De Groot, *Thermodynamics of Irreversible Processes*, North-Holland, Amsterdam (1951).
121. H. Holtan, Jr., *Proc. K. Ned. Acad. Wet.* **56B**, 498, 510 (1953).
122. H. Holtan, Jr., P. Mazur, and S. R. De Groot, *Physica* **19**, 1109 (1953).
123. R. E. Howard and A. B. Lidiard, in *Rep. Progr. Phys.* **27**, 161 (1964).
124. K. E. Johnson, S. J. Simi, and J. Dudley, *J. Electrochem. Soc.* **120**, 703 (1973).
125. S. Chandra, H. B. Lal, and K. Shahi, *J. Phys. D* **5**, 443 (1972).
126. H. F. Hunger, *J. Electrochem. Soc.* **120**, 1157 (1973).
127. B. M. Mogilevskii, V. N. Romanov, V. M. Reiterov, L. M. Trofimova, and A. F. Chudnovskii, *Fiz. Tverd. Tela* **15**, 1062 (1973).
128. K. Shahi and S. Chandra, *J. Phys. D* **9**, 3105 (1976).
129. K. Shahi, *Phys. Status Solidi* A **41**, 11 (1977).
130. W. Fischer, *Z. Naturforsch.* **22a**, 1575 (1967).
131. R. J. Ruka, J. E. Bauerle, and L. Dykstra, *J. Electrochem. Soc.* **115**, 497 (1966).
132. K. Goto, T. Ito, and M. Somono, *Trans. AIME* **245**, 1662 (1969).
133. S. Pizzini, C. Riccondi, and V. Wagner, *Z. Naturforsch.* **25a**, 559 (1970).
134. N. M. Tallan and I. Bransky, *J. Electrochem. Soc.* **118**, 345 (1971).
135. W. Fischer, *Z. Naturforsch.* **22a**, 1557 (1967).
136. W. A. Fischer and C. Pieper, *Arch. Bisenhuttenw.* **44**, 251 (1973).

137. C. Wagner, in *Progress in Solid State Chemistry*, Vol. 7, eds. H. Reiss and J. O. McCaldin, Pergamon Press, New York (1972), p. 1.
138. D. S. Tannhauser, *J. Electrochem. Soc.* **119**, 793 (1972).
139. A. S. Nowick and B. S. Berry, *Anelastic Relaxation in Crystalline Solids*, Academic Press, New York (1972).
140. R. G. Breckenridge, in *Imperfections in Nearly Perfect Crystals*, ed. W. Shockley, John Wiley, New York (1952), p. 219.
141. J. B. Wachtman, Jr., *Phys. Rev.* **131**, 517 (1963).
142. A. D. Franklin, A. Shorts, and J. B. Wachtman, Jr., *J. Res. Natl. Bur. Stand.* **68A**, 425 (1964).
143. K. W. Lay and D. H. Whitmore, *Phys. Status Solidi* **43**, 175 (1971).
144. H. B. Johnson, N. J. Tolar, G. R. Miller, and I. B. Cutler, *J. Am. Ceram. Soc.* **49**, 458 (1966).
145. J. B. Wachtman, Jr., and W. C. Corwin, *J. Res. Natl. Bur. Stand.* **69A**, 457 (1965).
146. M. S. Whittingham, in *Fast Ion Transport in Solids*, ed. W. Van Gool, North-Holland, Amsterdam (1973), p. 429.
147. W. L. Roth, H. S. Story, and D. Kline, quoted in M. S. Wittingham, in *Fast Ion Transport in Solids*, ed. W. Van Gool, North-Holland, Amsterdam (1973), p. 429.
148. R. E. Walstedt, R. Dupree, and J. P. Remeika, in *Superionic Conductors*, eds. G. D. Mahan and W. L. Roth, Plenum Press, New York (1976), p. 369.
149. J. B. Boyce, J. C. Mikkelsen Jr., and M. O'Keefe, *Solid State Commun.* **21**, 955 (1977).
150. L. D. Clark, M. S. Whittingham, and R. A. Huggins, *J. Solid State Chem.* **5**, 487 (1972).
151. M. S. Whittingham and R. A. Huggins, in *Fast Ion Transport in Solids*, ed. W. Van Gool, North-Holland, Amsterdam (1973), p. 645.
152. C. H. Hao, L. L. Chase, and G. D. Mahan, *Phys. Rev. B.* **13**, 4306 (1976).
153. M. J. Delaney and S. Ushioda, *Phys. Rev. B.* **16**, 1416 (1977).
154. G. Burns, F. H. Dacol, and M. W. Shafer, *Phys. Rev. B.* **16**, 1416 (1977).
155. R. J. Elliot, W. Hayes, W. G. Kleppmann, A. J. Rushworth, and J. F. Ryan, *Proc. R. Soc.* London **A360**, 317 (1978).
156. S. J. Allen, Jr. and J. P. Remeika, *Phys. Rev. Lett.* **33**, 1478 (1974).
157. H. R. Zeller, P. Bruesch, L. Pietronero, and S. Strassler, in *Superionic Conductors*, eds. G. D. Mahan and W. L. Roth, Plenum Press, New York (1976), p. 201.
158. J. F. Boilot, G. Collin, R. Comes, J. Thery, R. Collongues, and A. Guinier, in *Superionic Conductors*, eds. G. D. Mahan and W. L. Roth, Plenum Press, New York (1976), p. 243.
159. H. Fuess, K. Funke, and J. Kalus, *Phys. Status Solidi* B, **70**, 567 (1975).
160. C. Tubandt, H. Reinhold, and G. Liebold, *Z. Anorg. Allgem. Chem.* **197**, 225 (1931).
161. G. J. Landan and A. R. Ubbelohde, *Proc. R. Soc. London* **240**, 160 (1957).
162. B. B. Owens, *J. Electrochem. Soc.* **117**, 1536 (1970).
163. B. B. Owens, J. H. Christie, and G. T. Tiedeman, *J. Electrochem. Soc.* **118**, 1144 (1971).
164. M. L. Berardelli, C. Biondi, K. De Rossi, G. Fonseca, and M. Giomini, *J. Electrochem. Soc.* **119**, 114 (1972).
165. T. Takahashi, S. Ikeda, and O. Yamamoto, *J. Electrochem. Soc.* **119**, 477 (1972).
166. T. Takahashi, S. Ikeda, and O. Yamamoto, *J. Electrochem. Soc.* **120**, 647 (1973).
167. G. W. Mellors and D. V. Louzos, *J. Electrochem. Soc.* **118**, 846 (1971).
168. G. W. Mellors, D. V. Louzos, and J. A. Vanlier, *J. Electrochem. Soc.* **118**, 850 (1971).

169. O. Yamamoto and T. Takahashi, *Denki Kagaku* 35, 651 (1967).
170. T. Takahashi, K. Kuwabara, and O. Yamamoto, *Denki Kagaku* 35, 612 (1967); 36, 530 (1968).
171. T. Takahashi, K. Kuwabara, and Y. Yamamoto, *J. Electrochem. Soc.* 120, 1607 (1973).
172. W. L. Worrell, V. B. Tare, and F. J. Bruni, in *High Temperature Technology*, IUPAC, Butterworths, London (1969), p. 503.
173. W. L. Worrell, *Am. Ceram. Soc. Bull.* 53, 425 (1974).
174. L. Suchow and G. R. Pond, *J. Am. Chem. Soc.* 75, 5242 (1953).
175. J. N. Bradley and P. D. Greene, *Trans. Faraday Soc.* 63, 424 (1967).
176. T. Takahashi, O. Yamamoto, and S. Ikeda, *J. Electrochem. Soc.* 120, 1431 (1973).
177. H. A. Johansen and J. G. Cleary, *J. Electrochem. Soc.* 111, 100 (1964).
178. M. S. Whittingham and R. A. Huggins, in *Solid State Chemistry*, eds. R. S. Roth and S. J. Schneider, *Natl. Bur. Stand. (U.S.) Spec. Publ.* No. 364 (1972), p. 139.
179. C. Levine, *Power Sources Symp. Proc.* 25, 75 (1972).
180. T. H. Etsell and S. N. Flengas, *Chem. Rev.* 70, 341 (1970).
181. E. C. Subbarao, H. S. Maiti, and K. K. Srivastava, *Phys. Status Solidi* A 21, 9 (1974).
182. J. Stocker, *Bull. Soc. Chim. Fr.* 78 (1961).
183. K. Kiukkola and C. Wagner, *J. Electrochem. Soc.* 104, 379 (1957).
184. P. Duwez, F. Odell, and F. H. Brown, *J. Am. Ceram. Soc.* 35, 107 (1952).
185. D. W. Strickler and W. G. Carlson, *J. Am. Ceram. Soc.* 47, 122 (1964).
186. R. C. Garvie, *J. Am. Ceram. Soc.* 51, 553 (1968).
187. J. M. Dixon, L. D. La Grange, U. Merten, C. F. Miller, and J. T. Porter, *J. Electrochem. Soc.* 110, 276 (1963).
188. Y. A. Pyatenko, *Zh. Strukt. Khim.* 4, 708 (1963).
189. C. Delamarre and M. Perez Y. Jorba, *C.R. Acad. Sci., Paris* 261, 5128 (1965).
190. C. Delamarre and M. Perez Y. Jorba, *Rev. Hautes Temp. Refract.* 2, 313 (1965).
191. C. Wagner, *Naturwissenschaften* 31, 265 (1943).
192. H. Schmalzried, *Z. Elektrochem.* 66, 572 (1962).
193. Z. S. Volchenkova and S. F. Pal'guev, *Tr. Inst. Elektrochim. Akad. Nauk SSSR, Ural. Filial.* 2, 173 (1961).
194. A. Cocco and I. Barbariol, *Ric. Sci., Parte 2, Sez. A*, 296 (1962).
195. F. A. Kröger, *J. Am. Ceram. Soc.* 49, 215 (1966).
196. T. Y. Tien and E. C. Subbarao, *J. Am. Ceram. Soc.* 46, 489 (1963).
197. D. Michel, M. Perez Y. Jorba, and R. Collongues, *C.R. Acad. Sci. Paris* 266, 1602 (1968).
198. T. Y. Tien, *J. Am. Ceram. Soc.* 47, 430 (1964).
199. D. W. White, *Rev. Energ. Primaire* 2, 10 (1966).
200. D. Michel, *Rev. Int. Hauptes Temp. Refract.* 9, 225 (1972).
201. A. Kontopoulos and P. S. Nicholson, *J. Am. Ceram. Soc.* 54, 317 (1971).
202. A. W. Smith, F. W. Meszaros, and C. D. Amata, *J. Am. Ceram. Soc.* 49, 240 (1966).
203. W. Nernst, *Z. Elektrochem.* 6, 41 (1899).
204. E. Bauer and H. Preis, *Z. Elektrochem.* 43, 727 (1937).
205. P. Duwez, F. H. Brown, and F. Odell, *J. Electrochem. Soc.* 98, 356 (1951).
206. F. Hund, *Z. Elektrochem.* 55, 363 (1951).
207. K. K. Srivastava, R. N. Patil, C. B. Choudhary, K. V. G. K. Gokhale, and E. C. Subbarao, *Trans. and J. Brit. Ceram. Soc.* 73, 85 (1974).
208. D. W. Strickler and W. G. Carlson, *J. Am. Ceram. Soc.* 48, 286 (1965).
209. R. Ruh and H. J. Garrett, in *Thermal Analysis*, eds. R. F. Schwinkler and P. D. Garn, Academic Press, New York (1969), p. 851.
210. T. Noguchi, M. Mizuno, and T. Yamada, *Bull. Chem. Soc. Jpn.* 43, 2414 (1970).

211. F. K. Fan, A. K. Kuznetsov, and E. K. Keler, *Izv. Akad. Nauk SSSR, Otd. Khim. Nauk*, 1141 (1962).
212. D. K. Smith, *J. Am. Ceram. Soc.* **49**, 625 (1966).
213. S. P. Ray and V. S. Stubican, *Mat. Res. Bull.* **12**, 549 (1977).
214. D. T. Bray and U. Merten, *J. Electrochem. Soc.* **111**, 447 (1964).
215. J. E. Bauerle and J. Hrizo, *J. Phys. Chem. Solids* **30**, 565 (1969).
216. R. E. W. Casselton, *Phys. Stat. Solidi* A **2**, 571 (1970).
217. J. E. Bauerle, *J. Phys. Chem. Solids* **30**, 2657 (1969).
218. H. H. Moebius, H. Witzmann, and G. Proeve, *Z. Chem.* **4**, 195 (1964).
219. W. A. Fischer and D. Janke, *Z. Phys. Chem. Neue Folge* **69**, 11 (1970).
220. R. E. W. Casselton, in *Electromotive Force Measurements in High Temperature Systems*, ed. C. B. Alcock, Institute of Mining and Metals, London (1968), p. 151.
221. R. E. W. Casselton and J. C. Scott, *Phys. Lett.* **25A**, 264 (1967).
222. T. L. Markin, R. J. Bones, and R. M. Dell, in *Superionic Conductors*, eds. G. D. Mahan and W. L. Roth, Plenum Press, New York (1976), p. 15.
223. W. Baukal, *Electrochim. Acta* **14**, 1071 (1969).
224. K. W. Browall and R. H. Doremus, *J. Am. Ceram. Soc.* **60**, 262 (1977).
225. S. Geller, ed., *Solid Electrolytes*, Springer-Verlag, Berlin (1977).
226. D. Viechnicki and V. S. Stubican, *Nature* **206**, 1251 (1965).
227. C. F. Grain, *J. Am. Ceram. Soc.* **50**, 288 (1967).
228. D. Viechnicki and V. S. Stubican, *J. Am. Ceram. Soc.* **48**, 292 (1965).
229. T. W. Smoot and J. R. Ryan, *J. Am. Ceram. Soc.* **46**, 597 (1963).
230. S. F. Pal'guev, A. D. Neuimin, and V. N. Strekalovskii, in *Electrochemistry of Molten and Solid Electrolytes*, Vol. 6, ed. A. N. Baraboshkin, Consultants Bureau, New York (1968), p. 121.
231. M. Guillou, J. Millet, M. Asquiedge, N. Busson, M. Jacquin, and S. Palous, *C.R. Acad. Sci. Paris* **262**, 616 (1966).
232. F. Trombe and M. Foex, *C.R. Acad. Sci. Paris* **233**, 254 (1951).
233. F. H. Brown and P. Duwez, *J. Am. Ceram. Soc.* **38**, 95 (1955).
234. V. B. Glushkova, E. K. Keler, L. G. Shcherbakova, and E. Ya. Shcherbovskii, *Izv. Akad. Nauk SSSR, Neorg. Mater.* **7**, 2007 (1971).
235. R. S. Roth, *J. Res. Nat. Bur. Stand.* **56**, 17 (1956).
236. M. Perez, Y. Jorba, R. Collongues, and J. Lefevre, *C. R. Acad. Sci. Paris*, **249**, 1237 (1959).
237. M. Perez, Y. Jorba and R. Collongues, *Bull. Soc. Chim. Fr.* 1967 (1959); 1969 (1959); 70 (1961).
238. A. M. Anthony, F. Cabannes, and J. Renon, *Ann. Phys. (Paris)* **9**, 1 (1964).
239. Z. S. Volchenkova, *Izv. Akad. Nauk SSSR, Neorg. Mater.* **4**, 1975 (1968); *Inorg. Mater.* **4**, 1717 (1968).
240. Y. D. Tretyakov. *Izv. Akad. Nauk. SSSR, Neorg. Mater.* **2**, 501 (1966); *Inorg. Mater.* **2**, 432 (1966).
241. H. H. Moebius, H. Witzmann, and F. Zimmer, *Z. Chem.* **4**, 194 (1964).
242. F. M. Spiridonov, L. N. Popova, and R. Ya. Popil'skii, *J. Solid State Chem.* **2**, 430 (1970).
243. Z. S. Volchenkova and Z. N. Lakeeva, *Izv. Akad. Nauk SSSR Neorg. Mater.* **7**, 1076 (1971).
244. W. W. Barker, F. P. Bailey, and W. Garrett, *J. Solid State Chem.* **7**, 448 (1973).
245. R. Ruh, H. J. Garrett, R. F. Domagala, and V. A. Patel, *J. Am. Ceram. Soc.* **60**, 399 (1977).
246. W. E. Danforth, *J. Franklin Inst.* **266**, 483 (1955).
247. W. E. Danforth and J. H. Bodine, *J. Franklin Inst.* **260**, 467 (1955).

248. R, A. Rapp, in *Thermodynamics of Nuclear Materials (1967)* IAEA, Vienna (1968), p. 359.
249. J. M. Wimmer, L. R. Bidwell, and N. M. Tallan, *J. Am. Ceram. Soc.* **50**, 198 (1967).
250. J. B. Hardway III, J. W. Patterson, D. R. Wilder, and J. D. Shieltz, *J. Am. Ceram. Soc.* **54**, 94 (1971).
251. H. Ullman, *Z. Chem.* **9**, 39 (1969).
252. H. H. Moebius, *Z. Chem.* **4**, 81 (1964).
253. Z. S. Volchenkova and S. F. Palguev, *Trans. Inst. Electrochem.* **1**, 104 (1961).
254. H. S. Maiti and E. C. Subbarao, *J. Electrochem. Soc.* **123**, 1713 (1976).
255. H. S. Maiti and E. C. Subbarao, *J. Electrochem. Soc.* **123**, 1058 (1976).
256. H. Peters and H. Moebius, *Z. Phys. Chem.* **207**, 298 (1958).
257. A. M. Diness and R. Roy, *J. Mater. Sci.* **4**, 613 (1969).
258. F. Hund, *Ber. Deut. Keram. Ges.* **42**, 251 (1965).
259. D. M. Roy and R. Roy, *J. Electrochem. Soc.* **111**, 421 (1964).
260. G. Bauer and K. A. Gingerich, *J. Inorg. Nucl. Chem.* **16**, 87 (1960).
261. F. A. Kuznetzov, V. I. Belyi, and T. N. Rezukhina, *Dokl. Akad. Nauk. SSSR, Fiz-Khim* **139**, 1405 (1961).
262. D. J. M. Bevan and J. Kordis, *J. Inorg. Nucl. Chem.* **26**, 1509 (1964).
263. E. L. Holverson and C. J. Kevane, *J. Chem. Phys.* **44**, 3692 (1966).
264. P. Kofstad and A. Z. Hed, *J. Am. Ceram. Soc.* **50**, 681 (1967).
265. P. L. Land, *J. Phys. Chem. Solids* **34**, 1839 (1973).
266. R. N. Blumenthal, P. W. Lee, and R. J. Paulener, *J. Electrochem. Soc.* **118**, 123 (1971).
267. R. B. Blumenthal and R. L. Hofnaier, *J. Electrochem. Soc.* **121**, 126 (1974).
268. G. J. Van Handel and R. N. Blumenthal, *J. Electrochem. Soc.* **121**, 1198 (1974).
269. R. N. Blumenthal, *J. Solid State Chem.* **12**, 307 (1975).
270. R. N. Blumenthal and R. K. Sharma, *J. Solid State Chem.* **13**, 360 (1975).
271. R. J. Paulener, R. N. Blumenthal, and J. E. Garnier, *J. Phys. Chem. Solids* **36**, 1213 (1975).
272. R. N. Blumenthal and B. A. Pinz, *J. Appl. Phys.* **38**, 2376 (1967).
273. R. N. Blumenthal, F. S. Brugner, and J. E. Garnier, *J. Electrochem. Soc.* **120**, 1230 (1973).
274. H. L. Tuller and A. S. Nowick, *J. Electrochem. Soc.* **122**, 255 (1975).
275. J. M. Bevan, W. W. Barker, and T. C. Parks, *Proc. 4th Conf. on Rare Earth Res. Phoenix Ariz.*, ed. L. Eyring, Gordon and Breach, New York (1965), p. 441.
276. M. A. Seitz and T. B. Holliday, *J. Electrochem. Soc.* **121**, 122 (1974).
277. J. E. Garnier, R. N. Blumenthal, R. J. Paulener, and R. K. Sharma, *J. Phys. Chem. Solids* **37**, 369 (1976).
278. T. Kudo and H. Obayashi, *J. Electrochem. Soc.* **122**, 142 (1975).
279. T. Takahashi, H. Iwahara, and Y. Nagai, *J. Appl. Electrochem.* **2**, 97 (1972).
280. T. Takahashi and H. Iwahara, *J. Appl. Electrochem.* **3**, 65 (1973).
281. T. Takahashi, H. Iwahara, and T. Arao, *J. Appl. Electrochem.* **5**, 187 (1975).
282. T. Takahashi, T. Esaka, and H. Iwahara, *J. Appl. Elrctrochem.* **5**, 197 (1975).
283. T. Takahashi, T. Esaka, and H. Iwahara, *J. Solid State Chem.* **16**, 317 (1976).
284. T. Takahashi, T. Esaka, and H. Iwahara, *J. Appl. Electrochem.* **7**, 31 (1977).
285. T. Takahashi, T. Esaka, and H. Iwahara, *J. Appl. Electrochem.* **7**, 299 (1977).
286. T. Takahashi, H. Iwahara, and T. Esaka, *J. Electrochem. Soc.* **124**, 1563 (1977).
287. R. C. De Vries and W. L. Roth, *J. Am. Ceram. Soc.* **52**, 369 (1969).
288. J. H. Kennedy, in *Solid Electrolytes*, ed. S. Geller, Springer Verlag, Berlin (1977), p. 105.

289. P. Hagenmuller and W. Van Gool, eds., *Solid Electrolytes*, Academic Press, New York (1978).
290. R. Collongues, J. Thery, and J. P. Boilot, in *Solid Electrolytes*, eds. P. Hagenmuller and W. Van Gool, Academic Press, New York (1978), p. 253.
291. M. Bettman and L. T. Terner, *Inorg. Chem.* 10, 1442 (1971).
292. S. B. Hendricks and L. Pauling, *Z. Kristallogr.*, 64, 303 (1927).
293. W. L. Bragg, C. Gottfried, and J. West, *Z. Kristallogr.* 77, 255 (1931).
294. C. A. Beevers and M. A. S. Ross, *Z. Kristallogr.* 97, 59 (1937).
295. J. Felsche, *Naturwissenschaften* 54, 612 (1967); *Z. Kristallogr.* 127, 94 (1968).
296. C. R. Peters, M. Bettman, J. W. Moore, and M. D. Click, *Acta Crystallogr.* B27, 1826 (1971).
297. K. Toshiko and M. Giichi, *J. Solid State Chem.* 17, 61 (1976).
298. G. Collin, J. P. Boilot, A. Kahn, J. Thery, and R. Comes, *J. Solid State Chem.* 21, 283 (1977).
299. A. Kahn, J. P. Boilet, and J. Thery, *Mater. Res. Bull.* 11, 397 (1976).
300. G. Yamaguchi, *J. Jpn. Electrochem. Soc.* 11, 260 (1943).
301. J. Thery and D. Briancon, *C. R. Acad. Sci. Paris*, 254, 2782 (1962); *Rev. Hauptes Temp., Refract.* 1, 221 (1964).
302. G. Yamaguchi and K. Suzuki, *Bull. Chem. Soc. Jpn.* 41, 93 (1968).
303. M. Bettman and C. R. Peters, *J. Phys. Chem.* 73, 1774 (1969).
304. R. H. Radzilowski, *J. Am. Ceram. Soc.* 53, 699 (1970).
305. A. Imai and M. Harata, *Jpn. J. Appl. Phys.* 11, 180 (1972).
306. R. H. Radzilowski and J. T. Kummer, *Inorg. Chem.* 8, 2531 (1969).
307. R. H. Radzilowski, *Inorg. Chem.* 8, 994 (1969).
308. H. Saalfield, H. Matthies, and S. K. Dutta, *Ber. deut. Keram. Ges.* 45, 212 (1968).
309. N. A. Tarapov and M. M. Stukalova, *C. R. Akad. Nauk SSSR* 24, 459 (1939); 27, 974 (1940).
310. J. T. Kummer and N. Weber, *Trans. S.A.E.* 76, 1003 (1968).
311. W. L. Roth, in *Solid State Chemistry*, eds. R. S. Roth and S. J. Schneider, Natl. Bur. Stand. (U.S.) Spec. Publ. No. 364 (1972), p. 129.
312. P. D. Darnier and J. P. Remeika, *J. Solid State Chem.* 17, 245 (1976).
313. M. Harata, *Mat. Res. Bull.* 6, 461 (1971).
314. J. H. Kennedy and A. F. Sammells, *J. Electrochem. Soc.* 119, 1609 (1972).
315. J. H. Kennedy and A. F. Sammells, *Fast Ion Transport in Solids*, ed. W. Van Gool, North-Holland, Amsterdam (1973), p. 563.
316. D. Kline, H. S. Story ,and W. L. Roth, *J. Chem. Phys.*, 57, 5180 (1972).
317. W. L. Roth, F. Reidinger, and S. J. LaPlaca, in *Superionic Conductors*, eds. G. O. Mahan and W. L. Roth, Plenum Press, New York (1976), p. 273.
318. D. B. McWhan, S. M. Shapiro, J. P. Remeika, and G. Shirane, *J. Phys. C. Solid State Phys.* 8, 2487 (1975).
319. S. M. Shapiro in *Superionic Conductors*, eds. G. D. Mahan and W. L. Roth, Plenum Press, New York (1976), p. 261.
320. Y. Le Cars, R. Comes, L. Deschamps, and J. Thery, *Acta Crystallogr.* A30, 305 (1974).
321. R. D. Armstrong, P. M. A. Sherwood, and R. A. Wiggins, *Spectrochim. Acta* 30A, 1213 (1974).
322. D. J. M. Bevan, B. Hudson, and P. T. Moseley, *Mater. Res. Bull.* 9, 1073 (1974).
323. L. C. DeJonghe, *J. Mater. Sc.* 10, 2173 (1975); 11, 206 (1976).
324. D. Jerome and J. P. Boilot, *J. Phys. (Paris) Lett.* 35, 129 (1974).
325. I. Chung, H. S. Story, and W. L. Roth, *J. Chem. Phys.* 63, 4903 (1975).

326. W. Bailey, S. Glowinkowski, H. S. Story, and W. L. Roth, *J. Chem. Phys.* **64**, 4126 (1976).

327. Y. LeCars, J. Thery, and R. Collongues, *C. R. Acad. Sci. Paris* **274c**, 4 (1972).

328. R. H. Radzilowski and J. T. Kummer, *J. Electrochem. Soc.* **118**, 714 (1971).

329. W. Van Gool and A. G. Piken, *J. Mat. Sci.* **4**, 105 (1968).

330. L. L. Chase, in *Superionic Conductors*, eds. G. D. Mahan and W. L. Roth, Plenum Press, New York (1976), p. 299.

331. J. L. Sudworth, A. R. Tilley, and K. D. South, in *Fast Ion Transport in Solids*, ed. W. Van Gool, North-Holland, Amsterdam (1973), p. 581.

332. W. L. Fielder, H. E. Kautz, J. S. Fordyce, and J. Singer, *J. Electrochem. Soc.* **122**, 528 (1975).

333. M. S. Whittingham, in *Fast Ion Transport in Solids*, ed. W. Van Gool, North-Holland, Amsterdam (1973), p. 545.

334. V. V. Daniel, *Dielectric Relaxation*, Academic Press, New York (1967).

335. R. H. Radzilowski, Y. F. Yao, and J. T. Kummer, *J. Appl. Phys.* **40**, 4716 (1969).

336. L. Hsueh and D. N. Bennion, *J. Electrochem. Soc.* **118**, 1128 (1971).

337. R. Galli, F. A. Trapeano, P. Bazzarin, and U. Mirarchi, in *Fast Ion Transport in Solids*, ed. W. Van Gool, North-Holland, Amsterdam (1973), p. 573.

338. H. Sato and R. Kikuchi, in *Superionic Conductors*, eds. G. D. Mahan and W. L. Roth, Plenum Press, New York (1976), p. 135.

339. Y. Haven, in *Solid Electrolytes*, eds. P. Hagenmuller and W. Van Gool, Academic Press, New York (1978), p. 59.

340. W. van Gool and P. H. Bottelberghs, *J. Solid State Chem.* **7**, 59 (1973).

341. J. Antoine, D. Vivien, J. Livage, J. Thery, and R. Collongues, *Mater. Res. Bull.* **10**, 865 (1975).

342. J. P. Boilot and J. Thery, *Mater. Res. Bull.* **11**, 407 (1976).

343. J. P. Boilot, A. Kahn, J. Thery, R. Collongues, J. Antoine, D. Vivien, C. Chevrette, and D. Gourier, *Electrochim. Acta* **22**, 741 (1977).

344. R. W. Powers, in *Superionic Conductors*, eds. G. O. Mahan and W. L. Roth, Plenum Press, New York (1976), p. 351.

345. J. H. Kennedy, in *Superionic Conductors*, eds. G. O. Mahan and W. L. Roth, Plenum Press, New York (1976), p. 335.

346. D. B. McWhalen, S. J. Allen, J. P. Remeika, and P. D. Dernier, *Phys. Rev. Lett.* **35**, 953 (1975).

347. C. Tubandt and E. Lorenz, *Nernst's Festschrift*, W. Knapp Hatte (1912), p. 446; *Z. Phys. Chem.* **87**, 513 (1914).

348. M. L. Huggins, in *Phase Transformations in Solids*, eds. R. Smolukowski, M. J. Buerger, and W. A. Weyl, John Wiley, New York (1951), Chap. 8.

349. L. W. Strock and V. A. Brophy, *Am. Mineral.* **40**, 94 (1955).

350. J. W. Mansen, *J. Phys. Chem.* **60**, 806 (1956).

351. A. J. Majumdar and R. Roy, *J. Phys. Chem.* **63**, 1858 (1959).

352. G. Burley, *Am. Mineral.* **48**, 1266 (1963); *Acta Crystallogr.* **23**, 1 (1967).

353. S. Hoshino, *J. Phys. Soc. Jpn* **12**, 315 (1957).

354. R. J. Cava and B. J. Wuensch, in *Superionic Conductors*, eds. G. D. Mahan and W. L. Roth, Plenum Press, New York (1976), p. 218.

355. B. B. Owens, in *Advances in Electrochemistry and Electrochemical Engineering*, Vol. 8, ed. C. Tobias, Wiley Interscience, New York (1971), p. 1.

356. J. A. A. Ketalaar, *Z. Kristallogr.* **180**, 190 (1931); *Trans. Faraday Soc.* **34**, 874 (1938).

357. F. Hanic, J. S. Kasper and H. Wiedemeier, *J. Solid State Chem.* **10**, 20 (1974).

358. S. Hoshimo, *J. Phys. Soc. Jpn* **10**, 197 (1955).

359. J. S. Kasper and K. W. Browall, *J. Solid State Chem.* **13**, 49 (1975).
360. H. Wiederisch and S. Geller, in *The Chemistry of Extended Defects in Non-metallic Solids*, eds. L. Eyring and M. O'Keefe, North-Holland, Amsterdam, (1970), p. 629.
361. R. Weil and A. W. Lawson, *J. Chem. Phys.* **41**, 832 (1964).
362. J. N. Bradley and P. D. Greene, *Trans. Faraday Soc.* **62**, 2069 (1964).
363. B. B. Owens and G. R. Argue, *Science* **157**, 308 (1967); *J. Electrochem. Soc.* **117**, 898 (1974).
364. D. O. Raleigh, *J. Appl. Phys.* **41**, 1876 (1970).
365. S. Geller, *Science* **157**, 310 (1967).
366. F. P. Bundy, J. S. Kasper, and M. J. Moore, *High Temp.—High Pressure* **3**, 303 (1971).
367. S. Seetharaman and K. P. Abraham, *J. Sci. Ind. Res. (India)* **32**, 641 (1973).
368. T. Takahashi and O. Yamamoto, *J. Electrochem. Soc.* **117**, 1 (1970).
369. N. DeRossi, G. Pistoia, and B. Scrosati, *J. Electrochem. Soc.* **116**, 1642 (1969).
370. B. Scrosati, G. Germano, and G. Pistoia, *J. Electrochem. Soc.* **118**, 861 (1971).
371. B. Scrosati, *J. Electrochem. Soc.* **118**, 899 (1971).
372. D. M. Smyth, C. H. Tompkins, and S. D. Ross, *Proc. Ann. Power Sources Conf.* **24**, 24 (1970).
373. S. Geller and M. D. Lind, *J. Chem. Phys.* **52**, 5854 (1970).
374. S. Geller, *Science* **176**, 1016 (1972).
365. S. Geller and B. B. Owens, *J. Phys. Chem. Solids* **33**, 1241 (1972).
376. S. Geller, in *Solid Electrolytes*, ed. S. Geller, Springer-Verlag, Berlin (1977), p. 641.
377. B. Reuter and K. Hardel, *Naturwissenschaften* **48**, 161 (1961); *Z. Anorg. vllgem. Chem.* **340**, 158 (1965); *Ber. Bungenges Phys. Chem.* **70**, 87 (1966).
378. T. Takahashi and O. Yamamoto, *Denki Kagaku* **30**, 610 (1964); **33**, 346 (1965); *Electrochim. Acta* **11**, 779, 911 (1966).
379. B. B. Owens, G. R. Argue, J. J. Groce, and L. D. Hermo, *J. Electrochem. Soc.* **116**, 312 (1969).
380. B. B. Owens, in *Fast Ion Transport in Solids*, ed. W. Van Gool, North-Holland, Amsterdam (1973), p. 593.
381. E. A. Ukshe and N. G. Bukura, *Zh. Vses. Khim. Obshchest.* **16**, 658 (1971).
382. T. Takahashi, in *Solid Electrolytes*, eds. P. Hagenmuller and W. Van Gool, Academic Press, New York (1978), p. 201.
383. J. S. Kasper, in *Solid Electrolytes*, eds. P. Hagenmuller and W. Van Gool, Academic Press, New York (1978), p. 217.
384. R. D. Armstrong and T. Dickinson, in *Superionic Conductors*, eds. G. D. Mahan and W. L. Roth, Plenum Press, New York (1976), p. 65.
385. S. Geller, in *Superionic Conductors*, eds. G. D. Mahan and W. L. Roth, Plenum Press, New York (1976), 0. 171.
386. K. Funke, in *Superionic Conductors*, eds. G. D. Mahan and W. L. Roth, Plenum Press, New York (1976), p. 183.
387. T. Takahashi, in *Superionic Conductors*, eds. G. D. Mahan and W. L. Roth, Plenum Press, New York (1976), p. 379.
388. B. Scrosati, *J. Electrochem. Soc.* **120**, 519 (1973).
389. T. Matsui and J. B. Wagner, Jr., in *Solid Electrolytes*, eds. P. Hagenmuller and W. Van Gool, Academic Press, New York (1978), p. 237.
390. T. L. Markin in *Electromotive Force Measurements in High Temperature Systems*, ed. C. B. Alcock, Inst. Mining Met., London (1968), p. 125.
391. J. M. Reau and J. Portier, in *Solid Electrolytes*, eds. P. Hagenmuller and W. Van Gool, Academic Press, New York, (1978), p. 313.
392. E. Zintle and A. Udgard, *Z. Anorg. Allgem. Chem.* **290**, 150 (V939).

393. E. Barsis and A. Taylor, *J. Chem. Phys.* **45**, 1154 (1966).
394. H. Martze, *J. Mater. Sci.* **5**, 831 (1970).
395. M. Berard, *J. Am. Ceram. Soc.* **54**, 144 (1971).
396. H. Martze and R. Lindner, *Z. Naturforsch.* **A19**, 1178 (1964).
397. M. E. Baker and A. Taylor, *J. Phys. Chem. Solids* **30**, 1003 (1969).
398. J. M. Short and R. Roy, *J. Phys. Chem.* **68**, 3077 (1964).
399. D. L. Johnson, *J. Am. Ceram. Soc.* **55**, 327 (1972).
400. R. Scheidecker and M. V. Berard, *J. Am. Ceram. Soc.* **56**, 204 (1973).
401. E. Mollwo, *Nachr. Ges. Wiss. Göttingen, Matt. Phys. K.I., N.F.* **6**, 79 (1934).
402. C. Wagner, *J. Electrochem. Soc.* **115**, 933 (1968).
403. J. W. Patterson, in *Physics of Electronic Ceramics*, eds. L. L. Hench and D. B. Dove, Marcel Dekker, New York (1971), p. 131.
404. J. W. Hinze and J. W. Patterson, *J. Electrochem. Soc.* **120**, 96 (1973).
405. W. L. Phillips and J. E. Hanlon, *J. Am. Ceram. Soc.* **46**, 477 (1963).
406. D. V. Vechter, quoted by Y. O. Tretyakov and A. R. Kaul in *Physics of Solid Electrolytes*, ed. J. Hladik, Academic Press, London (1972), Chap. 14.
407. R. Benz and C. Wagner, *J. Phys. Chem.* **65**, 1308 (1961).
408. R. W. Taylor and H. Schmalzreid, *J. Phys. Chem.* **68**, 2444 (1964).
409. T. N. Rezhukina, V. A. Levitskii, and M. J. Frenkel, *Izv. Akad. Nauk SSSR, Neorg. Mater.* **2**, 325 (1966).
410. A. A. Vetcher and D. V. Vetcher, *Zh. Fiz. Khim.* **41**, 1288 (1967); **41**, 2916 (1967).
411. H. G. Schnering and P. Blackmann, *Naturwissenschaften* **52**, 53 (1965).
412. J. L. Holm, *Acta Chem. Scand.* **19**, 1512 (1965).
413. P. F. Werner, *Inorg. Chem.* **5**, 736 (1966).
414. L. E. Nagel and M. O'Keefe, in *Fast Ion Transport in Solids*, ed. W. Van Gool, North-Holland, Amsterdam (1973).
415. J. M. Reau, C. Lucat, G. Campel, J. Portier, and A. Hammou, *J. Solid State Chem.* **17**, 123 (1976).
416. A. K. Cheetham, B. E. F. Fender, D. Steele, R. I. Taylor, and B. T. M. Willis *Solid State Commun.* **8**, 171 (1970).
417. A. K. Cheetham, B. E. F. Fender, and M. J. Cooper, *J. Phys. C Solid State Phys.* **4**, 3107 (1971).
418. J. H. Kennedy and R. C. Miles, *J. Electrochem. Soc.* **123**, 1 (1976).
419. J. Schoonman, L. B. Ebert, C. H. Hsieuh, and R. A. Huggins, *J. Solid State Chem.* **17**, 123 (1976).
420. C. E. Darrington and M. O'Keefe, *Nature* **246**, 44 (1973).
421. C. E. Darrington, A. Navrotsky, and M. O'Keefe, *Solid State Commun.* **18**, 4 (1976).
422. C. Lucat, G. Campel, J. Claverie, J. Portier, J. M. Reau, and P. Hagenmuller, *Mater. Res. Bull.* **12**, 145 (1976).
423. U. Croatto and M. Bruno, *Gazz. Chem. Ital.* **78**, 95 (1948).
424. E. Zintle and U. Croatto, *Z. Anorg. Allgem. Chem.* **242**, 79 (1939).
425. E. W. Dewing, *Trans. AIME* **245**, 1829 (1969).
426. M. O'Keefe, in *Superionic Conductors*, eds. G. D. Mahan and W. L. Roth, Plenum Press, New York (1976), p. 101.
427. E. C. Subbarao, in *Non-Stoichiometric Compounds*, ed. L. Mandelcorn, Academic Press, New York (1964), Chap. 5.
428. J. Singer, H. E. Kautz, W. L. Fielder and J. S. Fordyce, in *Fast Ion Transport in Solids*, ed. W. Van Gool, North-Holland, Amsterdam (1973), p. 653.
429. P. H. Bottelberghs, in *Fast Ion Transport in Solids*, ed. W. Van Gool, North-Holland, Amsterdam (1973), p. 637.

430. R. S. Roth, H. S. Parker, W. S. Brower, and J. L. Waring, in *Fast Ion Transport in Solids*, ed. W. Van Gool, North-Holland, Amstercam (1973), p. 217.
431. A. D. Wadsley, *Rev. Pure Appl. Chem.* **5**, 165 (1965).
432. J. S. Dryden and A. D. Wadsley, *Trans. Faraday Soc.* **54**, 1574 (1958).
433. R. S. Roth, in *Fast Ion Transport in Solids*, ed. W. Van Gool, North-Holland, Amsterdam (1973), p. 663.
434. J. M. Reau, C. Delmas, and P. Hagenmuller, in *Solid Electrolytes*, eds. P. Hagenmuller and W. Van Gool, Academic Press, New York (1978), p. 381.
435. J. B. Goodenough, in *Solid Electrolytes*, eds. P. Hagenmuller and W. Van Gool, Academic Press, New York (1978), p. 393.
436. J. B. Goodenough, H. Y-P. Hong, and J. A. Kafalas, *Mater. Res. Bull.* **11**, 203 (1976).
437. D. Ravaine and J. L. Sonquet, in *Solid Electrolytes*, eds. P. Hagenmuller and W. Van Gool, Acacemic Press, New York (1978), p. 277.
438. M. Jozefowicz, in *Fast Ion Transport in Solids*, ed. W. Van Gool, North-Holland, Amsterdam (1973), p. 623.
439. W. Van Gool, in *Fast Ion Transport in Solids*, ed. W. Van Vool, North-Holland, Amsterdam (1973), p. 201.
440. R. D. Armstrong, R. S. Bulmer and T. Dickinson, in *Fast Ion Transport in Solids*, ed. W. Van Gool, North-Holland, Amsterdam (1973), p. 269.
441. S. Geller, in *Fast Ion Transport in Solids*, ed. W. Van Gool, North-Holland, Amsterdam (1973), p. 607.

Limiting Factors in Measurements Using Solid Electrolytes

T. A. Ramanarayanan

1. Introduction

Measurements using solid electrolytes are of two kinds: thermodynamic measurements and kinetic measurements. The use of solid electrolytes for thermodynamic measurements received much of its impetus from the work of Kiukkola and Wagner[1,2] and Peters et al.[3,4] They introduced the use of ZrO_2-base and ThO_2-base solid electrolytes for the determination of the standard free energy of formation of oxides. As early as 1943, Wagner had explained transport mechanisms in doped zirconia.[5] Since the work of Kiukkola and Wagner,[2] solid oxide and halide electrolytes have been extensively used in several thermodynamic investigations. Kinetic measurements using solid electrolytes fall into two categories. The first category includes the measurement of transport phenomena in solid electrolytes. The polarization technique devised by Hebb[6] and Wagner[7] may be used for this purpose. The second category includes the use of a solid electrolyte to study transport phenomena in an electrode or the kinetics of a phase boundary reaction. Detailed discussions on thermodynamic measurements and kinetic measurements are presented in Chapters 3, 4, and 5. Several excellent review articles have also appeared on the subject.[8-15]

In 1933, Wagner[16] derived an expression for the steady-state open-circuit voltage across a compact mixed-conductor scale. More detailed derivations when phases of locally variable composition are present have also been given

T. A. Ramanarayanan • Department of Metallurgical Engineering, Indian Institute of Technology, Kanpur 208016, India. Present address: Corporate Research Laboratories, Exxon Research and Engineering Company, Linden, New Jersey 07036.

by Wagner.[17] A brief derivation is given below to emphasize the difference in the equations to be used for thermodynamic and kinetic measurements and to draw attention to the origin and significance of certain parameters.

Consider an electrochemical cell made up of a solid compound, MO, in which oxygen ions, O^{2-}, excess electrons, e′, and electron holes, h·, are mobile, with oxygen chemical potentials μ'_{O_2} and μ''_{O_2} at the left-hand and right-hand electrode–electrolyte interfaces, respectively, as shown below:

$$\text{Pt, } \mu'_{O_2} \left| \text{ MO } \right| \mu''_{O_2}, \text{ Pt} \qquad\qquad [2.1]$$
$$\quad x=0 \quad x=L$$

The partial current density, I_i, of a mobile species, i, at any location within the electrolyte may be written as[7]

$$I_i = \frac{-\sigma_i}{z_i F} \frac{d\eta_i}{dx} \qquad\qquad (2.1)$$

where σ_i is the partial conductivity of species i, z_i is the valence of i, F is the Faraday constant, x is the distance coordinate, and η_i is the electrochemical potential of species i defined by

$$\eta_i = \mu_i + z_i F \varphi \qquad\qquad (2.2)$$

Here μ_i is the chemical potential of i and φ is the local electrostatic potential. In view of Eqs. (2.1) and (2.2), the sum of the current densities of the oxygen ions, excess electrons, and electron holes in MO may be written as

$$I_{O^{2-}} + I_{h·} + I_{e'} = \frac{-d\varphi}{dx}(\sigma_{O^{2-}} + \sigma_{h·} + \sigma_{e'}) + \frac{\sigma_{O^{2-}}}{2F}\frac{d\mu_{O^{2-}}}{dx}$$
$$+ \frac{\sigma_{e'}}{F}\frac{d\mu_{e'}}{dx} - \frac{\sigma_{h·}}{F}\frac{d\mu_{h·}}{dx} \qquad (2.3)$$

In thermodynamic measurements, one is interested in the open-circuit condition, which may be defined as the limiting case where the sum $I_{O^{2-}} + I_{h·} + I_{e'}$ tends to zero. Thus at open circuit, from Eq. (2.3),

$$d\varphi = \frac{t_{O^{2-}}}{2F}d\mu_{O^{2-}} + \frac{t_{e'}}{F}d\mu_{e'} - \frac{t_{h·}}{F}d\mu_{h·}. \qquad (2.4)$$

where $t_{O^{2-}}$, $t_{h·}$, and $t_{e'}$ are the transference numbers of oxygen ions, electron holes, and excess electrons, respectively, and are given by the ratios of the respective partial conductivities to the total conductivity, $\sigma_{O^{2-}} + \sigma_{h·} + \sigma_{e'}$. The following equilibria may be assumed at any location in MO:

$$\tfrac{1}{2}O_2 + 2e' = O^{-2} \qquad\qquad (2.5)$$

$$h· + e' = \text{null} \qquad\qquad (2.6)$$

From Eqs. (2.5) and (2.6),

$$d\mu_{O^{2-}} - 2d\mu_{e'} = \tfrac{1}{2}d\mu_{O_2} \tag{2.7a}$$

and

$$d\mu_{h\cdot} = -d\mu_{e'} \tag{2.7b}$$

Equations (2.4)–(2.7b), with the further condition that $t_{O^{2-}} + t_{h\cdot} + t_{e'} = 1$, give

$$d\varphi = \frac{t_{O^{2-}}}{4F}\, d\mu_{O_2} + \frac{d\mu_{e'}}{F} \tag{2.8}$$

Integrating from $x = 0$ to $x = L$,

$$\varphi'' - \varphi' = \frac{1}{4F}\int_{\mu'_{O_2}}^{\mu''_O} t_{O^{2-}}\, d\mu_{O_2} + \frac{1}{F}(\mu''_{e'} - \mu'_{e'}) \tag{2.9}$$

Since $\varphi'' - \varphi'$ equals the open-circuit voltage $(E_{o.c.})$ and $\mu'_{e'} = \mu^{Pt}_{e'} = \mu''_{e'}$, one has

$$E_{o.c.} = \frac{1}{4F}\int_{\mu'_{O_2}}^{\mu''_{O_2}} t_{O^{2-}}\, d\mu_{O_2} \tag{2.10}$$

The more general form of Eq. (2.10) is

$$E_{o.c.} = \frac{1}{4F}\int_{\mu'_{O_2}}^{\mu''_{O_2}} t_{ion}\, d\mu_{O_2} \tag{2.11}$$

However, $t_{ion} = t_{O^{2-}}$ in the oxide solid electrolytes. Equation (2.11) forms the basis for thermodynamic measurements using solid electrolytes. In most such measurements $t_{ion} \geq 0.99$.

In kinetic studies, one has a nonzero current through the electrochemical cell circuit. From Eq. (2.1) the oxygen ion current density is given by

$$I_{O^{2-}} = \frac{\sigma_{O^{2-}}}{2F}\frac{d\eta_{O^{2-}}}{dx} \tag{2.12}$$

From Eqs. (2.2), (2.7a), and (2.12),

$$I_{O^{2-}} = \frac{\sigma_{O^{2-}}}{2F}\left(\frac{1}{2}\frac{d\mu_{O_2}}{dx} + 2\frac{d\mu_{e'}}{dx} - 2F\frac{d\varphi}{dx}\right) \tag{2.13}$$

$I_{O^{2-}}$ is the current density of oxygen ions at any location in the electrolyte. This is related to the ionic current i_{ion} in the external measuring circuit through

$$I_{O^{2-}} = -\frac{i_{ion}}{A} \tag{2.14}$$

where A is the cross-sectional area of the electrolyte. Combining (2.13) and (2.14) and integrating from $x = 0$ to $x = L$, we obtain

$$E_{\text{cell}} = \frac{1}{4F} \int_{\mu'_{O_2}}^{\mu''_{O_2}} d\mu_{O_2} + i_{\text{ion}} \Omega_{\text{ion}} \tag{2.15}$$

where

$$\Omega_{\text{ion}} = \frac{1}{A} \int_{x=0}^{x=L} \frac{1}{\sigma_{O^{2-}}} \, dx$$

is the ionic resistance of the electrolyte and $E_{\text{cell}} = \varphi'' - \varphi'$.

Equation (2.15) forms the basis for kinetic measurements using solid electrolyte cells.

In using solid electrolyte cells for thermodynamic and kinetic measurements, several factors limit the applicability of the technique. In what follows, these factors are evaluated and discussed. Ramanarayanan and Worrell[18] have discussed the limitations in using solid state electrochemical cells for high-temperature thermodynamic studies. This paper emphasizes the limitations in the area of kinetic studies also.

2. Electronic Conduction in the Electrolyte

According to Eq. (2.11), the measured open-circuit voltage, $E_{\text{o.c.}}$, is a simple function of the oxygen chemical potentials or partial pressures at the electrolyte boundaries when $t_{\text{ion}} \geq 0.99$ over the experimental oxygen partial pressure range. When t_{ion} is less than 0.99, there is electronic conduction in the electrolyte and there may be a problem in maintaining desired chemical potentials at the electrode–electrolyte interfaces. In solid oxide electrolytes, there is a useful range of oxygen pressure and temperature where $t_{\text{ion}} \geq 0.99$. Outside this range, positive hole conduction occurs at high oxygen pressures and electron conduction at low oxygen pressures. The oxygen pressure dependence of positive hole conduction and electron conduction is based on the following equilibria:

$$\tfrac{1}{2}O_2 + V_O^{\cdot\cdot} + 2e' = O_O \tag{2.16}$$

$$\tfrac{1}{2}O_2 + V_O^{\cdot\cdot} = O_O + 2h^{\cdot} \tag{2.17}$$

Here the defect notation of Kröger and Vink[19,20] has been used. In oxide electrolytes, oxygen ion vacancies are the main charge carriers and their concentrations are fixed by doping. From the equilibrium constants for the reactions (2.16) and (2.17), it follows that $n \propto P_{O_2}^{-1/4}$ and that $p \propto P_{O_2}^{1/4}$, where n and p are the concentrations of excess electrons and electron holes, respectively. The partial conductivity, σ_i, of any species is given by

$$\sigma_i = C_i q_i^2 B_i \tag{2.18}$$

where q_i is the charge (coulombs/particle), C_i is the concentration (particles/cm^3), and B_i is the mobility (particles cm^2 sec^{-1} V^{-1} C^{-1}). In view of Eq. (2.18), for concentration-independent mobilities, one obtains $\sigma_h \propto P_{O_2}^{1/4}$, $\sigma_{e'} \propto P_{O_2}^{-1/4}$ and $\sigma_{ion} = \sigma_{V_{\ddot{O}}} =$ constant for any given temperature. Thus on a plot of $\log \sigma_t$ vs. $\log P_{O_2}$ (where σ_t is the total conductivity), one would expect an intermediate region of constant conductivity where $\sigma_t \simeq \sigma_{ion}$. At high P_{O_2}'s, a P_{O_2}-dependent positive hole conduction will occur. At low P_{O_2}'s, a P_{O_2}-dependent electronic conduction is expected.

Figure 2.1 shows the conductivity vs. P_{O_2} plot for $Zr_{0.85}$ $Ca_{0.15}$ $O_{1.85}$[21] and $Th_{0.85}$ $Y_{0.15}$ $O_{1.85}$[22] at 1000°C. Patterson, Bogren, and Rapp's[21] conductivity data seem to suggest that calcia-stabilized zirconia may be used for thermodynamic measurements down to a $P_{O_2} = 10^{-25}$ atm at 1000°C. However, several emf measurements indicate that the low P_{O_2} limit is much higher.[23,24] On the high P_{O_2} side, calcia-stabilized zirconia is useful up to even 1 atm oxygen pressure and there is no evidence for positive hole conduction. Yttria-doped thoria, on the other hand, starts to exhibit positive hole conduction around $P_{O_2} = 10^{-6}$ atm at 1000°C. However, on the low P_{O_2} side, yttria-doped thoria is ionic down to much lower oxygen pressures than calcia-stabilized zirconia. Down to a P_{O_2} of 10^{-25} atm at 1000°C, there is no evidence for electronic conduction. It has been suggested[21] that yttria-doped thoria may be ionic down to a P_{O_2} of 10^{-34} atm at 1000°C. The actual low P_{O_2} limit of yttria-doped thoria is yet to be established.

For the calcium fluoride electrolyte, Wagner[25] has provided an analysis to predict the onset of electronic conduction under reducing conditions. His analysis is based on the color center migration data of Mollwo.[26] Wagner has calculated that the calcium activity for which $t_{e'} = 10^{-2}$ equals 10^{-5} at 600°C and 6×10^{-6} at 840°C. At lower calcium activities, the electronic transference

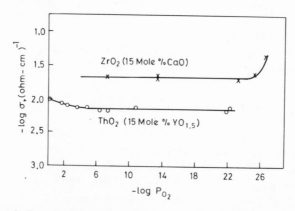

FIG. 2.1. Electrical conductivity of $ZrO_2(CaO)$[21] and $ThO_2(Y_2O_3)$[22] as a function of oxygen partial pressure at 1000°C (Reference 14).

number would be still less. Hinze and Patterson[27] have shown that calcium fluoride is ionic even in equilibrium with pure calcium. They studied the emf of the cell

$$Ca \mid CaF_2 \mid Th, ThF_4 \qquad\qquad [2.2]$$

Their results are in reasonable agreement with the expected thermodynamic emf's. However, recent polarization measurements by Delcet et al.[54] indicate that at temperatures in the range 800–950°C, n-type electronic conduction is present in single crystalline calcium fluoride at calcium activities higher than 10^{-3}.

Patterson[28] has suggested the use of chemical potential vs. $1/T$ diagrams in order to represent conduction domains in solid electrolytes. Such a diagram, as given by Patterson, is shown in Fig. 1.7 of Chapter 1. The low-temperature limits shown for calcia-stabilized zirconia and yttria-doped thoria are fixed by the requirement that a minimum conductivity is needed in order to use a compound as a solid electrolyte. According to Schmalzried,[11] this value is about 10^{-6} (ohm cm)$^{-1}$. The actual values of P_{O_2} at which electronic conduction is introduced in calcia-stabilized zirconia and yttria-doped thoria are probably lower than those represented in Fig. 1.7. Figure 1.7 also includes the electrolytic domain for other solid electrolytes.

The exact location of the domain boundaries shown in Fig. 1.7 depends on the purity of the electrolyte. For example, the oxygen pressure at which n-type electronic conduction becomes significant in high-purity yttria-doped thoria is probably much lower than that shown in Fig. 1.7. For a yttria-doped thoria electrolyte with 3 mol % of oxygen vacancies, it would take only 60 mol ppm of an impurity (which is compensated by excess electrons) to introduce significant electronic conduction (i.e., $t_{ion} < 0.99$), if the electronic mobility is 1000 times greater than the ionic mobility.[18] To maintain t_{ion} at 0.99 for the impure yttria-doped thoria, the electron concentration must be decreased by increasing the P_{O_2} in accordance with Eq. (2.16). In regions near the domain boundaries, it would take a smaller concentration of impurities to introduce electronic conductivity because a significant number of electronic defects are already present according to Eqs. (2.16) and (2.17).

2.1. Examples from Thermodynamic Measurements

The Cr–Cr_2O_3 equilibrium has an oxygen potential very near the low P_{O_2} electrolytic domain boundary for calcia-stabilized zirconia. Therefore impurities in the electrolyte may well determine whether reliable measurements can be made with the Cr–Cr_2O_3 electrode. The Cr–Cr_2O_3 equilibrium ($P_{O_2} = 10^{-22}$ at 1000°C) has been investigated by Tretyakov and Schmalzried[29] and by Pugliese and Fitterer[30] using a calcia-stabilized zirconia electrolyte. Recently this equilibrium has been reinvestigated by Mazandarany

Table 2.1. Calculated $t_{e'}$ in Calcia-Stabilized Zirconia Electrolytes[a]

Investigator	Cell investigated	emf (mV) at 1000°C	$t_{e'}$ at 1000°C
Tretyakov and Schmalzried	Cr, Cr_2O_3 | ZrO_2 + CaO | Co, CoO[b]	612.00	2.1×10^{-2}
Pugliese and Fitterer	Cr, Cr_2O_3 | ZrO_2 + CaO | Co, CoO	599.53	4.1×10^{-2}
Mazandarany and Pehlke	Cr, Cr_2O_3 | ThO_2 + Y_2O_3 | Co, CoO	625.13	
Hoch *et al.*	Nb, NbO | ZrO_2 + CaO | NbO, NbO_2	174.00	1.1×10^{-1}
Worrell	Nb, NbO | ThO_2 + Y_2O_3 | NbO, NbO_2	195.00	

[a] From Reference 18.
[b] Although an air reference electrode was used, the values in Table 2.1 were obtained by combining their data with the Co–CoO equilibrium data of Kiukkola and Wagner.[1]

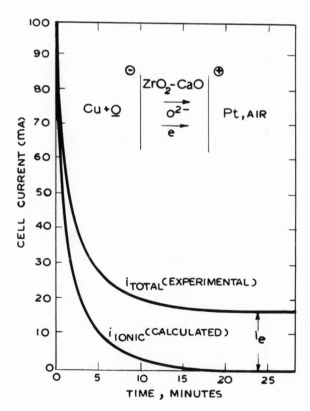

FIG. 2.2. Variation of current with time with change in applied emf (Reference 32).

and Pehlke[31] using a yttria-doped thoria electrolyte. Since electronic conductivity effects should be negligible in the yttria-doped thoria electrolyte, one can calculate the average electronic transference number ($t_{e'}$) in the calcia-stabilized zirconia electrolyte used by Tretyakov and Schmalzried[29] and by Pugliese and Fitterer.[30] Results of such calculations are shown in Table 2.1.

2.2. Examples from Kinetic Measurements

Rapp and co-workers[32–35] have used a potentiostatic electrochemical technique to determine the diffusivity of oxygen in liquid and solid metals. The method is discussed in Chapter 5, Section 3.1. The basic equation used in these studies is

$$E_{\text{appl}} = \frac{1}{4F} \int_{\mu'_{O_2}}^{\mu''_{O_2}} d\mu_{O_2} + i_{\text{ion}}\Omega_{\text{ion}} \qquad (2.19)$$

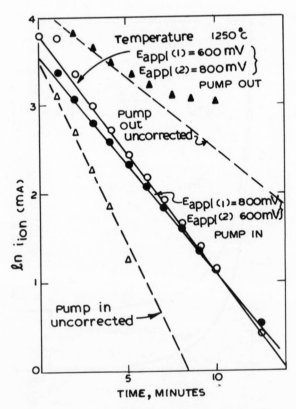

FIG. 2.3. Total and transient ionic currents (Reference 32).

which is a modified form of Eq. (2.15) with $E_{cell} = E_{appl}$, where E_{appl} is the applied voltage. In these diffusion experiments considerable electronic currents were observed in the stabilized zirconia electrolyte which was used, as shown in Fig. 2.2. The electronic currents must be subtracted from the total current to get the ionic current. If it is assumed that $i_{total} = i_{ion}$, there will be a considerable error in the calculated diffusivity as shown in Fig. 2.3. Thus, in potentiostatic diffusion measurements, always a proper correction must be made for the electronic current.

3. Maintaining a Desired Chemical Potential at the Electrode–Electrolyte Interface

3.1. Influence of Electronic Conductivity

If t_{ion} is less than unity, Eq. (2.11) can still be integrated if the variation of t_{ion} with μ_{O_2} is known.[36] But when $t_{ion} < 0.99$, there can be experimental difficulties, particularly with solid electrodes. There will be a short-circuiting current through the electrolyte and electrons will move through the electrolyte from the electrode with the lower oxygen chemical potential to that with the higher oxygen chemical potential. This electron flow must be balanced by an oxygen ion flow in the opposite direction. If diffusional processes cannot supply or remove oxygen rapidly enough at the electrode–electrolyte interfaces, there can be oxide formation at the low P_{O_2} interface and metal formation at the high P_{O_2} interface. Even when there is no layer formation, the chemical potential of oxygen at the interface can differ from that in the bulk of the electrode. If the magnitude of the short-circuiting current is very small, it is possible to obtain a stable emf which includes an overvoltage contribution. This overvoltage arises from a shift in the oxygen potential at the electrode–electrolyte interface from that established by the metal–metal-oxide equilibrium to that fixed by the steady state oxygen transfer. Recent measurements[37] with various metal–metal oxide electrodes have demonstrated that a steady-state overvoltage is obtained when extremely small currents (1–100 μA) are passed through symmetrical electrochemical cells of the type

$$M, MO \mid ZrO_2(CaO) \mid M, MO \qquad [2.3]$$

Results of measurements indicate that the major oxygen transfer at the electrode–electrolyte interface occurs between the electrolyte and the metal particles in the electrode. Results[37] are shown in Fig. 2.4 to illustrate the variation of the steady state overvoltage, η, vs. the imposed current, i, at 900°C for Ni–NiO, Fe–FeO, and Cu–Cu$_2$O electrodes. Cu–Cu$_2$O electrodes can adjust to small cell currents much more easily than either Fe–FeO or Ni–NiO electrodes.

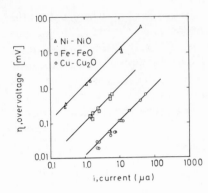

FIG. 2.4. Overvoltage vs. current plots for Cu, Cu₂O; Fe, FeO; and Ni, NiO electrodes at 900°C (Reference 37).

3.2. Influence of Chemical Reactions

Reactions between the electrode and the electrolyte can cause problems in precisely measuring thermodynamic parameters. Consider, for example, a cell of the type

$$A \mid AX \mid A\text{--}B \qquad [2.4]$$

where AX is a solid electrolyte and A–B is an alloy of metals A and B. The open-circuit emf of this cell is related to the activity of A in the alloy. However, at the AX | A–B interface, the following displacement reaction is possible:

$$B \text{ (alloy)} + AX = A \text{ (alloy)} + BX \qquad (2.20)$$

The reaction would lead to a change in the activity of A at the electrode-electrolyte interface, the extent of the change being a function of the initial composition, N_A°, of the alloy, the standard free-energy change for reaction (2.20), and the interdiffusion coefficients, D_{AX} and $D_{A\text{--}B}$ in the electrolyte and the alloy, respectively. Wagner and Werner[38] have estimated the change in the activity, $\varepsilon = \delta a_A / a_A^\circ$, for liquid systems. Schmalzried,[11] using Wagner's analysis, has estimated ε for solid systems as a function of D_{AX}, $D_{A\text{--}B}$, N_A°, and ΔG° for reaction (2.20). The calculations of Schmalzried[11] are presented in Fig. 2.5. It is clear from Fig. 2.5 that in order for ε to be very small, B should be much nobler than A.

Often, an electrode–electrolyte reaction can result in a reaction product which forms a layer between the electrode and the electrolyte. This case has been discussed by Schmalzried.[8] For example, consider the cell

$$\text{Pt, O}_2 \left| \text{ZrO}_2(\text{CaO}) \right| \text{M, MO, Pt} \qquad [2.5]$$
$$(\text{P}'_{\text{O}_2}) \qquad\qquad (\text{P}''_{\text{O}_2})$$

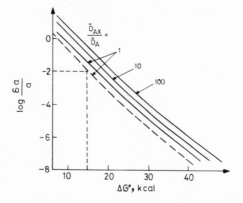

FIG. 2.5. Relative activity change resulting from the displacement reaction [B] + AX = [A] + BX at the electrolyte AX/alloy A–B phase boundary, plotted as a function of $\Delta G°$, the standard value of the change in Gibbs energy of the displacement reaction (Reference 11).

Under normal conditions, the emf of cell [2.5] is given by

$$E = \frac{RT}{4F} \ln \frac{P''_{O_2}}{P'_{O_2}} \tag{2.21}$$

when $t_{O^{2-}}$ is unity in the stabilized zirconia. However, if a reaction product such as $MZrO_3$ forms between the M–MO electrode and the calcia-stabilized zirconia electrolyte, then the emf is no longer given by Eq. (2.21) if the transport number of oxygen ions in $MZrO_3$ is less than unity. In studies with the cell

$$NbO_2, Nb_2O_{4.8} \mid ThO_2(Y_2O_3) \mid Fe, Fe_{0.95}O \tag{2.6}$$

Worrell[39] detected a $YFeO_3$ phase by x-ray diffraction which indicated that a reaction occurred between the Fe–$Fe_{0.95}O$ electrode and the thoria-base electrolyte. The observed emf of cell [2.6] was slightly lower than the expected value. According to Worrell, such a decrease in voltage is due to a reduction in the oxygen activity of the Fe–$Fe_{0.95}O$ electrode caused by the presence of $YFeO_3$.

Because of the high chemical stabilities of ZrO_2 and ThO_2, reactions between these electrolytes and electrode constituents can occur only when very stable ternary oxide compounds, intermetallic phases, or solid oxide solutions are formed as reaction products.

Reactions within the electrode can also lead to chemical potential variations in the electrode. When using alloy–oxide electrodes, one can prevent an oxide–alloy reaction by using only those alloys in which one metal oxide is at least 20 kcal/mol more stable than the other. If the two oxides can form solid solutions, then the difference in $\Delta G_f°$ of the oxides should be even greater. For example, electrochemical cell measurements of the Ta–Mo and Ta–W systems have been made,[40,41] whereas the Ta–V system cannot be investigated using an oxide solid electrolyte.

3.3. Influence of the Gaseous Atmosphere

The oxygen pressure of an inert gas can be reduced to as low as 10^{-15} atm by purification. Therefore it is preferable to make high-temperature electrochemical measurements in a flowing argon or helium atmosphere than in a vacuum. The oxygen in the inert gas is very rarely in equilibrium with the electrode. This is particularly true with metal–metal-oxide electrodes in which the oxide is a highly stable refractory oxide such as Ta–Ta$_2$O$_5$ and Nb–NbO. At oxygen pressures of the order of 10^{-15} atm in the inert gas, the number of oxygen molecules striking the electrode is negligible and the inert gas should not affect the chemical potential at the electrode–electrolyte interface.

Careful purification of the inert gas is essential especially when working with low-oxygen-activity electrodes. This is particularly true when the oxygen pressure fixed by the electrode is below about 10^{-20} atm. The inert gas may be purified by first passing through anhydrous calcium sulfate, magnesium perchlorate, or phosphorous pentoxide, and then through BASF catalyst (activated copper) maintained at 200°C or Ti–Zr chips maintained at 800°C. Finally an oxygen getter (i.e., tantalum or niobium foil) should be placed inside the reaction tube upstream from the cell. The metal foil will remove any residual oxygen or water vapor which may arise from internal sources such as degassing of the refractory tubes. When working with fluoride electrolytes, the removal of moisture and carbon dioxide from the inert gas is essential.

It is important to determine whether the inert gas flow rate has any effect on the measured emf. If the cell emf is independent of flow rate, it is unlikely that the gaseous atmosphere is influencing the measurements. If there is a flow rate effect, oxygen in the inert gas has established a mixed potential at the electrode–electrolyte interface. Improved electrode–electrolyte interfacial contact can be established by using pellets with prepolished smooth surfaces and by using springs or weights to promote intimate contact between electrodes and the electrolyte.

In some electrochemical cells, the difference between the oxygen pressures of the two electrodes may be so large that the inert gas atmosphere will tend to oxidize one electrode and reduce the other. To avoid this, one should choose a reference electrode which establishes an oxygen pressure which is within 10^5 atm of the estimated oxygen pressure of the unknown electrode. Rapp and Shores[10] have tabulated an extensive list of metal–metal-oxide reference electrodes. An extensive list is also given in Chapter 3. If a suitable reference electrode is not available, it is helpful to separate the electrodes into two compartments by using either an electrolyte tube or by sealing a refractory tube to the electrolyte pellet.

Solid oxide electrolytes are often used to measure oxygen activities in single-phase electrodes. To measure the change in the activity with composition, it is usual to change the oxygen content of such electrodes by a coulometric titration procedure. Since the oxygen concentrations involved in such measurements are very small, particular care must be taken to ensure that there is no oxygen input from the gaseous atmosphere.

3.4. Equilibration of the Electrode Constituents

Very often, problems in thermodynamic measurements are caused by the difficulty in attaining thermodynamic equilibrium with some electrodes. With single-phase electrodes of solid metals or nonstoichiometric oxides, equilibration is obviously limited by diffusional processes in the solid phase. In single-phase electrode studies, particular care is needed to ensure that the composition at the electrode–electrolyte interface is the same as in the bulk electrode. To avoid compositional gradients in the electrode at the electrode–electrolyte interface, intimate interfacial contact is essential.

In the study of two-phase electrodes such as metal–metal-oxide or alloy–metal-oxide mixtures, equilibration times may be limited by diffusional processes in the metal/alloy or the oxide. Difficulties may also arise because of the formation of a coherent oxide layer around the metal particles, thus preventing metal–oxide contact. This is particularly a problem with the $Cr–Cr_2O_3$ electrode. Pressed electrode pellets of small particles (-325 mesh) decrease the equilibration time by decreasing diffusion distances and increasing two-phase contact area. Singhal and Worrell[40,41] have observed equilibration times between 3 and 72 hr at temperatures between 800 and 1000°C for Ta–Mo, Ta_2O_5 and Ta–W, Ta_2O_5 electrodes.

With solid electrolytes, three-phase electrodes (and longer times for equilibration) are necessary to determine the thermodynamic properties of intermetallic compounds. For example, the thermodynamic properties of $ZrPt_3$[42,43] and $TiPt_3$[43] have been studied using (Zr–Pt) alloy, $ZrPt_3$, ZrO_2 and (Ti–Pt) alloy, $TiPt_3$, TiO_2 electrodes, respectively. Free energies of formation of Th–Co intermetallic compounds have been studied[44] using a CaF_2 electrolyte and a ThF_4 + Th–Co intermetallic compound + Co electrode.

3.5. Polarization Effects at the Electrode–Electrolyte Interface

Polarization at the electrode–electrolyte interface must be taken into account in several kinetic measurements. For example, Rickert and Steiner[45] studied the diffusivity of oxygen in solid silver using the cell

$$\text{Fe, FeO} \mid \text{ZrO}_2(\text{Y}_2\text{O}_3) \mid \text{Ag (O)} \qquad [2.7]$$

The current densities involved in their experiments varied from 500 $\mu\text{A/cm}^2$ to nearly 50 $\mu\text{A/cm}^2$. There is some polarization of the Fe–FeO electrode at these current densities (Fig. 2.5). In fact, a Cu–Cu$_2$O electrode would be more suitable.

Alcock and Belford have determined the solubility of oxygen in liquid lead[46] and liquid tin[47] using a stabilized zirconia and a doped thoria electrolyte, respectively. They used Ni/NiO as the reference electrode. In their experiments, the composition of the melt was changed by coulometrically titrating oxygen from the reference electrode into the melt. They observed that the Ni/NiO electrode polarized to a significant extent so that a separate measuring Ni/NiO electrode was found to be necessary for making thermodynamic measurements.

4. Effects of Porosity, Inhomogeneity, and Inadequate Mechanical Properties

The available literature on the effects of grain-size and porosity on the electrical conductivity of ZrO$_2$-base electrolytes has been reviewed by Möbius.[48] According to Tien,[49] grain-boundary conductivity contribution in Zr$_{0.84}$Ca$_{0.16}$O$_{1.84}$ is significant at 600°C, but not at 1000°C. A definite study of the effect of density on the partial conductivities of sintered compacts has not been made. However, sintered compacts with 95% theoretical density seem to be acceptable for most thermodynamic and kinetic applications. High-density electrolytes (above 98%) are usually transparent or translucent, while those with lower density are white or slightly colored. With low-density solid electrolytes, there is the possibility of gas transport through the electrolyte from one electrode compartment to the other. This will lead to unsteady emf's. In studies with the calcium fluoride electrolyte, it was found[50] that penetration of the metal fluoride electrode into the electrolyte occurred when pellets prepared from powdered CaF$_2$ were used as the electrolyte, resulting in unsteady emf. Better precision was obtained when plates cut from single crystals of calcium fluoride or CaF$_2$ doped with 3% YF$_3$ were used.

Inhomogeneities in the electrolyte can lead to regions where the electronic transference number is $> 10^{-2}$. When such regions are interconnected, mass transport through the electrolyte can occur under open-circuit conditions.

Solid oxide electrolyte tubes usually have poor thermal shock resistance. Rapp and co-workers[32–35] have found the commercial grader ZrO$_2$ (3%–3$\frac{1}{2}$% CaO) electrolyte tube to be less susceptible to cracking under temperature gradients. Hence these were used in oxygen diffusivity studies in liquid metals.

Even solidification of liquid metals in these tubes could be carried out. Tubes with higher CaO contents are susceptible to cracking under temperature gradients. Thermal cycling can also lead to microcracks.

5. Experimental Uncertainties

5.1. Limited Temperature Range of Study

In electrochemical cell measurements, the average uncertainty is ± 1 mV, which is 46 cal/g atom of oxygen for the Gibb's free-energy change, ΔG. Thus the accuracy of the measured Gibb's free-energy change is very high. However, determinations of the entropy changes (ΔS) and enthalpy changes (ΔH) involve the variation of emf with temperature, according to the expression

$$E = -\frac{\Delta H}{2F} + T\frac{dE}{dT} \qquad (2.22)$$

In view of the differential quotient, dE/dT, the error involved in the ΔH values obtained with the help of Eq. (2.22) is much higher than the corresponding error in the ΔG values. Using Eq. (2.22), Schmalzried has calculated the errors. Results of the calculation are shown in Fig. 2.6. Wagner[51] has pointed out that in spite of the above restriction, at times one obtains values of ΔH which are better than those obtained using calorimetric methods.

5.2. Instrumentation Problems

Measurements of open-circuit voltages of galvanic cells can be done using a potentiometer or an electrometer. When a potentiometer is used, only intermittent measurements are possible. Because the input impedance of potentiometers is small, the electrodes can get polarized. For continuous measurements of cell voltages, an electrometer with a high input impedance ($> 10^{12}$ ohms) must be used. While recording cell voltages, care must be taken to see that the recorder is not directly connected to the cell. Recorders usually have low input impedances and again electrode polarization can occur.

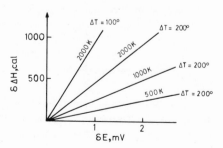

FIG. 2.6. Accuracy of ΔH values calculated with the help of Eq. (2.21). δE is the accuracy of the emf values. ΔT is the temperature range of emf measurements (Reference 11).

In potentiostatic experiments where an applied voltage, E_{appl}, is used to fix the concentration or activity at an electrode–electrolyte interface, some experimental difficulties have been encountered. It follows from Eq. (2.19) that

$$E_{appl} = \frac{RT}{4F} \ln \frac{P''_{O_2}}{P'_{O_2}} + i_{ion}\Omega_{ion} \qquad (2.23)$$

where P'_{O_2} is the reference P_{O_2}, P''_{O_2} is the oxygen pressure (or activity), which is desired to be fixed at the electrode–electrolyte interface, and i_{ion} is the ionic current, which is measured as a function of time. Since i_{ion} is varying with time, P''_{O_2} cannot be fixed by a constant applied voltage. Errors are minimized, however, by choosing applied voltages such that the $i_{ion}\Omega_{ion}$ term is small in comparison with the chemical term[31-34] [first term in Eq. (2.23)].

Electrical pickup leading to stray emf's is a problem in electrochemical measurements. To eliminate electrical pickup, the resistance furnace should be noninductively wound. A grounded tube of Pt foil or nichrome should surround the combustion tube. All electrical lead wires should be properly shielded and all measuring instruments should be grounded.

It is essential that the electrochemical cell should be placed within a constant temperature zone in which the temperature variation is less than $1°C$. The temperature variation across a cell makes a contribution to the measured emf of the cell. This contribution, for an oxide electrolyte cell, can be written as[52]

$$E = \frac{1}{4F}[\mu_{O_2}(T_2, P''_{O_2}) - \mu_{O_2}(T_1, P'_{O_2})] + \alpha(T_2 - T_1) \qquad (2.24)$$

where T_2 and T_1 are the temperatures at the two electrode–electrolyte interfaces and P'_{O_2} and P''_{O_2} are the corresponding oxygen pressures. α is a term related to the partial entropies and heats of transfer of the electrolyte and the electrode. When Pt electrodes were used, the α values were found to be equal to $0.095 \pm 0.005 \, mV/°C$ and $0.05 \pm 0.005 \, mV/°C$ for zirconia–lime and thoria–lime electrolytes, respectively.[52]

In electrochemical measurements, a suitable temperature controller must be used to provide temperature control within $\pm 1°C$ or better. A simple solid state circuit, using field effect transistors, has been developed, which, when used in conjunction with on–off controllers, provides temperature control within $\pm 0.2°C$ at $1000°C$.[53]

6. Concluding Remarks

The foregoing considerations show that many limiting factors influence the precise determination of thermodynamic and kinetic parameters using the solid electrolyte technique. Depending on the particular investigation, some of these factors will have greater significance. Preliminary investigations can usually determine the factors which influence a given study. A general precaution in thermodynamic measurements involving coexistence electrodes is to pass a small polarization current for a few seconds and to see whether the cell emf returns to the original value. In both thermodynamic and kinetic measurements, the voltage (or the current) at a given temperature must be reproducible irrespective of whether this temperature is reached by heating or by cooling.

References

1. K. Kiukkola and C. Wagner, *J. Electrochem. Soc.* **104**, 308 (1957).
2. K. Kiukkola and C. Wagner, *J. Electrochem. Soc.* **104**, 379 (1957).
3. H. Peters and H. Möbius, *Z. Phys. Chem.* **209**, 298 (1958).
4. H. Peters and G. Mann, *Z. Electrochem.* **63**, 244 (1959).
5. C. Wagner, *Naturwissenschaften* **31**, 265 (1943).
6. M. H. Hebb, *J. Chem. Phys.* **20**, 185 (1952).
7. C. Wagner, in *International Committee of Electrochemical Thermodynamics and Kinetics, Proceedings of the Seventh Meeting, Lindau*, 1955, Butterworths, London (1957), p. 361.
8. H. Schmalzried, *Thermodynamics*, Vol. 1, IAEA, Vienna, (1966), p. 97.
9. B. C. H. Steele, in *Electromotiveforce Measurements in High Temperature Systems*, ed. C. B. Alcock, The Institution of Mining and Metallurgy, London (1968), p. 3.
10. R. A. Rapp and D. A. Shores, in *Physicochemical Measurements in Metals Research*, ed. R. A. Rapp, Wiley-Interscience, New York (1970), p. 123.
11. H. Schmalzried, *Proceedings of the International Symposium on Metallurgical Chemistry*, Brunel University, England, 1971, p. 39.
12. T. H. Etsell and S. N. Flengas, *Chem. Rev.* **70**, 339 (1970).
13. H. Schmalzried and A. D. Pelton, *Ann. Rev. Mater. Sci.* **2**, 143 (1972).
14. W. L. Worrell, *Am. Ceram. Soc. Bull.* **53**, 425 (1974).
15. H. Rickert, in *Electromotive Force Measurements in High Temperature Systems*, ed. C. B. Alcock, The Institution of Mining and Metallurgy, London (1968), p. 59.
16. C. Wagner, *Z. Phys. Chem.* **B21**, 25 (1933).
17. C. Wagner, in *Advances of Electrochemistry and Electrochemical Engineering*, Vol. IV, ed. P. Delahay, Interscience, New York (1966), p. 2.
18. T. A. Ramanarayanan and W. L. Worrell, *Proceedings of the Symposium on Physico-Chemical Techniques at High Temperatures*, Vol. 7, IUPAC, Baden Near Vienna (1973), p. 240; *Can. Met. Quart.* **13**, 325 (1974).
19. F. A. Kröger and H. J. Vink, *Solid State Phys.* **3**, 307 (1956).
20. F. A. Kröger, *The Chemistry of Imperfect Crystals*, North-Holland, Amsterdam (1964).
21. J. W. Patterson, E. C. Bogren, and R. A. Rapp, *J. Electrochem. Soc.* **114**, 752 (1967).
22. M. F. Lasker and R. A. Rapp, *Z. Phys. Chem. N.F.* **49**, 198 (1966).
23. R. Baker and J. M. West, *J. Iron Steel Inst.* **204**, 212 (1966).
24. H. Schmalzried, *Z. Electrochem* **66**, 572 (1962).
25. C. Wagner, *J. Electrochem. Soc.* **115**, 933 (1968).
26. E. Mollwo, *Nachr. Gesellsch. Wissensch. Gottingen Math. Phys. Al. N.F.* **6**, 79 (1934).
27. J. W. Hinze and J. W. Patterson, *J. Electrochem. Soc.* **120**, 96 (1973).

28. J. W. Patterson, *J. Electrochem. Soc.* **118**, 1033 (1971).
29. J. D. Tretyakov and H. Schmalzried, *Ber. Bunsenges. Phys. Chem.* **69**, 396 (1965).
30. L. A. Pugliese and G. R. Fitterer, *Met. Trans.* **2**, 1997 (1970).
31. F. N. Mazandarany and R. D. Pehlke, *J. Electrochem. Soc.* **121**, 711 (1974).
32. K. E. Oberg, L. M. Friedman, W. M. Boorstein, and R. A. Rapp, *Met. Trans.* **4**, 61 (1973).
33. K. E. Oberg, L. M. Friedman, R. Szwarc, W. M. Boorstein, and R. A. Rapp, *J. Iron Steel Inst.* **210**, 359 (1972).
34. R. Szwarc, K. E. Oberg, and R. A. Rapp, *High Temp. Sc.* **4**, 347 (1972).
35. T. A. Ramanarayanan and R. A. Rapp, *Met. Trans.* **3**, 3239 (1972).
36. H. Schmalzried, *Z. Phys. Chem.* (*Frankfurt a. Mm*) **38**, 87 (1963).
37. W. L. Worrell and J. L. Iskoe, in *Fast Ion Transport in Solids*, ed. W. van Gool, IUPAC, North-Holland, Amsterdam (1973), p. 513.
38. C. Wagner and A. Werner, *J. Electrochem. Soc.* **110**, 326 (1963).
39. W. L. Worrell, *Thermodynamics*, Vol. 1, IAEA, Vienna (1966), p. 131.
40. S. C. Singhal and W. L. Worrell, in *Metallurgical Chemistry*, ed. O. Kubaschewski, Her Majesty's Stationery Office, London (1972), p. 65.
41. S. C. Singhal and W. L. Worrell, *Met. Trans.* **4**, 895 (1973).
42. P. J. Meschter and W. L. Worrell, in *Proceedings of the Third International Conference on Chemical Thermodynamics*, IUPAC, Baden Near Vienna, Austria (1973).
43. P. J. Meschter, Ph.D. dissertation, University of Pennsylvania (1974).
44. W. H. Skelton, N. J. Magrani, and J. F. Smith, *Met. Trans.* **2**, 473 (1971).
45. H. Rickert and R. Steiner, *Z. Phys. Chem.* **49**, 127 (1966).
46. T. N. Belford and C. B. Alcock, *Trans. Faraday Soc.* **60**, 822 (1964).
47. T. N. Belford and C. B. Alcock, *Trans. Faraday Soc.* **61**, 443 (1965).
48. H. H. Möbius, *Silikattechnik* **17**, 358, 385 (1966).
49. T. Y. Tien, *J. Appl. Phys.* **35**, 122 (1964).
50. N. L. Lofgren and E. J. McIver, quoted by T. L. Markin, in *Electromotive Force Measurements in High Temperature Systems*, ed. C. B. Alcock, The Institution of Mining and Metallurgy, London (1968), p. 91.
51. C. Wagner, in *Physico Chemical Measurements in Metals Research*, Part 1, ed. R. A. Rapp, Interscience, New York (1970), p. 1.
52. K. S. Goto and W. Pluschkell, in *Physics of Electrolytes*, Vol. 2, ed. J. Hladik, Academic Press, New York (1972), p. 539.
53. J. G. Burt, *J. Electrochem. Soc.* **117**, 267 (1970).
54. J. Delcet, R. J. Heus, and J. J. Egan, *J. Electrochem. Soc.* **125**, 755 (1978).

3

Thermodynamic Studies of Alloys and Intermetallic Compounds

M. S. Chandrasekharaiah, O. M. Sreedharan, and G. Chattopadhyay

1. Introduction

1.1. General

The thermodynamic description of alloy systems has been and still is in terms of phenomenological thermodynamics. The chemical potential, μ_i^α, of a component i in the alloy phase α as a function of composition and temperature is in principle a sufficient property to characterize the alloy system. Hence the development of experimental methods specific to alloy thermodynamics has been in obtaining more reliable chemical potential data over wider temperature and composition ranges. No single experimental method has been proved suitable for the study of the varieties of alloys encountered in practice.

Calorimetry would yield the necessary heat effects directly if suitable calorimeters with the desired accuracy over the temperature and composition ranges of interest could be designed. At present, however, various practical difficulties limit direct calorimetric measurements to a few alloys only. The information for a large number of alloys continues to be obtained from the measurements of dissociation or reaction equilibria and from the emf of suitable galvanic cells. The revival of the solid electrolyte galvanic cell method in the past two decades in the investigation of the free energy of formation of materials at high temperature has added much valuable data to alloy thermo-

M. S. Chandrasekharaiah, O. M. Sreedharan, and G. Chattopadhyay • Chemistry Division, Bhabha Atomic Research Centre, Trombay, Bombay 400085, India. Dr. Sreedharan's present address: Reactor Research Centre, Kalpakkam 603102, India.

chemistry. A brief survey of the application of this method to the study of the thermodynamics of alloy systems is presented in this chapter. The other aspects of solid electrolyte galvanic cells are covered in the other chapters of this book. For a general survey of the recent progress in alloy thermochemistry, the excellent review of Kubaschewski and Slough[1] should be consulted.

1.2. Solid Electrolyte Galvanic Cell Method

The recorded emf between the two terminals of an isothermal galvanic cell is thermodynamically well defined and can be related to the relevant chemical reactions occurring at the two electrodes by means of Nernst's equation provided the electrode processes satisfy certain conditions. The factors that affect these processes determine the choice of the electrolyte and the electrodes for a particular application. The solid electrolyte galvanic cell is no exception to this. Several types of solid electrolytes have come into vogue in alloy studies, viz., oxygen-ion-conducting electrolytes, fluoride-ion-conducting electrolytes, and metal-ion-conducting electrolytes like substituted β-alumina.

1.2.1. Oxide Electrolyte Galvanic Cells

When a solid oxide electrolyte having $t_{O^{2-}} \simeq 1.00$ is interposed between two electrodes of well-defined oxygen potentials, the electromotive force E established across the two electrode–electrolyte interfaces yield the oxygen chemical potential difference of the two electrode systems. Besides several other chapters of this book, many excellent reviews[2-11] have appeared recently dealing with the theory and applications of oxide electrolyte galvanic cell measurements. Hence, only specific points regarding their application to alloy study are outlined here.

Galvanic cells incorporating solid oxide electrolytes can all be considered as oxygen concentration cells and can be represented formally as follows:

$$\text{Pt, O}_2(\mu'_{O_2}) \mid \text{Solid oxide electrolyte} \mid \text{O}_2(\mu''_{O_2}), \text{Pt} \qquad [3.1]$$

The quantities μ'_{O_2} and μ''_{O_2} represent the relevant chemical potentials of oxygen established at the two electrodes, respectively, by appropriate electrode reactions. Provided local thermodynamic equilibrium is maintained at the interfaces between the various phases, the emf E of this cell is given by

$$E = \frac{-1}{4F}(\mu'_{O_2} - \mu''_{O_2}) + \frac{1}{4F}\int_{\mu''_{O_2}}^{\mu'_{O_2}} t_e d\mu_{O_2} \qquad (3.1)$$

where $\mu'_{O_2} < \mu''_{O_2}$. In Eq. (3.1), F is the Faraday constant and t_e is the electronic transference number in the electrolyte. In general, it is not possible to evaluate the integral in this expression with sufficient accuracy owing to the paucity of necessary information. Hence every attempt is made in the setting up of experiments to eliminate this correction term or at least reduce it to an insignificant magnitude. This is possible when t_e is less than 0.01 in the oxygen chemical potential range under study. The dissolution of electrode components may effect the local thermodynamic equilibrium as well as modify the kinetics of the relevant electrochemical processes at the electrolyte–electrode interfaces.

A typical cell configuration suitable for alloy activity measurement employing oxide electrolytes is

$$\text{Pt, A, AO} \mid \text{oxide electrolyte} \mid \text{AO, [A]}_B, \text{Pt} \qquad [3.2]$$

A is the less noble metal of the alloy [A]$_B$ and AO is the oxide of A in equilibrium with the metal A in the temperature and oxygen potential range of measurement. The following are the electrochemical reactions at the electrodes:

$$\text{A} + \text{O}^{2-} = \text{AO} + 2e^- \qquad (3.2)$$

$$\text{AO} + 2e^- = [\text{A}]_{\text{alloy}} + \text{O}^{2-} \qquad (3.3)$$

When the coexisting oxide phase, AO, is essentially stoichiometric at both electrodes, and other sources of mixed potentials are absent, the measured emf E of the cell directly gives the activity of A in the alloy:

$$-\ln a_A = \frac{2FE}{RT} \qquad (3.4)$$

If a fluoride-ion-conducting electrolyte replaces the oxide electrolyte in cell [3.2], the resulting emf would again yield the activity data upon substituting the corresponding metal fluoride for the metal oxide AO.

1.2.2. Reference Electrodes

In cell [3.2], the left-hand electrode is the reference electrode. An electrode system should satisfy several conditions in order to be selected as the reference.

(a) The reversibility of all electrode processes with minimum polarization under the measuring conditions is the prime consideration in the selection of reference electrodes. It may be necessary to establish the degree of reversibility in each case separately. According to Worrell and Iskoe,[12] the order of increasing polarizability among the common metal–metal-oxide electrodes is

$$\text{Cu, Cu}_2\text{O} < \text{Fe, Fe}_{0.95}\text{O} < \text{Ni, NiO}$$

This aspect is discussed in greater detail in Chapter 2. The concentration polarization generally diminishes with the increase in temperature. For example, the Cr, Cr_2O_3 electrode is unsuitable at $1272°K$ because of its polarizability. But at liquid steel temperatures ($\sim 1875°K$), Turkdogan and Fruehan[13] recommend the use of this electrode as the reference. At present, there is no detailed investigation available on the polarizability of metal fluoride electrodes.

(b) The change in composition of coexisting metal oxide phase with temperature should be small so that the concentration equilibrium is readily attained with the change of temperature.[14] It will be an advantage if the coexisting metal oxide (or metal fluoride) phase is a compound of essentially stoichiometric composition. In this regard, the fluoride system has an edge over the oxide system.

(c) The equilibrium oxygen or fluorine potential of the reference electrode should lie in the ionic conducting region (electrolytic domain) of the oxide or fluoride electrolyte (see Fig. 1.7).

(d) The oxygen or fluorine potential of the reference should be preferably within a few orders of magnitude ($\ln p_{O_2}^R/\ln p_{O_2} \simeq 2\text{--}3$) of that of the test electrode.[15]

1.2.3. Some General Observations about Experimental Procedures

Solid electrolyte galvanic cell measurements present a number of experimental problems many of which are specific to the particular system only. But there are a few which are common to all solid electrolyte galvanic cell measurements and in Chapter 2, these are considered in detail. A brief mention of some of the problems is made here.

An essentially inert gaseous environment around each electrode is a necessity and is usually achieved by continuously flushing the electrode chambers with purified argon or helium. The gas phase interaction, particularly when pellets are stacked in an open cell configuration, should be eliminated in all the measurements. It is preferred to separate physically the two electrode chambers by introducing a wall impervious to gas, especially in those measurements involving gaseous reference electrodes. Several ingenious ways of achieving this separation have been reported,[3,6,9] each with its own added problems. Such separation is essential in some measurements involving molten systems.[9,13,16]

The solid electrolytes and many electrode components are either insulators or semiconductors[4,6] and generally have large thermoelectric coefficients.[4] The contribution from the thermoelectric voltages to the measured emf of the cell is very difficult to estimate and should be eliminated experimentally. This problem is further complicated in measurements where the lead wire on the reference electrode is different from that on the other elec-

Table 3.1. *Estimate of Thermoelectric Voltage*
for Pt–Metal Couple in Liquid Steel[a]

Lead wire[b]	$E(\text{mV}) = A + Bt$ (°C)	$E(\text{mV})$ at 1600°C
Mo	$\begin{cases} -8.64 + 0.0372t \\ -12.26 + 0.0406t \end{cases}$	51 53
W	$-14.86 + 0.0429t$	54
Fe	—	26–27

[a] From Reference 16.
[b] In all cases, the lead wire for the reference electrode side was platinum.

trode, as in the case of molten metals. This correction becomes large and estimates are not always reliable. To illustrate the magnitude of this correction, an estimate as given by Gatellier *et al.*[16] of the voltages that arise in the investigation of liquid steel is presented in Table 3.1. Therefore, particular attention should be given to the temperature uniformity inside the furnace, especially where the cell assembly is located.

In any solid electrolyte galvanic cell arrangement, several solid–solid surface contacts are involved, and ensuring good electrical contacts with insignificant contact resistances is not always easy. Enough care should therefore be exercised to ensure good electrical contacts. The nonreacting electronic conducting lead wires are part of the cell arrangement. Platinum has been the first choice for this purpose but there are many occasions (e.g., liquid metals, alloys etc.) where other lead materials become indispensable. Suitability of the lead materials has to be proved under the operating conditions before accepting any new lead wires.

1.2.4. Experimental Limitations of Fluoride Cells

A distinct advantage of solid electrolyte galvanic cells with a fluoride-ion-conducting electrolyte in alloy studies is the fact that the transition metal fluorides exhibit much smaller nonstoichiometry compared to the corresponding oxides. Galvanic cells with CaF_2 as the electrolyte, besides providing information about the activity of the less noble component of an alloy, enable the calculation of the ΔG_f° of many metal fluorides,[7,17–19] carbides,[20,21] phosphides,[22] etc.

Although from a theoretical consideration, fluoride electrolyte cells might offer certain advantages in alloy investigations, there are a few additional experimental limitations in realizing these advantages. Firstly, the presently available free-energy data for transition metal fluorides are limited and are less reliable than the corresponding data for oxides. In addition, most of the metal fluorides of interest have considerable volatility. Thus, no suitable nonvolatile fluorides are available for the second- and the third-period transition metals.

The second difficulty arises in the preparation of anhydrous metal fluorides. The preparation poses special problems involved in the handling of anhydrous HF and/or fluorine gas. To complicate the matter, the severe corrosive nature of the fluorides limits the choice of container materials for the cells as well.

Lastly the high volatility of fluorides precludes the use of a simple, single-gas assembly and requires an elaborate two-gas cell design.

1.3. Limitations of Data Derived from emf Measurements

Some of the limitations of data obtainable from emf measurements have been discussed in Chapter 2 and elsewhere.[23] In alloy thermochemistry, only the activity data of the less noble metal is obtained from the measured emf. Even in the most favorable case of large difference in the free energy of formation of the two compounds AO_x, BO_y or AF_x, BF_y, it is very difficult to obtain reliable emf data at the limiting terminal solution range. So the activity information of the more noble metal derived from Gibbs–Duhem relation would be even less reliable. This factor should never be ignored.

Another point worth noting about data treatment is that when the emf of a galvanic cell with solid electrolyte is measured in a limited temperature range, and the data are treated by the method of least squares to obtain an emf–temperature relation, a comparison of the individual coefficients of the expression alone for judging the agreement between different measurements may lead to wrong conclusions. For example, the least-squares fit of the emf data for the cell

$$\text{Pt} \,\Big|\, \text{Co, CoO} \,\Big|\, \begin{array}{c} \text{calcia-stabilized zirconia} \\ \text{or yttria-doped thoria} \end{array} \,\Big|\, \text{NiO, Ni} \,\Big|\, \text{Pt} \qquad [3.3]$$

measured between 1100 and 1375°K was found to be

$$E(\text{mV}) \pm 1.0 = 31.4 + 6.21 \times 10^{-2}\, t(°C) \qquad (3.5)$$

and

$$E'(\text{mV}) \pm 0.63 = 24.89 + 6.83 \times 10^{-2}\, t(°C) \qquad (3.6)$$

by two different investigators.[24,25] In both these expressions, t refers to temperature in °C. A cursory examination of the corresponding individual coefficients of these two expressions would suggest a very poor agreement between the two measurements. But a comparison of the emf values calculated separately from each expression shows an excellent agreement (Table 3.2).

Table 3.2. *Comparison of emf Values of cells [3.3] at Various Temperatures*

emf	A	900°C		1000°C		1100°C	
		Bt	A + Bt	Bt	A + Bt	Bt	A + Bt
E (mV) $\pm$ 1.00	31.4	55.9	87.3	62.1	93.5	68.2	99.6
E' (mV) $\pm$ 0.63	24.89	61.48	86.37	68.31	93.20	75.14	100.03

2. Metallic Alloy Systems

2.1. Introduction

Phase equilibrium diagrams of metallic systems are usually established by thermal methods. The slow rates of attaining composition equilibria in solids can cause errors in the equilibrium diagram obtained by such methods and the equilibrium diagram constructed from free-energy data may be more accurate. Kubaschewski[26] has discussed at length the advantages of constructing equilibrium diagrams from free-energy data. The solid electrolyte galvanic cell method yields directly the free energy of mixing and hence is finding increasing application in alloy thermochemistry. The general configuration of a solid electrolyte galvanic cell for activity determination in an alloy can be given as (cf. [3.2])

$$\text{Pt} \mid \text{A, AX}_n \mid \text{X}^--\text{ion-conducting electrolyte} \mid \text{AX}_n, [\text{A}]_\text{B} \mid \text{Pt} \qquad [3.4]$$

where A is the less noble metal of the alloy A–B. AX_n is the compound of A coexisting with pure metal A. The measured emf of the cell is related to the chemical potential difference of the anionic species, X_2, between the two electrodes. Under circumstances where the alloy composition $[\text{A}]_\text{B}$ does not significantly alter during the measurement, and the chemical displacement reaction

$$n[\text{B}]_\text{alloy} + m\text{AX}_n = n\text{BX}_m + m[\text{A}]_\text{alloy} \qquad (3.7)$$

is insignificant, the observed emf E of the cell directly gives the activity a_A of the alloy

$$E = -\frac{RT}{nF} \ln a_\text{A} \qquad (3.8)$$

In deriving Eq. (3.8), several assumptions are made all of which affect the reliability of the data. The first factor is the neglect of the contribution from reaction (3.7). The relative stability of AX_n and BX_m determines the extent of the reaction. This relative stability, viz., $(1/n)\Delta G_f^\circ \langle \text{AX}_m \rangle - 1/m\Delta G_f^\circ \langle \text{BX}_m \rangle$ determines to a great extent the composition range over which the measurement can be made. Rapp and Maak[27] and Barbi[28] have reported on activity measurements in Cu–Ni alloys by the oxide electrolyte method. Probably this represents the limiting case. The oxygen potentials of the Ni–NiO and Cu–Cu_2O electrodes at 1273°K calculated from the free energy of formation data leads to

$$\frac{[P_{O_2}]_{Cu_2O}}{[P_{O_2}]_{NiO}} \simeq 10^4$$

Even in this case it is very difficult to assess unambiguously the contribution owing to the displacement reaction.

The uncertainty introduced by the displacement reaction (3.7) progressively increases as the terminal solution region of the more noble metal [B] is reached, even in the case of alloys with a large difference in the nobility of the components. Davies and Smeltzer[29] have recently reported on activity measurements in Fe–Ni alloys. The results were complicated by the significant extent of the reaction

$$[Ni]_{alloy} + Fe_{1-x}O = [Ni-Fe]O + [Fe] \tag{3.9}$$

at $X_{Ni} \to 1$. Wüstite coexists with alloy of composition up to 79 at % of nickel only. But for alloys of higher nickel content, this displacement reaction resulted in a spinel phase gradually replacing the wüstite and considerable uncertainty exists regarding the a_{Fe} data.

It has been assumed that the activity a_A in contact with AX_n is essentially unity throughout the measurement. This is not true in many oxide systems. Many metals like Ag, U, Zr, Ti, Co, etc. dissolve appreciable amounts of oxygen at high temperatures, thereby altering a_A from unity. This would complicate the analysis of the data. Though accurate data are lacking, metals generally exhibit very little solubility for fluorine. The existence of nonstoichiometry in many metal oxides further complicates the data analysis. Sometimes the coexisting phase itself may change with change in the alloy composition. For example, Kubik and Alcock[30] have reported that in the Fe–Au system, the wüstite phase changes to Fe_3O_4 when the iron content of the alloy is less than 1%. So, caution is necessary in the study of such systems.

2.2. Binary Solid Metallic Alloys

A number of binary alloys have been investigated with the help of oxide electrolytes. The majority of the alloy systems reported have iron, cobalt, nickel, or copper as the less noble metal. In general, the activity of the less noble component was measured as a function of composition and temperature and that of the more noble component was then calculated with the help of the Gibbs–Duhem relation. Recently alloys containing metals other than the four mentioned above as the less noble metal have been investigated.[31,32] A summary of the binary metallic alloys studied is presented in Table 3.3.

Table 3.3. Solid Binary Metallic Alloys Investigated by this Method

System	Temperature range (°K)	Remarks and references
Cu-Ni	973-1273	a_{Ni} over complete composition (27, 28, 187)
Pt-Ti	1173-1373	2-25 at Ti (181)
Co-W	1150-1400	$\Delta_f G_T^\circ$ Co$_3$W (168)
Co-Cu	1073-1323	Precision ± 10 cal per g-ion oxygen (142)
Co-Pd	1023-1548	$X_{Co} = 0.55$ to -0.90 (144)
	1273-1473	Complete composition (143)
Co-Mo	1073-1373	$X_{Mo} = 0.03$ to 0.40 (145)
Co-Au	1170-1470	Complete composition (30)
Ni-Ga	1023-1273	Ga-Ni field (33, 188)
Ni-Pd	1108-1173	Complete composition (146) (143)
Ni-Pt	973-1473	Complete composition (143)
Ni-Au	1273-1473	Complete composition (147)
Ag-In	1048-1208	$X_{In} = 0.043$ to 0.190 (148)
Ag-Sn		
Bi-Mg	900	CaF$_2$ electrolyte (165)
Cu-Al	933-1033	CaF$_2$ electrolyte (166)
Al-Sb	778-895	CaF$_2$ electrolyte (167)
Ni-Cr	1073-1488	$X_{Cr} = 0.25$ to 0.98 (45)
Ni-Fe	1073-1373	$X_{Ni} = 0.6$ to 1.0 (29)
Ni-Mo	1273-1373	(187)
Co-Ga	823-1073	(189)
Cu-Pt	900-1200	$X_{Pt} = 0.06$ to 0.5 (34)
Cu-Pd	1098-1173	Complete composition (151)
Cu-Au	973-1273	$X_{Cu} = 0.20$ to 0.98 (152)
Fe-Pd	973-1273	$X_{Fe} = 0.2$ to 0.9 (153)
Fe-Pt	1023-1273	$X_{Fe} = 0.2$ to 0.8 (153)
Fe-Au	1300-1700	$X_{Fe} = 0.2$ to 0.9 (30)
Fe-Cr	900-1200	Complete composition (154)
Pd-Sn	1050-1300	0-5.5 wt % Sn (35)
Ta-W	1050-1300	Complete (31)
Ta-Mo	1050-1300	Complete (157)
Nb-Mo	1050-1300	Complete (32)
Ag-Pd	825-1150	β-alumina (159)
Pd-In	1000	(160)
Cu-In	1000	$0 < X_{In} < 0.1$ (184)
Cu-Ga	1000	$0.01 < X_{Ga} < 0.35$ (184)
Cu-Ge	1000	$0.015 < X_{Ge} < 0.24$ (184)
Cu-Sn	1000	$0.015 < X_{Sn} < 0.22$ (184)
Co-Cr	1000-1473	Complete composition (185)
Ni-Cr	1173, 1273, 1373	$0.028 < X_{Cr} < 0.882$ (186)
Na-Pb	623-773	$X_{Na} = 0.044$ to 0.78 (199) β-alumina

Binary Intermetallics

In the case of a binary metallic alloy system investigated over a range of composition by an isothermal galvanic cell, a discontinuity in the emi vs. composition plot indicates a possible intermetallic phase or a miscibility gap. If the formation of the intermetallic phase is confirmed by another independent measurement, its standard free energy of formation can be obtained directly from the measured emf. If a number of intermetallic phases exist in the system, by appropriate choice of the electrode systems, the free-energy information of all of them could be determined. For example, in the Ni–Ga,[33] Cu–Pt,[34] and Pd–Sn[35] systems, the standard free energy of formation of intermetallics has been reported by the solid oxide electrolyte method.

Application of fluoride electrolyte cells to investigate metallic alloy systems is very limited. A few systems reported so far refer to intermetallics

Table 3.4. Intermetallics by Solid Electrolyte Galvanic Cells

System	Electrolyte	References
Co–W	Calcia-stabilized zirconia/ yttria-doped thoria	168
Co–Mo	Calcia-stabilized zirconia/ yttria-doped thoria	176
Co–Nb	Calcia-stabilized zirconia/ yttria-doped thoria	177
Fe–Nb	Calcia-stabilized zirconia/ yttria-doped thoria	178
Ta–Pt	Calcia-stabilized zirconia/ yttria-doped thoria	178a ($TaPt_2$, $TaPt_3$)
Zr–Pt	Yttria-doped thoria	179 ($ZrPt_3$, $ZrPt_5$ phases)
Hf–Pt	Yttria-doped thoria	179 ($HfPt_3$)
Ti–Pt	Yttria-doped thoria	181 ($TiPt_3$, $TiPt_8$ phases)
Th–Ru (1020–1170°K)	CaF_2	180
Th–Re	CaF_2	38
Y–Re		
Th–Ni	CaF_2	36
Th–Co	CaF_2	37
Th–Fe	CaF_2	182
Th–Rh	CaF_2	192
Ba–Pt		
Ce–Ir	CaF_2	183
La–Ir		
U–Fe		
U–Co	CaF_2	191
U–Cu		
U–Ru	CaF_2	191a

rather than single-phase alloys. Th–Ni,[36] Th–Co,[37] Th–Re,[38] and Y–Re[38] are some for which data are available. Table 3.4 includes the thermodynamic data for some intermetallic compounds obtained using the emf method.

2.3. Binary Liquid Alloys

All the considerations mentioned for the solid alloys apply equally well to liquid alloys. In addition, a few other aspects must be considered in measurements with liquid alloys.

The selection of pure solid phase of the less noble component of the solid alloy in equilibrium with its compound as the state of unit activity was in general sufficiently accurate because of the limited solubility of $\langle AX_n \rangle$ in $\langle A \rangle$. This approximation may not be adequate in liquid alloys, which some-times show a tendency to dissolve the compound of the less noble component of the alloy.[39] The change in alloy composition owing to dissolution of electronically conducting lead material in solid alloys is generally insignificant and Pt is, therefore, employed. But in liquid alloy systems an acceptable, non-reacting conductor wire has to be found for each case separately.

The preferential volatility of the more volatile component of the alloy phase would alter the composition and hence an additional uncertainty would be introduced. Because higher temperatures must be used in the study of liquid alloys, the vapor pressures of the component metals are higher. There are a few instances reported in the literature (e.g., Cu–Zn,[40] Cu–Ag,[41] Ga–In,[42] Pb–Sn,[43] Fe–Cr,[44] Ni–Cr[45]) where vapor loss has been significant enough to necessitate the addition of the volatile component to compensate for the loss due to vaporization during the measurement. Another point which is specific to liquid alloys is the choice of the standard state for the less noble metal when the alloy is liquid but the pure metal is yet a solid. This problem arose in the investigation of the Fe–Cr system[44] with the cell

$$\text{Cr, Cr}_2\text{O}_3 \mid \text{oxide electrolyte} \mid \text{[Fe–Cr], Cr}_2\text{O}_3 \qquad [3.5]$$

at 1600°C. The alloy was liquid and the reference (Cr) was solid. The activity of chromium, a_{Cr}, in the alloy was calculated with reference to supercooled metastable liquid chromium. Table 3.5 presents a summary of the data on liquid alloys from galvanic cell measurements.

2.4. Conclusion

The majority of metallic alloys reported so far are for binary systems. There too the activity of less noble component of the alloy is measured as a function of composition, the other being evaluated by the Gibbs–Duhem relation. Though ΔG^M data in general are reliable, the other derived quantities

Table 3.5. Liquid Alloys by Solid Electrolyte Galvanic Cells

Alloy	Temperature range (°K)	Composition range of X_i (mole fraction)	Electrical contact leads	Remarks and references
Ni–Pb	973–1393	$0.01 < X_{Ni} < 0.11$	NiO-coated Ni wire	Ni(s) saturated with Pb(l) as standard state (50)
Pb–Sn	873–1100	$0.1 \leqslant X_{Sn} \leqslant 0.9$	Cr-Al$_2$O$_3$ cermet and nichrome	near regular solution (43, 161)
Ni–Cr	1873	$0.109 \leqslant X_{Cr} \leqslant 0.54$	Pt-wire, Cr$_2$O$_3$ tube	slightly negative deviation for Cr (162)
Fe–Si	1773 and 1873	0.002 to 0.75 at 1873; 0.06 to 0.65 at 1773	80% Mo and 20% Al$_2$O$_3$ cermet	(51)
Fe–Cr	1873	$0.11 \leqslant X_{Cr} \leqslant 0.52$	80% Mo, 20% Al$_2$O$_3$ Pt or Pt–Rh	Small deviation in Cr (44)
Cu–Ag	1330	$0.15 \leqslant X_{Cu} \leqslant 0.907$	Cr + Al$_2$O$_3$ cermet	Nearly regular solution (41)
Cu–Zn	1268	$0.2 \leqslant X_{Zn} \leqslant 0.4$	Ta	40
Ag–Sn	825–1100	$0.2 \leqslant X_{Sn} \leqslant 0.9$	—	At 900°K, Sn exhibit both negative and positive deviations, but only negative at 1100°K (50)
Ga–In	1073–1223	$0.1 \leqslant X_{In} \leqslant 0.9$		Moderate positive deviation (42)
Cu–Ni	1373–1473	$X_{Cu} = 0.23$ to 1.0	Chromium cermet	163
Cu–Fe	1673	$X_{Fe} = 0.08$ to 0.1	Chromium cermet	164
In–Sb	833–1073	$X_{In} = 0.16$ to 0.50	W	194
Pb–Sn–Ag	1073	—	Cr + Al$_2$O$_3$ cermet	193

like the entropy and the enthalpy of mixing are less reliable. The normal practice for presenting the data is to expand the $\ln \gamma_j$ in terms of interaction coefficients ε_j^i and ρ_j^i [46-48] or in terms of quadratic formalism of Darken.[49] The presently available accuracy does not warrant consideration of the higher-order terms.

Complete data covering the entire composition range and over a signifi-cant temperature range are rather limited. Among those reported, the Ga–In[42] system has an almost ideal solution behavior. Das and Ghosh[43] have made a comprehensive study of the Pb–Sn system, while Ghosh et al.[50] have reported data on Ag–Sn. The studies of Cu–Zn alloy system was limited to the region of practical brass making compositions of 20–40 at % of Zn.[40] The investigation by Fruehan[51] of Si in molten iron is of significance in ferrous metallurgy. He had investigated Fe–Si in the concentration of X_{Si} up to 0.75 at 1873°K and up to $X_{Si} = 0.686$ at 1773°K. He has reported that the quad-ratic expression is sufficient to represent the data.

Hardly any ternary alloy system had been investigated by the solid electrolyte galvanic cell method. One example of a solid ternary alloy system that had been reported over a limited composition range is by Jacobi et al.[190] on the Ni–Cu–Ga system. The method offers a unique opportunity for further systematic investigation of ternary alloy systems. The application of solid electrolyte galvanic cell measurement to the study of liquid metals has only just begun and it is too early to state exactly the extent of its contribution. It is obvious that under favorable conditions, it would give better results than other equilibrium measurements.

3. Oxygen Dissolved in Metals and Alloys

3.1. Introduction

The thermodynamics of oxygen dissolved in metals is vital to metallur-gists because the physical properties of many metals and alloys are greatly influenced by the amount of oxygen present. Solid oxide electrolyte galvanic cells are very suitable to determine quantitatively the dissolved oxygen in metals and alloys. A cell of the type

$$\text{Pt} \mid \text{reference electrode} \mid \text{electrolyte} \mid [\text{O}]_M \mid \text{Pt} \qquad [3.6]$$

is employed in the determination of dissolved oxygen in metal M. The emf E of the cell is given by

$$E = \frac{RT}{2F} \ln \frac{a_O}{(p_{O_2})^{1/2}_{\text{ref}}} \qquad (3.10)$$

The activity of oxygen in the metal, a_O, is related to the atom fraction of oxygen, X_O, through

$$a_O = \gamma_O \cdot X_O \qquad (3.11)$$

when γ_O is the activity coefficient of oxygen.

3.2. Dilute Alloys of Oxygen in Metals

The oxygen solubility in most metals and alloys is small. In addition to the convenience of continuous monitoring in the metallurgical processing, the galvanic cell method differs from the other analytical methods of determining oxygen content of metals in one important way. Vacuum fusion, inert gas fusion and extraction, activation analysis, etc., all yield the total oxygen present irrespective of whether all or only a part of it is in the dissolved state. The solid electrolyte method gives the activity of the dissolved oxygen only.

For an unambiguous determination of oxygen content of a metal (solid or liquid) by the solid electrolyte galvanic cell method, the following pertinent factors have to be observed.

(i) In the determination of oxygen activity, the measured emf corresponds to the interface concentration of $[O]_M$ alloy at the electrolyte surface. If the oxygen diffusion rates are slow, then the actual measured activity, a_O, may significantly differ from the bulk activity a'_O.[52]

(ii) The oxygen potential (p_{O_2}) of the dissolved oxygen should be within the purely oxygen-ion-conducting region of the electrolyte at that temperature. Calcia-stabilized zirconia has been the most extensively investigated electrolyte.[4,6,13] Still there is some ambiguity regarding the oxygen potentials at which it exhibits significant electronic conductance.[53-56]

(iii) The oxygen content should not be affected by interaction with the surrounding atmosphere.

In addition, in calculating the solubility of oxygen from the measured oxygen activity data, it is assumed that the corresponding activity of the host metal remains virtually near unity. If it is changed, the change is small enough to be calculated by Raoultian approximation. The limited data reported so far indicate the reasonableness of the assumption. Belford and Alcock[57,58] reported the influence of dissolved oxygen on the activity of liquid lead and tin. Oxygen solubility in lead increased from 0.0074% at 783°K to 0.136% at 973°K and in tin from 0.0013% at 809°K to 0.020% at 973°K. But their observed change in the relative partial molar free energy of metal (in both) was less than 6 cal mol^{-1} and correction based on Raoultian behavior was sufficient. A similar trend in the case of liquid copper[59] and liquid silver[60] was also reported. Table 3.6 shows a summary of the data on oxygen activity in metals. The literature in this area is indeed extensive.[86-141] Measurements of oxygen activity in solid metals are relatively few. Examples are oxygen activity determinations in niobium[32,138,158] and tantalum.[311,158,173]

Table 3.6. *Dissolved Oxygen in Liquid Metals by Solid Electrolyte Galvanic Cells*

Molten metal	Temperature (°K)	Contact leads	Reference electrodes and electrolyte	Remarks and references
Lead	783–973	Ir wire	Cu, Cu_2O and calcia-stabilized zirconia	Henry's law obeyed (57)
	773–1373	Pt wire	Ni, NiO and yttria-doped thoria	Diffusion studies (63–68)
Tin	800–1023	Osmium	Calcia-stabilized zirconia, Ni, NiO	58
Silver	1253–1433		Yttria-doped thoria	14, 69
	1073–1373	Pt	Calcia-stabilized zirconia	60
			Calcia-stabilized zirconia and Cu, Cu_2O	67
	1273–1493	Ir	Calcia-stabilized zirconia	83
	1273	Stainless steel	Calcia-stabilized zirconia	85
Copper	1423	Mo	Calcia-stabilized zirconia	Henry's law obeyed (80)
	1383–1703	Pt	Calcia-stabilized zirconia	81
	1388–1573	Cr_2O_3	Calcia-stabilized zirconia	82
	1373–1473	Ta	Calcia-stabilized zirconia	59
	1373–1523	Mo	Calcia-stabilized zirconia	83
	1873	Mo	Calcia-stabilized zirconia and Cr, Cr_2O_3 reference electrode	81
Iron and steel	1873	Mo–Al_2O_3 cermet	Yttria-doped thoria	99, 100
	1273–1823	Pt–Rh alloy	Yttria-doped thoria	119
	1873	Mo, Mo–Al_2O_3 cermet	Yttria-doped thoria and Cr, Cr_2O_3 reference electrode	114–117
	1533–1913	Pt–18% Rh	Yttria-doped thoria and Cr, Cr_2O_3 ref.	112
Sodium	1800	—	(Al_2O_3–SiO_2) electrolyte	110, 134, 139
	600–800	Stainless steel	Calcia-stabilized zirconia and yttria-doped thoria	200, 201

3.3. Effect of Third Alloying Element on the Oxygen Activity

In the traditional metallurgical practice, in addition to the host metal and oxygen, there is often a third element present either by intentional addition or as an impurity. Information about the influence of this third element on the activity of oxygen dissolved in the host metal is very useful in process metallurgy. A quantitative prediction of the change in the activity coefficient of oxygen is not possible except in the limiting case where both the solutes are present in such low concentrations that dilute limiting solution expressions are valid. The available information on the change in ln γ_0 upon the addition of third element is limited, yet some general trends are discernible.

The usual cell configuration for the study of this effect is similar to the cell arrangement for the oxygen solubility in liquid metals, viz.,

$$\text{Pt} \mid [\text{O}]_{A,B} \mid \text{electrolyte} \mid \text{MO, M} \mid \text{Pt} \qquad [3.7]$$

where (MO, M) is the reference electrode. From the emf values, interaction parameters can be evaluated and the effective concentrations of oxygen can be calculated. The data for ferrous alloys is quite extensive. Fruehan[54,55,61] and Schwerdtfeger[56] have reported on the changes of the activity of oxygen in liquid iron on adding Ti, Al, Si, Cr, V, B, and Mn. Of course, the presence of carbon alters the oxygen activity and this effect also has been studied in some detail. Among the nonferrous systems for which there are sufficient data are the alloys of copper,[59,62-64] lead,[65-68] and tin.[69] These investigations have firmly established the potentiality of the solid oxide electrolyte method for the study of dissolved oxygen in molten alloys.

For the concise presentation of the change in the activity of the dissolved oxygen in a metal upon the addition of the small quantities of a third alloying element, j, ln γ_0 has been selected as the representative quantity. There has been no theoretical model for the metallic solutions that explains satisfactorily the concentration dependence of the activity or yield an analytic functional form for ln γ_i. Several semiempirical formalisms exist[46-49] and among them, polynomial formalism[156] combines simplicity with quantitative representation. Expressing ln γ_0 in polynomial form one obtains the following expression[156]:

$$\ln \gamma_0 = \ln \gamma_0^0 + \sum_{j=1}^{n} \varepsilon_0^j X_j + \sum_{i,j} \rho_0^{i,j} X_i X_j + \cdots \qquad (3.12)$$

In Eq. (3.12) ln γ_0^0 is the limiting activity coefficient of oxygen in the liquid solvent metal when the mole fraction of the solvent approaches unity. In other words, γ_0^0 is the Henry's law constant for oxygen in the binary solution of oxygen and the host metal. ε_0^j and $\rho_0^{(i,j)}$ are, respectively, called the first- and the second-order interaction coefficients. They are expressed as follows:

$$\varepsilon_0^0 = \left(\frac{\partial \ln \gamma_0}{\partial X_0} \right)_{X_s \to 1}, \qquad \varepsilon_0^j = \left(\frac{\partial \ln \gamma_0}{\partial X_j} \right)_{X_s \to 1} \qquad (3.13)$$

ε_0^0 is termed the binary coefficient of the first order and represents the influence of added oxygen on γ_0. ε_0^j is the ternary coefficient of the first order and is a measure of how the presence of j affects $\ln \gamma_0$. $\rho_0^{(0)}$ is the binary interaction coefficient of the second order, given by the expression

$$\rho_0^{(0)} = \left(\frac{\partial^2 \ln \gamma_0}{\partial X_0^2}\right)_{X_s \to 1} \tag{3.14}$$

$\rho_0^{(i,j)}$ and $\rho_0^{(j)}$ are, respectively, the two ternary coefficients of the second order and are likewise defined.

In metallic systems, it is common to use the 1-wt % standard state for the solute. When this standard state is used, Eq. (3.12) takes the form

$$\log f_0 = \sum e_0^j w_j + \sum_{j,i} \rho_0^{\prime(i,j)} w_j w_i + \cdots \tag{3.15}$$

where w's denote wt %, f_0 is the activity coefficient of oxygen based on the 1-wt % standard state, and e_0^j, $\rho_0^{\prime(i,j)}$ are the corresponding interaction coefficients. It is possible to derive the relations between the two sets of coefficients.[70,71] For example,

$$e_0^j = \frac{0.2425}{M_j} \varepsilon_0^j \tag{3.16}$$

where M_j is the molecular weight of the added element. These relations in principle can be extended to other ternaries also. It must be emphasized that the above formalisms apply only to dilute solutions.

Attempts have been made to interpret the influence of the alloying element, j, on a_0 in terms of quasichemical models with limited success. Alcock and Richardson[72] assumed a random distribution in the alloy, while Belton and Tankins[73] included both s-O- and j-O-type bondings in calculating $\Delta \bar{G}_0$. Jacob and Alcock[74] assumed a partial distortion of the electronic distribution around the metal atom bonded to oxygen in their calculations. None of these models is sufficiently general and precise to give quantitative agreement.

The investigation of dissolved oxygen in molten metals by the solid electrolyte method has yielded some interesting results. Fruehan[54] has drawn the following conclusions based on the available data primarily of molten steels.

(a) $\ln \gamma_0/\gamma_0^0$ vs. X_j or $\log f_0/f_0^0$ vs. w_j plots show that at high dilution, ε_0^j or the corresponding e_0^j coefficients are adequate to represent the data. But deviations from linearity at finite concentrations is the rule. Higher-order coefficients of the type $\rho_0^{(i,j)}$ then become necessary. Fruehan and Turkdogan[13] have evaluated $\log f_0$ as a function of the wt % of added solute j to the molten iron at 1873°K. Their plot of $\log f_0$ vs. wt % of j is shown in Fig. 3.1.

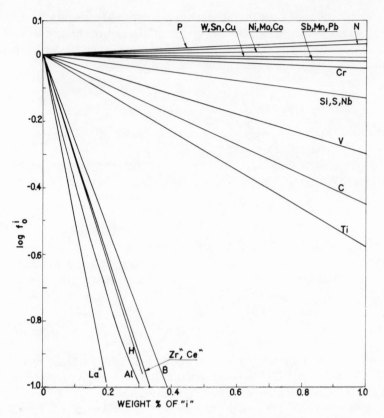

FIG. 3.1. Effect of alloying elements (i) on the activity coefficient of oxygen in liquid iron at 1600°C (Reference 198, and for asterisked elements, Reference 202).

(b) ε_O^j becomes more negative as the oxide stability of the added solute, j, increases. The effective oxygen solubility when plotted vs. wt % of added solute exhibits a minimum in Fe–j–O systems (Fig. 3.2). In Table 3.7, some of the interaction coefficients ε_O^j in molten steel at 1873°K are presented.

In spite of several investigations, the data are not always consistent. Sigworth and Elliott[198] have discussed the thermodynamic implication of the data of the iron-based alloys. Among the nonferrous systems, copper has been the most investigated one. A plot of ln γ_O vs. X_j for molten copper at 1473°K is presented in Fig. 3.3 It can be seen that the experimental scatter from one group of investigators to another is not always small.

As an illustration of the present state of knowledge about dissolved oxygen in molten metals, oxygen in liquid copper is briefly discussed. As can be seen from the Table 3.6, several sets of data are available in the literature.

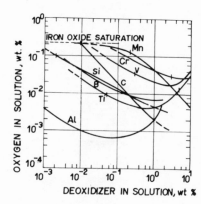

FIG. 3.2. Deoxidation equilibria in liquid iron alloys at 1600°C (Reference 13).

Nanda and Geiger[63,64] have recently studied the solubility and the activity of oxygen in copper and copper–tin alloys and evaluated a value of γ_O^0 (activity coefficient at infinite dilution) of 0.12 at 1373°K in good agreement with the corresponding data by Wilder[59] viz., $\gamma_O^0 = 0.12$ at 1373°K and 0.205 at 1473°K. Using values of 0.023 and 0.067 at % for the saturation solubility of oxygen in liquid copper at 1373 and 1473°K, respectively, and assuming that Henrian behavior is valid up to the saturation limit of oxygen, Wilder[59] obtained $\Delta_f G°$ of $Cu_2O(s)$ in reasonable agreement with the data of Kiukkola and Wagner[2] at 1373°K and lower by 3 kJ at 1473°K. If the error is attributed solely to the assumption that $\gamma_O = \gamma_O^0$ throughout the concentration range, then γ_O must have increased with X_O. This is in contradiction with the gas equilibration measurement reported by Dompas and Van Melle[84] and the emf results of Osterwald,[82] who have reported a decrease of activity coefficient of oxygen with concentration between 1302 and 1573°K.

Table 3.7. *Interaction Parameters for Oxygen in Molten Steel on the Addition of Third Element*

Alloying element	Al	Ti	B	V	Cr
$\Delta_f G°_{M_2O_3}/3$ kcal mol^{-1}	−85.8	−81.6	−66.2	−62.0	−51.9
Minimum [O] (ppm)	6	40	80	180	270
Mole fraction of M at min [O]	0.0002	0.01	0.03 to 0.05	0.03	0.07
Weight fraction of M at min [O]	0.09	0.9	0.67	3	7
Experimental e_O^M	−3.90	−1.12	—	—	−0.037
ε_O^M	−433	−222	−115	−29.0	−8.0
$\ln \gamma_O^0$	0.021	0.038	0.083	−1.0	—
Reference state for $\ln \gamma_O^0$	liquid Al	solid Ti	solid B	solid V	solid Cr

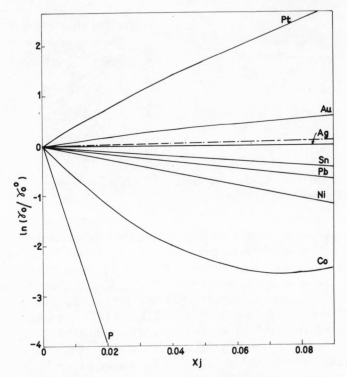

FIG. 3.3. In (γ_0/γ_0^0) vs. X_j in copper at 1200°C (Reference 197).

The agreement among the other derived quantities is not any better. ElNaggar and Parlee[90] have tabulated the partial molar enthalpies and entropies of solution of oxygen in liquid copper. It shows a spread of -163 to -18 kJ mol^{-1} for the partial molar enthalpies and a -61.9 J K^{-1} mol^{-1} to $+32.2$ J K^{-1} mol^{-1} for the partial molar entropies. Sigworth and Elliott[197] have recently published a detailed account of dissolved oxygen in liquid copper. They give the following equation:

$$\Delta\bar{G}_{O_2}^M \text{ (kJ mol}^{-1}) = -85.35 + 45.2 \times 10^{-3}T \text{ (°K)} \quad [1373 \leqslant T \text{ (°K)} \leqslant 1573]$$
(3.17)

Their evaluated $\Delta\bar{S}_{O_2}^M$ of -45.2 J K^{-1} mol^{-1} appears to be reasonable and in good agreement with that reported by Osterwald.[82] However, they state that the estimate of errors was difficult and errors were not always small. Many other investigators[87,89,92,93,97,163,164] seem to agree with their observation.

The application of solid electrolyte galvanic cells to the investigation of

Table 3.8. Oxygen Activity in Ternary Systems

System	Temperature (°K)	Interaction parameter	Reference
Fe–Cr–O	1873	e_O^M (1873°K) = -0.037	44
Fe–Si–O	1873	-0.14	51, 56, 124, 133
Fe–Si–O	1773		56
Fe–Ti–O	1873	-1.15	55, 136
Fe–V–O	1873	-0.4	61
Fe–B–O	1873	-2.6	61
Fe–Al–O	1873	-3.9	55, 136, 137
Fe–Mn–O	1873		56, 98
Fe–Zr–O	1873	-3.0	136, 202[a]
Fe–Cu–O	1873	-0.016	112, 203
Fe–Ni–O	1833–1973	0.006	112, 203
Fe–Ce–O	1873	-3.0	136, 202[a]
Fe–P–O	1873	0.07	81
Fe–S–O	1873	0.091	81
Co–Ni–O	1833–1973		112
Cu–Au–O	1473	ε_O^M (1473°K) = 23	90
Cu–Pt–O	1473	21	90
Cu–Ag–O	1473	13	90, 62[b], 150
Cu–Ni–O	1473	9	90
Cu–Ni–O	1473	-7.1	135, 203
Cu–Co–O	1473	-63	135, 203
Cu–Fe–O	1473	-553	135, 203
Cu–Sn–O	1373 and 1473	ε_O^M (1373°K) = -10.5	62
Cu–Sn–O	1373	-4.6	69
Cu–Pb–O	1373		65
Pb–Sn–O	1373		65
Fe–La–O	1953	-5.0	202[a]
Cu–Se–O	1473		149

[a] Measurement at 1953°K.
[b] Measurement at 1373°K also.

dissolved oxygen in molten metals is only a few years old, but the potentiality of the method is now firmly established. Table 3.8 presents a summary of the oxygen activity in molten metals in the presence of a small amount of a third element.

4. Fluoride Cells to Study Carbides, Borides, Phosphides, and Sulfides

Calcium fluoride has been employed in the thermodynamic investigations of carbides of Th, U, Mn, and Cr, and borides and phosphides of thorium.

The typical cell configurations are

$$\text{Ta} \mid \text{Th, ThF}_4 \mid \text{CaF}_2 \mid \text{ThF}_4, \text{ThC}_2, \text{C} \mid \text{Ta} \qquad [3.8]$$

$$\text{Ta} \mid \text{Th, ThF}_4 \mid \text{CaF}_2 \mid \text{ThF}_4, \text{ThC}, \text{ThC}_2 \mid \text{Ta} \qquad [3.9]$$

$$\text{Ta} \mid \text{Th, ThF}_4 \mid \text{CaF}_2 \mid \text{ThF}_4, \text{ThC}_{0.7}, \text{Th}_\alpha \mid \text{Ta} \qquad [3.10]$$

where Th_α refers to thorium saturated with carbon.

Similar cells have been employed to measure the free energy of formation of ThB_6, ThB_4, and thorium saturated with boron.

Free energies of formation of ThP and thorium saturated with phosphorous, Th_7S_{12}, Th_2S_3, ThS, and thorium saturated with sulfur were determined in conjunction with the calorimetric data on ThS_2 and Th_3P_4.

Aronson[75] and Satow[76] each studied the nonstoichiometric thorium mono- and dicarbide phases by employing the following types of cells:

$$\text{W} \mid \text{Th, ThF}_4 \mid \text{CaF}_2 \mid \text{ThF}_4, \text{ThC}_{2-x} \mid \text{W} \qquad [3.11]$$

$$\text{W} \mid \text{Th, ThF}_4 \mid \text{CaF}_2 \mid \text{ThF}_4, \text{ThC}_{1-x} \mid \text{W} \qquad [3.12]$$

Aronson's values were nearly 30 mV lower than Satow's. Satow attributes this to the formation of a conducting deposit along the edge of the electrolyte pellet and thus partially shorting out the cell.

Storms[21] has reviewed the data on the Th–C system and critically assessed those from solid electrolyte galvanic cell and vapor pressure methods.

Table 3.9. Fluoride Cells to Study Carbides, Borides, Phosphides, or Sulfides

System	Temperature range (°C)	Precision in emf measurement (mV)	Electrical contacts	Reference
Th–C	973–1243		Mo	169, 170
	1073–1273	±5	Ta	169, 170
	1073–1223		Ta	75
	993–1223		W	76
U–C	973–1173	±4	Mo	171
Th–B	1073–1223	±5	Ta	172
Th–P	1073–1223	±30		22
Th–S	1098–1198			174
Mn–C	900–1100	±1.8	Ta	77, 78
			Carbided Mo	
	900–1073		W	155
Cr–C	880–1100		Pt	175
Fe–Mn–C	963–1333			196
U–Fe–C	998–1084		W	195
U–W–C	981–1098		W	195

Free energies of formation of Mn_7C_3, Mn_5C_2, $Mn_{23}C_6$, Cr_3C_2, Cr_7C_3, were also determined by this method.[77-79]

In all these measurements equilibration was sluggish and required days before the cells gave steady emf's. Table 3.9 summarizes the systems studied, the precision of emf measurement, and the temperature range of investigation.

5. Conclusion

The solid electrolyte galvanic cell method has established itself as a sensitive and versatile experimental method in high-temperature alloy thermochemistry. The inherent high precision of the galvanic cell measurement could be fruitfully utilized in obtaining reliable free-energy data for metallic systems.

ACKNOWLEDGMENT

The authors would like to offer their appreciation to Dr. M. D. Karkhanavala, Head, Chemistry Division, Bhabha Atomic Research Centre, for his invaluable guidance during this work.

References

1. O. Kubaschewski and W. Slough, *Progr. Mater. Sci.* **14**, 1 (1969).
2. K. Kiukkola and C. Wagner, *J. Electrochem. Soc.* **104**, 308, 379 (1957).
3. B. C. H. Steele, in *Electromotive Force Measurements in High Temperature Systems*, ed. C. B. Alcock, Institution of Mining and Metallurgy, London (1968), p. 1.
4. T. H. Etsell and S. N. Flengas, *Chem. Rev.* **70**, 339 (1970).
5. H. Schmalzried, ed., *Thermodynamics*, Vol. I, IAEA, Vienna, (1966), p. 97.
6. R. A. Rapp and D. A. Shores, in *Techniques in Metals Research*, Vol. IV, part 2 ed. R. A. Rapp, Interscience, New York (1971), p. 122.
7. T. L. Markin, R. J. Bones, and V. J. Wheeler, *Proc. Brit. Ceram. Soc.* **8**, 175 (1967).
8. B. C. H. Steele and C. B. Alcock, *Trans. AIME* **233**, 1359 (1965).
9. M. Sato, in *Research Techniques for High Pressure and High Temperature*, ed. G. C. Ulmer, Springer-Verlag, Berlin (1971), p. 43.
10. D. O. Rayleigh, in *Progress in Solid State Chemistry*, Vol. 3, ed. H. Reiss, Pergamon Press, New York, (1970), p. 83.
11. D. O. Rayleigh, in *Electroanalytical Chemistry*, Vol. 6, ed. A. J. Bard, Marcel Dekker, New York, (1972), p. 87.
12. W. L. Worrell and J. L. Iskoe, in *Fast Ion Transport in Solids*, ed. W. Van Gool, North-Holland, Amsterdam (1973), p. 513.
13. E. T. Turkdogan and R. J. Fuehan, *Can. Met. Quart.* **11**, 371 (1972).
14. R. N. Blumenthal and D. H. Whitmore, *J. Electrochem. Soc.* **110**, 92 (1963).
15. D. A. Shores and R. A. Rapp, *J. Electrochem. Soc.* **118**, 1107 (1971).
16. C. Gatellier, K. Torssell, M. Olette, M. Meysson, M. Chastant, A. Rist, and P. Vicens, *Rev. Met.* **66**, 673 (1969).

17. W. H. Skelton and J. W. Patterson, *J. Less Common Met.*, **31**, 47 (1973).
18. Chattopadhyay and M. S. Chandrasekharaiah, unpublished work.
19. N. L. Lofgren and E. I. McIver, U.K. AEA-AERE-R 5169 (1965).
20. C. E. Holley, Jr., and E. K. Storms, in *Thermodynamics of Nuclear Materials*, IAEA Vienna, (1967), p. 397.
21. E. K. Storms, *The Refractory Carbides*, Academic Press, New York (1967).
22. K. Gingerich and S. Aronson, *J. Phys. Chem.* **70**, 2517 (1966).
23. O. Kubaschewski, *High Temp. High Press.* **4**, 1 (1972).
24. O. M. Sreedharan, M. S. Chandrasekharaiah, and M. D. Karkhanarala, *High Temp. Sci.*, **9**, 109 (1977).
25. C. B. Alcock and G. P. Stavropoulos, *J. Am. Ceram. Soc.* **54**, 436 (1971).
26. O. Kubaschewski, *Naturwissenschaften* **55**, 525 (1968).
27. R. A. Rapp and F. Maak, *Acta Met.* **10**, 63 (1962).
28. G. B. Barbi, *Ann. Chim. (Rome)* **56**, 62 (1966).
29. H. Davies and W. W. Smeltzer, *J. Electrochem. Soc.* **119**, 1362 (1972).
30. A. Kubik and C. B. Alcock, *Met. Sci. J.* **1**, 19 (1967).
31. S. C. Singhal and W. L. Worrell, *Met. Trans.* **4**, 895 (1973).
32. S. C. Singhal and W. L. Worrell, *Met. Trans.* **4**, 1125 (1973).
33. A. U. Seybolt, *J. Electrochem. Soc.* **111**, 697 (1964).
34. L. R. Bidwell, W. J. Schultz, and R. K. Sues, *Acta Metall.* **15**, 1143 (1967).
35. A. W. Bryant, W. G. Bugden, and J. N. Pratt, *Acta Metall.* **18**, 101 (1970).
36. W. H. Skelton, N. J. Magnani, and J. C. Smith, *Met. Trans.* **1**, 1833 (1970).
37. W. H. Skelton, N. J. Magnani, and J. C. Smith, *Met. Trans.* **2**, 473 (1971).
38. T. N. Rezukhina and B. S. Pokarev, *J. Chem. Thermodyn.* **3**, 369 (1971).
39. C. R. Cavanaugh and J. F. Elliott, *Trans. AIME* **230**, 633 (1964).
40. T. C. Wilder and W. E. Galin, *Trans. AIME* **245**, 1287 (1969).
41. U. V. Chowdary and A. Ghosh, *J. Electrochem. Soc.* **117**, 1024 (1970).
42. K. A. Kleinedinst, M. V. Rao, and D. A. Stevenson, *J. Electrochem. Soc.* **119**, 1261 (1972).
43. S. K. Das and A. Ghosh, *Met. Trans.* **3**, 803 (1972).
44. R. J. Fruehan, *Trans. AIME* **245**, 1215 (1969).
45. L. A. Pugliese and G, R. Fitterer, *Met. Trans.* **1**, 1997 (1970).
46. C. H. P. Lupis, "Thermodynamic Formalism of Metallic Solutions," in *Liquid Metals*, ed. S. Z. Beer, Marcel Dekker, New York (1972), p. 1.
47. C. H. P. Lupis and J. F. Elliott, *Acta Metall.* **14**, 529 (1966); **15**, 265 (1967).
48. C. H. P, Lupis and J. F. Elliott, *Trans. AIME* **233**, 829 (1965).
49. L. S. Darken, *Trans. AIME* **239**, 90 (1967).
50. P. J. Roychowdhuri and A. Ghosh, *Met. Trans.* **2**, 2171 (1971).
51. R. J. Fruehan, *Met. Trans.* **1**, 865 (1970).
52. T. A. Ramanarayanan and R. A. Rapp, *Met. Trans.* **3**, 3239 (1972).
53. R. Baker and J. M. West, *J. Iron Steel Inst.* **204**, 12 (1966).
54. R. J. Fruehan, I. J. Martonik, and E. T. Turkdogan, *Trans. AIME* **245**, 1501 (1969).
55. R. J. Fruehan, *Met. Trans.* **1**, 3403 (1970).
56. K. Schwerdtfeger, *Trans. AIME* **239**, 1276 (1967).
57. T. N. Belford and C. B. Alcock, *Trans. Faraday Soc.* **60**, 822 (1964).
58. T. N. Belford and C. B. Alcock, *Trans. Faraday Soc.* **61**, 443 (1965).
59. T. C. Wilder, *Trans. AIME* **236**, 1035 (1966).
60. C. M. Diaz, C. R. Masson, and F. D. Richardson, *Trans. Inst. Min. Met.* **75**, C183 (1966).
61. R. J. Fruehan, *Met. Trans.* **1**, 2083 (1970).
62. R. J. Fruehan and F. D. Richardson, *Trans. AIME* **245**, 1721 (1966).

63. C. R. Nanda and G. H. Geiger, *Met. Trans.* 1, 1235 (1970).
64. C. R. Nanda and G. H. Geiger, *Met. Trans.* 2, 1101 (1971).
65. K. T. Jacob and J. H. E. Jeffes, *Trans. Inst. Min. Met.* 80, C32, C181 (1971).
66. G. Bandopadhyay and H. S. Ray, *Met. Trans.* 2, 3055 (1971).
67. K. Goto, M. Sasable, and M. Someno, *Trans. AIME* 242, 1757 (1968).
68. S. Honma, N. Sano, and Y. Matsushita, *Met. Trans.* 2, 1494 (1971).
69. K. T. Jacob, S. K. Seshadri, and F. D. Richardson, *Trans. Inst. Min. Met.* 79, C275 (1970).
70. C. H. P. Lupis and J. F. Elliott, *Trans. AIME* 233, 829 (1965).
71. J. F. Elliott, *Trans. AIME* 236, 130 (1966).
72. C. B. Alcock and F. D. Richardson, *Acta Met.* 8, 882 (1960).
73. G. R. Belton and E. S. Tankins, *Trans. AIME* 233, 1892 (1963).
74. K. T. Jacob and C. B. Alcock, *Acta Metall.* 20, 221 (1972).
75. S. Aronson and J. Sadofsky, *J. Inorg. Nucl. Chem.* 27, 1769 (1965).
76. J. Satow, *J. Nucl. Mat.* 21, 249, 255 (1967).
77. F. Moattar and J. S. Anderson, *Trans. Faraday Soc.* 67, 2303 (1971).
78. H. Tanaka, Y. Kishida, A. Yamaguchi, and J. Moriyama, *J. Jpn. Inst. Met.* 35, 997 (1971).
79. W. A. Fischer, *Chem. Anal. (Warsaw)* 16, 975 (1971).
80. W. Plushkell and H. J. Engell, *Z. Metallkd.* 56, 450 (1965).
81. W. A. Fischer and W. Ackermann, *Arch. Eisenhuttenwes.* 36, 643, 695 (1965); 37, 43 (1966).
82. J. Osterwald, *Z. Phys. Chem. (N.F.)* 49, 138 (1966).
83. H. Rickert and A. A. Elmiligy, *Z. Metallkd.* 59, 635 (1968).
84. J. M. Dompas and J. Van Melle, *J. Inst. Met. (London)* 98, 304 (1970).
85. C. M. Diaz and F. D. Richardson, *Trans. Inst. Min. Met.* 76, C196 (1967); 75, C183 (1966).
86. C. M. Diaz and F. D. Richardson, in *Electromotive Force Measurements in High Temperature Systems*, ed. C. B. Alcock, Institution of Mining and Metallurgy, London (1968), p. 29.
87. J. M. Dompas and L. Hens, Belgium Non Ferrous Metal Research Association Conference, Liege, BNFMRA, Paper No. 9, October 1971.
88. J. M. Dompas and P. C. Lockyer, *Met. Trans.* 3, 2597 (1972).
89. J. M. Dompas and J. Van Melle, *J. Met.* 26, 443 (1972).
90. M. M. El Naggar and N. A. I. Parlee, *Met. Trans.* 1, 2975 (1970); 2, 909 (1971).
91. T. H. Etsell and S. N. Flengas, *J. Electrochem. Soc.* 119, 198 (1972).
92. A. D. Kulkarni, R. E. Johnson, and G. W. Perbix, *J. Inst. Met.* 99, 15 (1971).
93. T. Oishi, A. Yamaguchi, and J. Moriyama, *Nippon Kogyo Kaishi* 88, 103 (1972).
94. J. Osterwald, *Z. Metallkd.* 59, 573 (1968).
95. W. P. Thompson and P. Tarassoff, *Can. Met. Quart* 10, 315 (1971).
96. I. Taukshara, *Nippon Konzokku Gakkaishi* 34, 679 (1970).
97. H. Rickert and H. Wagner, *Electrochim. Acta* 11, 83 (1966).
98. M. G. Fröhberg and P. M. Mathew, *Schweiz. Arch.* 38, 251 (1972).
99. M. Olette, C. Gatellier, and F. Torssell, *Berg Huettenmaenn. Monatsch.* 113, 484 (1968).
100. C. Gatellier and M. Olette, C.R. *Acad. Sci. Ser. C* 266, 1133 (1968).
101. Von H. Boeck and E. Teichert, *Stahl Eisen* 89, 61 (1969).
102. L. Brhacek, T. Myslivec, and A. Golonks, *Hutn. Listy* 26, 87 (1971).
103. P. Catoul, P. Tyou and A. Hans, *Rep. C.N.R.M. (Liege)*, No. 11, 57 (1967).
104. P. Catoul and A. Hans, *C.N.R.M. (Brussels)*, 27 (1971).
105. P. Catoul and A. Hans, *C.N.R.M. (Liege)*, 4 (1969).

106. D. A. Dukelow, J. M. Steltzer, and G. F. Koons, *J. Met.* **23** (12), 22 (1971).
107. E. J. Turkdogan and R. J. Fruehan, General Meeting A.I.S.I. May 1968, A.I.S.I. Yearbook 279–301 (1968).
108. H. J. Engell, W. Esche, and E. Schulte, *Hoesch Ber.* **4/67**, 146 (1967).
109. W. A. Fischer and M. Haussmann, *Arch. Eisenhuttenwes.* **37**, 959 (1966).
110. W. A. Fischer and D. Janke, *Arch. Eisenhuttenwes.* **39**, 89 (1968).
111. W. A. Fischer, *Berg Huetenmaen. Monatsch.* **113** (3), 141 (1968).
112. W. A. Fischer, D. Janke, and W. Ackermann, *Arch. Eisenhuttenwes.* **41**, 361 (1970).
113. W. A. Fischer and D. Janke, *Arch. Eisenhuttenwes.* **41**, 1027 (1970); **44**, 15 (1973).
114. G. R. Fitterer, *J. Met.* **18**, 961 (1966).
115. G. R. Fitterer, C. D. Cassler, and V. L. Vierbicky, *J. Met.* **21** (8), 46 (1969); **20** (6), 74 (1968).
116. G. R. Fitterer, C. D. Cassler, and J. I. Nurminen, NASA Grant NGR 39-011-067, 1970.
117. G. R. Fitterer, *Instrum. Iron Steel Ind.* **20**, 48 (1970).
118. E. Forster and H. Richter, *Arch. Eisenhuttenwes.* **40**, 475 (1969).
119. K. Goto and Y. Matsushita, *Tetsu To Hagane* **50**, 1818 (1964).
120. K. Goto and Y. Matsushita, *Tetsu To Hagane* **50**, 1821 (1964).
121. B. Korousic, *Rud-Met. Zr.* **1**, 43 (1970).
122. Von L. Logdandy, E. Forster, W. Klapdar, and H. Richter, *Stahl Eisen* **89**, 704 (1969).
123. W. Loscher, *Arch. Eisenhuttenwes.* **40**, 479 (1969).
124. M. Macozek and Z. Buzek, *Hutn. Listy* **27**, 394 (1972).
125. Y. Matsushita and K. Goto, *Trans. I.S.I.J.* **6**, 131 (1966).
126. M. Ohtani and K. Sambongi, *Tetsu To Hagane* **49**, 22 (1963).
127. J. K. Pargeter, *Can. Met. Quart.* **6**, 21 (1967).
128. J. K. Pargeter, *J. Met.* **20** (10), 27 (1968).
129. J. K. Pargeter and D. K. Faurchou, *J. Met.* **21**, 46 (1969).
130. S. R. Richards, D. A. J. Swinkels, and J. B. Henderson, paper presented at the International Conference on Science and Technology, Iron and Steel, Tokyo, September 7–11, 1970.
131. C. K. Russell, R. J. Fruehan, and R. S. Rittiger, *J. Met.* **23** (11), 44 (1971).
132. K. Sambongi, M. Ohtani, Y. Omori, and H. Inove, *Tetsu To Hagane* **50**, 1823 (1964); *Iron Steel Inst. Jpn. I.S.I.J.* **6**, 76 (1966).
133. K. Schwerdtfeger and H. J. Engell, *Arch. Eisenhuttenwes.* **35**, 533 (1964).
134. K. H. Ulrich and K. Borowski, *Arch. Eisenhuttenwes.* **39**, 259 (1968).
135. K. P. Abraham, *Trans. Indian Inst. Met.* **22**, 5 (1969).
136. Z. Buzek and A. Hatla, *Freiberg. Forschungsch. Met.* **117**, 59 (1969).
137. F. W. Euler and W. Loscher, *Hoesch. Ber.* **3**, 15 (1966).
138. R. Kirchin, Dissertation, Techn. Hochschule, 1971.
139. B. Shuh, B. Korousic, and B. Marincek, *Schweiz, Arch.* **34**, 380 (1968).
140. Y. Matsushita and K. Goto, in *Thermodynamics*, Vol. I, ed. H. Schmalzried, IAEA, Vienna (1966), p. 111.
141. R. J. Fruehan, *Trans. AIME* **242**, 2007 (1968).
142. W. A. Dench and O. Kubaschewski, *High Temp. High Press.* **1**, 357 (1969).
143. K. Schwerdtfeger and A. Muan, *Acta Metall.* **13**, 509 (1965).
144. L. R. Bidwell, F. E. Rizzo, and J. V. Smith, *Acta Metall.* **18**, 1013 (1969).
145. V. N. Drobyshev and T. N. Rezukhina, *Russ. J. Phys. Chem.* **39**, 75 (1970).
146. L. R. Bidwell and R. Speiser, *Acta Metall.* **13**, 61 (1965).
147. C. M. Sellars and F. Maak, *Trans. AIME* **236**, 457 (1966).
148. C. B. Alcock, K. T. Jacob, and T. Palmutau, *Acta Metall.* **21**, 1003 (1973).

149. L. Staffanson, L. Bentell, and I. Svenson, *Scand. J. Met.* **3**, 153 (1974).
150. K. T. Jacob and J. H. E. Jeffes, *J. Chem. Thermodyn.* **5**, 365 (1973).
151. W. G. Bugden and J. N. Pratt, *J. Chem. Thermodyn.* **1**, 353 (1969).
152. J. Troudsen and P. Bolsaitis, *Met. Trans.* **1**, 2023 (1970).
153. C. B. Alcock and A. Kubik, *Acta Metall.* **17**, 437 (1969).
154. P. C. Lidster and H. B. Bell, *Trans. AIME* **245**, 2273 (1969).
155. H. Tanaka, Y. Kishida, T. Kotani, and J. Moriyama, *J. Jpn. Inst. Met.* **37**, 568 (1973).
156. C. Wagner, *Acta Metall.* **21**, 1297 (1973).
157. S. C. Singhal and W. L. Worrell, Metallurgical Chemistry Symposium 1971 held at Brunel University, ed. O. Kubaschewski, Her Majesty's Stationery Office, London (1972), p. 65.
158. E. Fromm, *J. Less Common Met.* **22**, 139 (1970).
159. D. C. Bartosik, P. K. Raychoudhuri, and D. H. Whitmore, 3rd International Conference on Chemical Thermodynamics, Baden, September 1973, paper 9/4.
160. J. N. Pratt, J. M. Bird, and A. W. Bryant, 3rd International Conference on Chemical Thermodynamics, Baden, September 1973, paper 9/7.
161. Y. Matsushita and K. Goto, *J. Faculty Eng. Tokyo Univ.* **27**, 217 (1964).
162. R. J. Fruehan, *Trans. AIME* **242**, 2007 (1968).
163. A. D. Kulkarni and R. E. Johnson, *Met. Trans.* **4**, 1723 (1973).
164. A. D. Kulkarni, *Met. Trans.* **4**, 1713 (1973).
165. R. J. Heus and J. J. Egan, *Z. Phys. Chem. N.F.* **74**, 108 (1971).
166. Samir Abu Ali, V. V. Samokhval, U. S. Geidrikh, and A. A. Vecher, *Russ. J. Phys. Chem.* **46**, 139 (1972).
167. V. V. Samokhval and A. A. Vecher, *Russ. J. Phys. Chem.* **42**, 340 (1968).
168. T. N. Rezukhina and Z. V. Proshina, *Russ. J. Phys. Chem.* **36**, 333 (1962).
169. J. J. Egan, W. McCoy, and J. Bracker, in *Thermodynamics of Nuclear Materials*, IAEA, Vienna (1962), p. 163.
170. J. J. Egan, *J. Phys. Chem.* **68**, 978 (1964).
171. W. K. Bek and J. J. Egan, *J. Electrochem. Soc.* **113**, 396 (1966).
172. S. Aronson and A. Auskern, in *Thermodynamics*, Vol. I, IAEA, Vienna (1966), p. 165.
173. W. Nickerson and C. Altstetter, *Scripta Met.* **7**, 377 (1973),
174. S. Aronson, *J. Inorg. Nucl. Chem.* **29**, 1611 (1967).
175. A. Kleykamp. *Ber. Bunsenges. Phys. Chem.* **73**, 354 (1969).
176. V. N. Drobyshev, T. N. Rezukhina, and L. A. Tarasova, *Russ. J. Phys. Chem.* **39**, 70 (1965).
177. V. N. Drobyshev and T. N. Rezukhina, *Russ. J. Phys. Chem.* **39**, 75 (1965).
178. V. N. Drobyshev and T. N. Rezukhina, *Izv. Akad. Nauk SSSR, Metall.*, 156 (1966).
178a. S. K. Lau, Ph.D. thesis, University of Pennsylvania, 1978.
179. P. J. Meschter and W. L. Worrell, *Met. Trans.* **8A**, 503 (1977).
180. H. Kleykamp and M. Murabayashi, 3rd International Conference on Chemical Thermodynamics, Baden, paper 9/8a (1973).
181. P. J. Meschter and W. L. Worrell, *Met. Trans.* **7A**, 299 (1976).
182. W. H. Skelton, N. J. Magnani, and J. F. Smith, *Met. Trans.* **4**, 917 (1973).
183. T. N. Rezukhina, 3rd International Conference on Chemical Thermodynamics, Baden, Paper 9/6, 1973.
184. B. Predel and U. Schallner, *Mater. Sci. Eng.* **10**, 249 (1972).
185. H. B. Bell, J. P. Hafra, F. H. Pullard, and P. J. Spencer, *Met. Sci. J.* **7**, 185 (1973).
186. H. Davies and W. W. Smeltzer, *J. Electrochem. Soc.* **121**, 543 (1974).
187. I. Katayama, H. Shimatani, and Z. Kozuka, *J. Jpn. Inst. Met.* **37**, 509 (1973).

188. I. Katayama, S. Igi, and Z. Kozuka, *J. Jpn. Inst. Met.* **38**, 332 (1974).
189. I. Katayama, N. Kemori, and Z. Kozuka, *Trans. Jpn. Inst. Met.* **16**, 423 (1975).
190. Von H. Jacobi, D. Stockel, and H. Leo Lukas, *Z. Metallkd.* **62**, 305 (1971).
191. M. Kanno, *J. Nucl. Mater.* **51**, 24 (1974).
191a. H. Holleck and A. Kleykamp, *J. Nucl. Mater.* **35**, 158 (1970).
192. M. Murabayashi and H. Kleykamp, *J. Less Common Met.* **39**, 235 (1975).
193. K. P. Jagannathan and A. Ghosh, *Met. Trans.* **4**, 1577 (1973).
194. D. Chatterjee and J. V. Smith, *J. Electrochem. Soc.* **120**, 770 (1973).
195. H. Tanaka, Y. Kishida, and J. Moriyama, *J. Jpn. Inst. Met.* **37**, 564 (1973).
196. R. Benz, *Met. Trans.* **5**, 2217 (1974).
197. G. K. Sigworth and J. F. Elliott, *Can. Met. Quart.* **13**, 455 (1974).
198. G. K. Sigworth and J. F. Elliott, *Met. Sci. J.* **8**, 298 (1974).
199. D. J. Fray and B. Savory, *J. Chem. Thermodyn.* **7**, 485 (1975).
200. M. Kolodney, B. Minushkin, and H. Sternmetz, *Electrochem. Technol.* **3**, 214 (1965).
201. H. S. Isaacs, *J. Electrochem. Soc.* **119**, 455 (1972).
202. R. J. Fruehan, *Met. Trans.* **5**, 345 (1974).
203. W. A. Fischer and D. Janke, *Z. Metallkd.* **62**, 747 (1971).

4

Thermodynamic Properties of Oxide Systems

S. Seetharaman and K. P. Abraham

1. Introduction

The high-temperature solid electrolyte galvanic cells have recently come into prominence on account of their numerous applications in many thermodynamic and kinetic investigations. The growing demand for materials for application at high temperatures stimulated interest in a systematic investigation of the thermodynamic properties of many oxide systems. Solid state galvanic cells with a ceramic electrolyte have been found to be highly suitable for these investigations. A significant amount of thermodynamic data pertaining to oxide systems has been collected by this technique. The technique has made possible the accurate study of a number of new oxide systems and nonstoichiometric compounds. The applications of these galvanic cells to the study of oxide systems is briefly surveyed and an account is given of the systems investigated.

The electrochemical method can be used to determine the standard free-energy change of a reaction, provided a galvanic cell can be experimentally realized in which the reaction occurs electrochemically. Moreover, the galvanic cell must be so designed as to eliminate all the side reactions, so that the reversible cell potential, E, is related to the Gibbs free-energy change, ΔG, for the virtual chemical reaction of the cell by the relationship

$$\Delta G = -nFE \qquad (4.1)$$

where n is the number of equivalents of charge passed through the external circuit and F is the Faraday constant. Further, the following relationship has

S. Seetharaman · Metallurgisk Kemi, Institution for Metallurgy, Kungl Teckniska Hogskolan, Stockholm 70, Sweden. K. P. Abraham · Department of Metallurgy, Indian Institute of Science, Bangalore 560012, India.

been derived in Chapter 2 for the cell potential E in terms of the ionic transport number, t_{ion}, of the solid electrolyte and the oxygen potential μ'_{O_2} and μ''_{O_2} at the electrodes

$$E = \frac{1}{4F}\int_{\mu'_{O_2}}^{\mu''_{O_2}} t_{ion}\, d\mu_{O_2} \tag{4.2}$$

However, the presence of even a small electronic contribution to the conductivity will give rise to internal short circuiting, thereby disturbing the thermodynamic equilibrium existing at the electrode–electrolyte interface. A value of t_{ion} of 0.99 or above is considered to be suitable for most thermodynamic measurements.

The limitations arising out of a value of $t_{ion} < 1$ and problems arising out of electrode–electrolyte reactions are discussed in detail in Chapter 2. The limits of applicability of the commonly used solid electrolytes have been established by a number of workers and the results are summarized in Fig. 2.1 of Chapter 2. A comprehensive account of the various solid electrolytes, the conduction mechanism, an assessment of the errors involved in the emf measurements, and the applications have been given by Schmalzried.[1]

2. Comparison of the Solid Electrolyte Method with Other Methods for Thermodynamic Measurements

The complete thermodynamic characterization of oxide systems can be done by the measurement of the enthalpies of formation by calorimetric methods and the determination of the free energies of formation, either by the equilibration of the system with a gas mixture of known oxygen partial pressure or by the emf method. The equilibration method has been in use for a long time, but has the disadvantage that it involves the chemical and/or physical analysis of the equilibrated phases after quenching. The possibility of phase changes while cooling from the state of equilibrium introduces errors in the measured data. The emf method is much quicker than the equilibration method and permits direct measurement of thermodynamic potentials at the temperature of the experiment and no quenching from high temperatures is involved. Moreover, the emf method can be employed for thermodynamic studies at slightly lower temperatures than possible by equilibration methods. Comparing the uncertainties involved in the two methods, an uncertainty of 1% in the measurement of the equilibrium constant (K) in the equilibration method, leads to an uncertainty of 20 cal in ΔG° at 1000°K. However, 1% error in K is rarely realized in high-temperature measurements and the uncertainty involved is generally much more, and can be of the order of 100–200 cal. On the other hand, an uncertainty of 1 mV in the measurement of emf will bring in an uncertainty in ΔG°, e.g., of 92 cal for a reaction involving 1 mol of oxygen. This accuracy can be achieved by careful experimentation and by ensuring that the cell reactions are well defined and truly reversible. For any given configuration of electrodes and electrolytes, it is commonly found that emf's are reproducible to within ± 1 mV.

3. Types of Solid Electrolytes Used in the Study of Oxide Systems

The types of solid electrolytes used in high-temperature galvanic cells and their properties have been given in Chapter 1. Both oxide and fluoride solid electrolytes have been successfully employed for the thermodynamic studies on oxide systems. Schmalzried[1] and Pizzini and Bianchi[2] have given a list of all possible solid electrolytes which can be used. The possible applications of fluoride solid electrolytes have been described elsewhere.[3]

The applications of the oxide and fluoride solid electrolytes for thermodynamic studies on oxide systems are discussed separately in Sections 4 and 5, respectively.

4. Thermodynamic Measurements of Oxide Systems with Cells Involving Oxide Solid Electrolytes

The use of solid electrolyte galvanic cells for thermodynamic measurements in oxide systems implies that the electrolyte should always have a lower free energy of formation than any of the electrode components. The possibility of any exchange reaction between the constituents of the electrode material and interaction between electrode material and electrolyte due to mutual solid solubility or eutectic formations should be examined. The volatility of any of the cell components at the operating temperature is another factor to be taken into consideration. Moreover, the electrolyte must be so chosen that in the oxygen partial pressure ranges involved in the measurements, there is practically no electronic conduction in the electrolyte. The obvious choice is to use the zirconia-based electrolytes for moderate oxygen partial pressures and thoria-based electrolytes for low oxygen partial pressures.[43]

4.1. Apparatus

4.1.1. Cell Design

In the earliest form of the cell, Kiukkola and Wagner[4] kept the solid electrolyte pellet sandwiched between the reference and measuring electrodes. The outer surface of the electrode in each case was kept in contact with a platinum foil connected to the two electrical leads of the galvanic cell, as

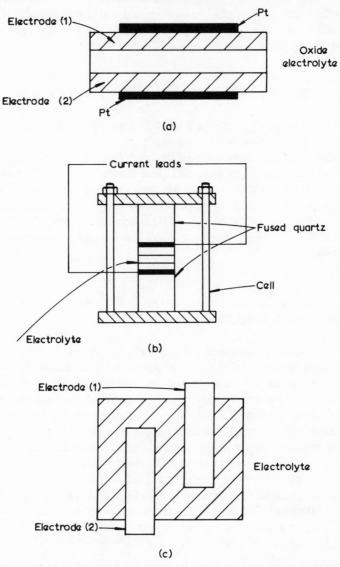

FIG. 4.1. Cells without separate electrode compartments (References 5 and 7).

shown in Fig. 4.1a. This type of cell assembly is adequate for systems with high equilibrium oxygen pressures. The electrode–electrolyte contact was normally improved by polishing the surfaces of the electrodes and the electrolyte. Spring-loaded supports from the top and bottom ensure good electrical contact. A simple but elegant way of ensuring good contact was devised by Gerasimov *et al.*[5] They used a clamp that can be tightened to give the best contact, insulated well from the electrodes as shown Fig. 4.1b. Schmalzried,[6,7] however, embedded the electrodes in the electrolyte as shown in Fig. 4.1c. A similar cell was used by Rapp.[8]

It was realized that though the transfer of oxygen between the two electrodes via the gas phase is small, the accuracy of the data could be improved by isolating the two electrodes from each other. A cell design which permits the isolation of the two electrodes is due to Steele[9] and is shown in Fig. 4.2. This cell assembly has provision to keep an "oxygen getter" in the immediate vicinity of the electrodes.

A more sophisticated apparatus which permits the isolation of the electrodes and ensures good contact between the electrodes and the electrolyte was used by Charatte and Flengas[10] and is shown in Fig. 4.3. The maintenance of true thermodynamic equilibrium between the electrodes and their respective gaseous environments is ensured by keeping a static inert atmosphere in each of the sealed electrode compartments. The standard free energies of formation of various oxides reported by these authors have been widely accepted as highly reliable data.

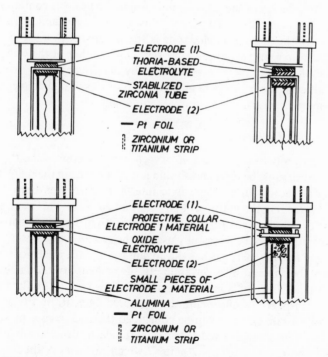

FIG. 4.2. Cells with separate electrode compartments (Reference 9).

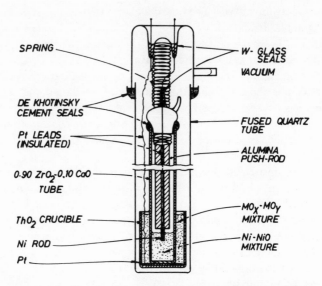

FIG. 4.3. Closed cell for emf measurements (Reference 10).

A unique design of the cell assembly where the possible contamination of the electrolyte material can be prevented is due to Taylor and Schmalzried.[11] In this design the measuring electrode is equilibrated with a static gas phase which has an oxygen partial pressure very close to that of the electrode and the partial pressure of oxygen is measured by the galvanic cell. Though there are difficulties in the measurement with a cell of this design as mentioned by the authors themselves, the advantages are obvious.

4.1.2. The Inert Environment

For the proper functioning of oxygen concentration cells, the environment of the cell must be inert and should not participate in the redox reactions occurring at the electrode–electrolyte interface. There has been much discussion[12] regarding the nature of this environment as to whether it should be a vacuum or a flowing stream of an inert gas. Both methods of controlling the environment have been successfully used in the case of oxide systems. A setup employing a flowing inert gas has the advantage that the electrode–electrolyte contact can be checked by ascertaining the dependence of the cell emf on the gas flow rate. If the emf is independent of the flow rate, it is unlikely that the gaseous atmosphere is influencing the measured data. If, on the other hand, there is a flow rate effect, it can be concluded that the oxygen present in the inert gas is establishing a mixed potential at the electrode–electrolyte interface.

The importance of using a "getter" which can remove the last traces of oxygen in the vicinity of the electrolyte has been stressed by many authors.[12]

On the other hand, Goto and Matsushita[13] recommend that the oxygen partial pressure of the gas phase should be very close to that of the two electrodes for successful measurement. If the difference between the oxygen pressures of the two electrodes is very large, separation of the electrodes into two compartments is recommended by using a solid electrolyte tube. Too large a difference in oxygen pressure would mean that the inert gas atmosphere may be oxidizing to one electrode and reducing to the other. This difficulty can be avoided by choosing a reference electrode which establishes an oxygen pressure which is within 10^5 atm of the estimated equilibrium oxygen pressure of the unknown electrode.

The oxygen pressure of the inert gas can be reduced to 10^{-15} atm by arranging a train of suitable purifying agents. It should be pointed out that it will be difficult to establish equilibrium between the oxygen pressure of the inert gas and that fixed by the electrode. However, the inert gas should not influence the oxygen potential at the electrode interface, because the number of oxygen molecules from the gas phase striking the interface will be negligible if the inert gas phase is purified to an oxygen pressure of 10^{-15} atm.

4.1.3. Emf Measuring Device

A potentiometer can normally be used for measuring the emf developed by the cell. However, the use of an electrometer with input impedance of the order of 10^{14} ohms or more, like the Vibron capillary electrometer, is recommended[15] for most purposes and would be highly suitable for these electrode systems with high electrical resistance. Some authors[256] have used Solartron or Keithley digital voltmeters for the measurement of the cell emf.

An account of the experimental errors and precautions to be adopted in the measurements is given in Chapter 2. An analysis of the errors arising out of temperature gradients in the cell, oxygen transfer through solid electrolyte, and reaction between electrode material and solid electrolyte has been made by Goto and Pluschkell.[16] Rapp and Shores[14] have made some recommendations concerning the most acceptable cell geometry. The need for ensuring the truly reversible nature of the measured emf cannot be overemphasized and has been dealt with in Chapter 2. The application of an ac ripple with an amplitude of 50 mV is found[278] to shorten the time required for equilibration.

4.2. Determination of the Free Energy of Formation of Oxides, Spinels, Silicates, and Other Compounds

4.2.1. Oxides

Accurate values of the standard free energies of formation of metal oxides have been obtained from emf measurements, using a galvanic cell with a suitable reference electrode of accurately known oxygen potential at a given temperature. Such a cell can be represented by

$$A, AO \mid solid\ electrolyte \mid BO, B \qquad\qquad [4.1]$$
$$p'_{O_2} \qquad\qquad p''_{O_2}$$

where AO and BO are oxides of the metals A and B that can exist in equilibrium with the respective metals, without dissolution or disproportionation. In this particular case the reference electrode BO is assumed to be more stable than AO. The partial pressure of oxygen in equilibrium with A and AO (p'_{O_2}) is greater than that at the B, BO electrode (p''_{O_2}), as determined by the relative stabilities of the two oxides. Hence oxygen tends to be transferred across the "oxide membrane" from left to right. The emf E of the cell is related to the two oxygen potentials by the relationship

$$E = -\frac{RT}{4F} \ln \frac{p''_{O_2}}{p'_{O_2}} \qquad\qquad (4.3)$$

Hence from a knowledge of the oxygen pressure at the reference electrode, the oxygen pressure in equilibrium with A and AO at the experimental temperature is known and hence its standard free energy of formation. The standard free energies of formation of a number of oxides have been determined by this method with a high degree of accuracy and the data obtained have been listed in Table 4.1. The uncertainty limits in their results have been estimated by many authors and these are also indicated in the table. Even though in most cell arrangements, the emf's are reproducible to ± 1 mV, corresponding to an uncertainty of ± 50 cal/g atom of oxygen, the uncertainty limits reported in some cases are much higher. For example, Rizzo, Bidwell, and Frank[106] have measured the free energy of formation of WO_2 and they showed from their results that the values obtained for materials of different purity, with impurities at 100 ppm level, could vary by as much as ± 300 cal. Moreover, they found that between their results and results obtained from other sources, there was an uncertainty of ± 300 cal/g atom of oxygen. However, as can be seen from Table 4.1, the limits of uncertainty in many other cases were well within the experimental reproducibility of results.

4.2.2. Spinels

The oxygen concentration cells involving solid oxide electrolytes have also been used for the determination of the standard free energies of formation of a number of compounds between metal oxides. The cell used is of the type

$$A, B_2O_3, AB_2O_4 \mid oxide\ electrolyte \mid AO, A \qquad\qquad [4.2]$$

The cell reaction is the formation of the spinel AB_2O_4 from the metal A and oxide B_2O_3.

Table 4.1. The Standard Free Energies of Formation of Metal Oxides by Solid State Galvanic Cell Measurements

Reaction oxygen gas (1 atm)	Temperature range (°K)	$\Delta G°$ (cal/mol)	Ref.
$2Al(s) + 1\frac{1}{2}O_2 = Al_2O_3(s)$	930	$-319,500$	21
$2Bi(l) + 1\frac{1}{2}O_2 = Bi_2O_3(s)$	773–973	$-150,480 + 79.94T \pm 230$	22
$2Bi(l) + 1\frac{1}{2}O_2 = \alpha\text{-}Bi_2O_3(s)$	885–991	$-143,640 + 75.34T$	223
$2Bi(l) + 1\frac{1}{2}O_2 = \beta\text{-}Bi_2O_3(s)$	991–1095	$-133,290 + 64.89T$	223
$2Bi(l) + 1\frac{1}{2}O_2 = Bi_2O_3(l)$	1095–1223	$-119,420 + 52.05T$	223
$2Ce(s) + 1\frac{1}{2}O_2 = Ce_2O_3(s)$	298–1003	$-434,189 + 0.88T \log T$ $-3.83 \times 10^{-3}T^2 + 0.40 \times 10^{-6}T^3$ $-1.41/T \times 10^{-5} + 74.02T$	23
$Ce(s) + O_2 = CeO_2(s)$	298–1003	$-259,845 - 1.54T \log T$ $-1.16 \times 10^{-3}T^2 + 0.20 \times 10^{-6}T^3$ $-0.94 \times 10^5/T + 54.90T$	23 (also 24)
$Co(s) + \frac{1}{2}O_2 = CoO(s)$	1000–1500	$-57,600 + 18.61T \pm 100$	25
	850–1250	$-55,697 + 16.96T \pm 300$	26
	850–1250	$-55,693 + 16.99T \pm 180$	26
		$-57,120 + 17.86T \pm 200$	27 (also 1, 28–30)
$Co(s) + \frac{1}{2}O_2 = CoO(s)$	1073–1673	$-55,200 + 16.40T$	224
	1173–1373	$-56,420 + 17.05T$	225
$3CoO(s) + \frac{1}{3}O_2 = Co_3O_4(s)$	800–1200	$-47,623 + 39.38T \pm 250$	26
	800–1200	$-47,052 + 38.81T \pm 300$	26 (also 27)
$2Cu(s) + \frac{1}{2}O_2 = Cu_2O(s)$	973–1273	$-40,200 + 17.37T \pm 100$	31
	850–1250	$-39,735 + 16.96T \pm 200$	26
		$-40,080 + 17.45T \pm 300$	32
	1000–M.P.	$-42,000 + 17.9T \pm 200$	25
	845–1270	$-42,600 - 5.69T \log T + 36.76T$	33
	773–1356	$-35,360 + 16.68T(°C) \pm 100$	34
	900–1300	$-39,925 + 17.08T \pm 280$	9
	924–1328	$-39,855 + 17.041T \pm 65$	10 (also 35–37)
$2Cu(s) + \frac{1}{2}O_2 = Cu_2O(s)$	1095–1273	$-40,020 + 16.70T$	225
	1000–1273	$-41,717 + 18.29T$	226
$2Cu(l) + \frac{1}{2}O_2 = Cu_2O(s)$	1338–1573		38
$Cu(s) + \frac{1}{2}O_2 = CuO(s)$	973–1273	$-35,750 + 19.97T \pm 100$	31
	845–1270	$-38,800 - 6.55T \log T + 42.50T$	33
	773–1173	$-33,640 + 23.66T(°C) \pm 300$	34 (also 36)
$Cu_2O(s) + \frac{1}{2}O_2 = 2CuO(s)$	1073–1273	$-61,090 + 43.90T$	225
$Cu_2O(s) + \frac{1}{2}O_2 = 2CuO(s)$	892–1320	$-31,347 + 22.643T \pm 65$	10

—continued overleaf

Table 4.1 (continued)

Reaction oxygen gas (1 atm)	Temperature range (°K)	$\Delta G°$ (cal/mol)	Ref.
$2Cr(s) + 1\frac{1}{2}O_2 = Cr_2O_3(s)$	1000–1500	$-258,600 + 55.2T \pm 300$	25 (also 29, 39–41)
$2Cr(s) + \frac{3}{2}O_2 = Cr_2O_3(s)$	1550–1725	$-257,400 + 55.53T$	227
	1150–1540	$-266,600 + 59.78T$	228
	1073–1448	$-263,590 + 59.60T$	229
$xFe(s) + \frac{1}{2}O_2 = Fe_xO(s)$	1000–1500	$-63,050 + 15.75T \pm 100$	25
	773–1423	$-62,580(\pm100) + 15.25 (\pm0.05)T$	42
	773–	$-57,110 + 14.13T(°C) \pm 300$	34
	923–1273	$-63,240 + 15.63T \pm 130$	9
	903–1540	$-62,952 + 15.493T \pm 100$	10
	813–1473	$-62,452\ (\pm46) + 15.127 (\pm0.05)T$	44
	1000–1600	$-63,390 + 15.66T$	45 (also 4, 30, 36, 46–50)
$Fe(\gamma) + \frac{1}{2}O_2 = FeO(l)$	1684	$-56,750 + 11.2T$	51
$3FeO(s) + \frac{1}{2}O_2 = Fe_3O_4(s)$	949–1273	$-74,538 + 29.373T \pm 85$	10
	1173–1473	$-(75,985 \pm 194) + (30.550 \pm 0.15)T$	44 (also 30, 52)
$3Fe(s) + 2O_2 = Fe_3O_4(s)$	773– Trans. Pt.	$-236,600 + 63.35T(°C) \pm 1000$	34 (also 25, 49, 53)
$2Fe_3O_4(s) + \frac{1}{2}O_2 = 3Fe_2O_3(s)$	967–1373	$-59,047 + 33.681T \pm 120$	10
	1099–1321	$-51,930 + 35.39T(°C) \pm 850$	54 (also 25, 30, 52, 55)
$2Ga(l) + \frac{3}{2}O_2 = \beta\text{-}Ga_2O_3(s)$	873–1273	$-252,400 + 70.2T \pm 400$	56
		$-252,200 + 71.1T \pm 400$	56 (also 56a)
$2Ga(l) + \frac{3}{2}O_2 = Ga_2O_3(s)$	1073–1273	$-260,760 + 78.6T$	230
$2Ga(l) + \frac{3}{2}O_2 = Ga_2O_3(s)$		$-260,500 + 78.9T$	231
$Ga_2O + O_2 = Ga_2O_3(s)$	1022–1107	$+204,000 - 108.0T$	232
$Ge(s) + O_2 = GeO_2(s)$	933–1103	$-135,470 + 44.59T \pm 130$	233
Hexagonal	823–973	$-132,900 + 41.45T$	232
$Ge(l) + O_2 = GeO_2(s)$	1223–1423	$-137,332 + 44.865T$	234
Tetragonal	922–1096	$-141,360 + 48.93T \pm 60$	233
$H_2(g) + \frac{1}{2}O_2 = H_2O(g)$	1073–2023		57
$2In(l) + \frac{3}{2}O_2 = In_2O_3(s)$	725–968	$-(222,700 \pm 1000) + (78.1 \pm 0.7)T$	58
	969–1233	$-218,780 + 75.34T \pm 700$	35
	900–1073	$-219,400 + 76.03T \pm 200$	59
	823–1073	$-215,550 + 72.63T \pm 450$	60
	873–1073	$-218,700 + 75.6T \pm 400$	60a

—continued

Table 4.1 (continued)

Reaction oxygen gas (1 atm)	Temperature range (°K)	$\Delta G°$ (cal/mol)	Ref.
$2In(l) + \frac{3}{2}O_2 = In_2O_3(s)$	959–1284	$-219,050 + 76.07T \pm 150$	233
$2In(l) + \frac{3}{2}O_2 = In_2O_3(s)$	823–973	$-218,235 + 75.53T$	232
$2In(l) + \frac{3}{2}O_2 = In_2O_3(s)$		$-220,600 + 76.8T$	231
$Ir(s) + O_2 = IrO_2(s)$	298–1397	$-57,554 - 3.56T \log T$ $+ 52.415T$	61
	945–1125	$-57,484 + 40.76T \pm 400$	62
	950–1170	$-57,945 + 41.41T \pm 200$	63
	950–1170	$-57,984 + 41.30T \pm 400$	63
$Ir(s) + O_2 = IrO_2(s)$	950–1170	$-56,760 + 40.41T$	239
$Mn(s) + \frac{1}{2}O_2 = MnO(s)$	923–1273	$-92,940 + 18.24T \pm 150$	64, 53 (also 41)
$Mn(l) + \frac{1}{2}O_2 = MnO(s)$	1553–1823	$-97,940 + 21.37T$	65
$3MnO(s) + \frac{1}{2}O_2 = Mn_3O_4(s)$	992–1393	$-53,157 + 26.585T \pm 80$	10
	1061–1324	$-46,395 + 26.81T(°C)$	54 (also 66)
$2Mn_3O_4(s) + \frac{1}{2}O_2 =$ $3Mn_2O_3(s)$	884–1126	$-27,112 + 22.012T \pm 190$	10 (also 53)
$Mo(s) + O_2 = MoO_2(s)$	1023–1323	$-137,500 + 40.0T$	8
	1260–1360	$-137,580 + 40.08T$	67 (also 66)
$Mo(s) + O_2 = MoO_2(s)$	1739–1933	$-117,280 + 28.28T$	227
$Mo(s) + O_2 = MoO_2(s)$		$-137,560 + 40.31T$	231
$2Mo_9O_{26}(s) + O_2 = 18MoO_3(s)$	773–1023	$-84,300 + 46.3T(\pm 1700)$	68
$Nb(s) + \frac{1}{2}O_2 = NbO(s)$	1000–1400	$-100,200 + 22.1T$	69
	1050–1300	$-100,350 + 21.42T$	19
	298–1346	$-98,450 - 0.564T \log T$ $- 0.63 \times 10^{-3}T^2$ $-0.08 \times 10^5 T^{-1} + 22.10T$	70
	1073–1373	$-98,830 + 20.57T$	71
	1244–1378	$-97,790(\pm 950)$ $+ 19.88(\pm 0.70)T$	72 (also 73)
$Nb(s) + \frac{1}{2}O_2 = NbO(s)$	1196–1291	$-97,525 + 19.55T$	225
	1089–1426	$-99,870 + 21.51T$	236
$2Nb(s) + 2.4O_2 = Nb_2O_{4.8}(s)$	1050–1300	$-430,480 + 93.17T$	206 (also 74)
$Nb(s) + O_2 = NbO_2(s)$	1000–1400	$-188,150 + 40.05T$	75 (also 11)
$Ni(s) + \frac{1}{2}O_2 = NiO(s)$	1000–1500	$-56,050 + 20.4T \pm 100$	25
		$-56,450 + 20.75T$	76
	923–1173	$-55,965 + 20.29T \pm 140$	9
	911–1376	$-55,844 + 20.29T \pm 50$	10
		$-56,650 + 20.75T$	77 (also 4, 8, 30, 37, 49, 53, 56, 66, 78–81)

continued overleaf

Table 4.1 (continued)

Reaction oxygen gas (1 atm)	Temperature range (°K)	$\Delta G°$ (cal/mol)	Ref.
$Ni(s) + \frac{1}{2}O_2 = NiO(s)$	1023–1373	$-56,100 + 20.28T$	225
	1000–1273	$-58,322 + 21.98T$	226
	973–1723	$-55,130 + 19.81T$	237
$2Na(l) + \frac{1}{2}O_2 = Na_2O(s)$	714–934	$-193,890 + 61.66T \pm 800$	82
	593–823	$-189,800 + 56.60T$	83
$Os(s) + O_2 = OsO_2$	298–1200	$-69,800 + 42.5T \pm 1500$	84
$Pb(l) + \frac{1}{2}O_2 = PbO(s)$	720–1070	$-52,430 + 24.12T \pm 310$	32
	773–1160	$-45,710 + 23.47T(°C) \pm 100$	34
	772–1160	$-51,400 + 23.037T \pm 100$	10
$Pb(l) + \frac{1}{2}O_2 = PbO(s)$	1073–1143	$-51,770 + 23.43T$	238
	923–1152	$-51,400 + 23.25T$	237
$Pb(l) + \frac{1}{2}O_2 = PbO(l)$	1160–1323	$-40,840 + 17.82T(°C) \pm 100$	34
	1160–1371	$-45,560 + 17.96T \pm 40$	10
$Pb(l) + \frac{1}{2}O_2 = PbO(l)$	1143–1373	$-45,080 + 17.58T$	238
	1152–1323	$-44,150 + 16.95T$	237
$Pd(s) + \frac{1}{2}O_2 = PdO(s)$	1000–1140	$-27,460 + 23.9T \pm 280$	85
$Pr(s) + 1\frac{1}{2}O_2 = Pr_2O_3(s)$	823–1473	$-380,000 + 88.5T(°C)$	86
$PrO_{1.5}(s) + 0.107O_2 = PrO_{1.714}$	823–1473	$-53,600 + 5.4T(°C)$	
		$- 1.0 \times 10^{-2}T^2(°C)$	86
Plutonium oxides			87, 24
$Re(s) + O_2 = ReO_2(s)$	850–1130	$-104,840 + 43.20T \pm 530$	88
$Ru(s) + O_2 = RuO_2(s)$	780–1040	$-79,150 + 48.0T$	89
	723–1473	$-(73,350 \pm 350)$	
		$+ (41.880 \pm 0.06)T$	90
	873–1273	$-71,440 + 38.97T \pm 120$	91
$2Sb(l) + \frac{3}{2}O_2 = Sb_2O_3(s)$	962–1121	$-166,330 + 59.81T \pm 70$	233
$Si(s) + O_2 = SiO_2(s)$	873–1273		92
$Sn(l) + \frac{1}{2}O_2 = SnO(s)$	505–1273	$-69,670 + 18.37T$	
		$- 1.50 \times 10^{-3}T^2$	
		$-10,000T^{-1} + 3.06T\log_{10} T$	93
$Sn(l) + \frac{1}{2}O_2 = SnO(l)$	1350–1420	$-66,242 + 22.19T$	235
$Sn(l) + \frac{1}{2}O_2 = SnO(l)$	1300–1425	$-64,300 + 21.4T$	94
	770–980	$-70,090 + 25.76T \pm 280$	93
	1173–1373	$-67,350 + 23.45T$	95
$\frac{1}{2}Sn(l) + \frac{1}{2}O_2 = \frac{1}{2}SnO_2(s)$	823–1023	$-70,000 + 25.17T$	232
	1046–1653	$-68,869 + 24.74T$	240
	990–1371	$-67,475 + 23.72T$	241
	773–1173	$-68,725 + 24.74T$	241a
$2Ta(s) + 2\frac{1}{2}O_2 = Ta_2O_5(s)$	1000–1300	$-483,100 + 102.85T$	96
	1073–1473	$-413,800 + 70.66T(°C) \pm 300$	34
	1050–1300	$-481,600 + 97.20T$	19
	1073–1373	$-480,100 + 97.65T$	71
			(also 97, 41)

continued

Table 4.1 (continued)

Reaction oxygen gas (1 atm)	Temperature range (°K)	$\Delta G°$ (cal/mol)	Ref.
$16Ti_7O_{13}(s) + O_2 = 14Ti_8O_{15}(s)$	1148–1273	$-162,200 + 41.4T$ $+ [37.0 \times 10^{-1} + 26.0$ $\times 10^{-5} \times (T - 1193)^2$ $+ 58.3 \times 10^{-1}$ $+ 14.5 \times 10^{-5}(T - 1342)^2]^{1/2}$	17
			(also 41, 98–100)
Uranium oxides			24, 77, 101–104
$\frac{2.5}{6}VO_{1.26}(s) + \frac{1}{2}O_2 = \frac{2.5}{12}V_2O_3$	832–1073	$-92,900 + 24.2T(\pm 200)$	105
$3V_2O_3 + \frac{1}{2}O_2 = 2V_3O_5$	973–1373	$-53,640 + 18.6T \pm 150$	105
$4V_3O_5(s) + \frac{1}{2}O_2 = 3V_4O_7(s)$	873–1273	$-52,320 + 21.6T \pm 100$	105
$5V_4O_7(s) + \frac{1}{2}O_2 = 4V_5O_9(s)$	973–1173	$-47,560 + 20.0T \pm 60$	105
$6V_5O_9(s) + \frac{1}{2}O_2 = 5V_6O_{11}(s)$	860–1000	$-51,630 + 21.6T \pm 50$	105
	1000–	$-48,870 + 21.9T \pm 50$	105
$7V_6O_{11}(s) + \frac{1}{2}O_2 = 6V_7O_{13}$	873–1153	$-43,810 + 18.5T \pm 50$	105
$V_7O_{13}(s) + \frac{1}{2}O_2 = 7VO_2(s)$	1120–	$-42,560 + 17.5T \pm 150$	105
$V_2O_3(s) + \frac{1}{2}O_2 = 2VO_2(s)$	1120–	$-48,860 + 19.14T \pm 200$	105
$\frac{1}{2}W(s) + \frac{1}{2}O_2 = \frac{1}{2}WO_2(s)$	973–1173	$-68,600 + 20.31T \pm 300$	106
	943–1230	$-68,542 - 7.21T \log T$ $+ 1.26 \times 10^{-3}T^2 - 0.47$ $\times 10^{-5}T^{-1} + 40.62T$	107
	1180–1340	$-69,525 + 21.26T$	108
			(also 5, 53, 235)
$W + O_2 = WO_2$		$-136,470 - 41.42T$	231
$\frac{1}{0.72}WO_2(s) + \frac{1}{2}O_2 = \frac{1}{0.72}WO_{2.72}$	973–1273	$-59,640 + 15.01T \pm 300$	106
	902–1158		5
	900–1173	$-65,308 - 7.21T \log T$ $+ 1.26 \times 10^{-3}T^2 - 0.47$ $\times 10^5 T^{-1} + 39.33$	107
$\frac{1}{0.18}WO_{2.72}(s) + \frac{1}{2}O_2$ $= \frac{1}{0.18}WO_{2.90}(s)$	973–1273	$-67,860 + 24.21T \pm 300$	106
$\frac{1}{0.10}WO_{2.90}(s) + \frac{1}{2}O_2$ $= \frac{1}{0.10}WO_3(s)$	973–1273	$-66,765 + 26.76T \pm 500$	106 (also 109)
$Zn(l) + \frac{1}{2}O_2 = ZnO(s)$	793–1168	$-84,770 + 25.75T$	110

Schmalzried[111] measured the free energies of formation of a number of compounds having the spinel structure by the solid state galvanic cell method and reported that the standard free-energy change of the reaction

$$AO + B_2O_3 = AB_2O_4 \qquad (4.4)$$

at 1000°C lies between -5000 and -6000 cal. Apart from this a number of workers have measured the thermodynamic data for the spinel systems by the emf method and the results obtained are listed in Table 4.2.

A recent compilation by Kubaschewski[129] gives the thermodynamic data for a number of double oxides.

4.2.3. Silicates and Other Compounds

The standard free energies of formation of a few silicates have been measured by the solid state galvanic cell method with oxide electrolytes. The silicates studied include Cu_2O–SiO_2,[114,115] $2Cu_2O$–SiO_2,[115–130] $2CoO$–SiO_2,[224–246] $2FeO$–SiO_2,[11,121,246,247] and $2NiO$–SiO_2.[11,244,246,248,249] Apart from spinels and silicates, the standard free energies of formation of a number of oxide compounds have been measured by the emf method incorporating solid electrolytes. These are listed in Table 4.3.

The main difficulty experienced with cells of the above type has been the sluggish equilibration both in the oxide phase mixtures and with respect to the electrode reaction. In some cases, as already mentioned, it was found necessary to use catalysts at the electrodes to speed up the electrode equilibria.

4.3. Activity Measurements in Oxide Systems

Although the gas equilibration method has been an established technique for determining the thermodynamic activities in oxide systems, the solid state galvanic cell method is sometimes preferred in view of its simplicity. The activity of the least stable oxide can be measured in binary and multicomponent systems at temperatures as low as 700°C. A number of binary oxides of rock salt as well as spinel type and silicate systems have been investigated by the emf method.

4.3.1. The Cell Reaction

For the measurement of activity a_{AO} of the constituent oxide in a binary oxide solution (A, B)O, the following typical galvanic cell can be used:

$$A, AO \left| \text{Solid electrolyte} \right| (A, B)O, A \qquad [4.3]$$
$$\qquad p'_{O_2} \qquad\qquad\qquad p''_{O_2}$$

Table 4.2. ΔG° of Spinel Formation by the Solid-Electrolyte Method

Reaction (all solids)	Temperature range (°K)	ΔG° (cal/mol)	Ref.
$CoO + Al_2O_3 = CoAl_2O_4$	1000–1500	$-10,200 + 2.67T(\pm 500)$	25, 111
$CoO + Cr_2O_3 = CoCr_2O_4$	1000–1500	$-19,360 + 5.77T(\pm 200)$	25, 111
	1270–1470		29
$CoO + Fe_2O_3 = CoFe_2O_4$	1173–1700	$-5,400 - 3.20T(\pm 400)$	25, 112
$CoO + La_2O_3 = CoLa_2O_4$	1100–1325		113
$CoO + CoTiO_3 = Co_2TiO_4$	1073–1473		11
$CuO + Al_2O_3 = CuAl_2O_4$	1023–1273		114
$CuO + Al_2O_3 = CuAl_2O_4$	1373–1473	$4,403 - 4.97T$	242
$CuO + Cr_2O_3 = CuCr_2O_4$	1000–1500	$-12,270 + 1.845T(\pm 400)$	25
	1073–1273	$-10,100(\pm 200)$ $-1.36(\pm 0.39)T$	116
Copper ferrite (nonstoichiometry)	1273–1373		117
$CuO + Ga_2O_3 = CuGa_2O_4$	1073–1273		118
$FeO + (\alpha)Al_2O_3 = FeAl_2O_4$	1235–1323	$-10,800 + 4.085T(\pm 35)$	119
$FeO + Cr_2O_3 = FeCr_2O_4$	1000–1500	$-13,750 + 3.93T(\pm 200)$	25
	1300–1433		120
$2Fe(s) + O_2 + 2Cr_2O_3(s)$ $= 2FeCr_2O_4(s)$	1023–1809	$-151,400 + 34.7T (\pm 300)$	256
$2Fe(l) + O_2 + 2Cr_2O_3(s)$ $= 2FeCr_2O_4(s)$	1809–1973	$-158,000 + 38.4T (\pm 300)$	256
$Fe(l) + 2Cr(l) + 4O = FeCr_2O_4$	1873	$-79,800 (\pm 1500)$	121
$FeO + Fe_2O_3 = Fe_3O_4$	1100–1700	$-4,030 - 2.76T(\pm 250)$[a]	25
$FeO + FeTiO_3 = Fe_2TiO_4$	1073–1473		11
$2Fe(s) + O_2 + 2V_2O_3(s)$ $= 2FeV_2O_4(s)$	1023–1809	$-138,000 + 29.8T (\pm 300)$	256
$2Fe(l) + O_2 + 2V_2O_3(s)$ $= 2FeV_2O_4(s)$	1809–1973	$-144,600 + 33.45T \pm 300$	256
Lithium ferrite	1173–1473		122
$MgO + Cr_2O_3 = MgCr_2O_4$	1000–1500	$-10,270 + 1.7T(\pm 500)$	25
$MgO + Fe_2O_3 = MgFe_2O_4$	1100–1700	$-5,750 + 0.32T(\pm 400)$	25, 112
$MgO + Fe_2O_3 = MgFe_2O_n$ (not balanced; $n = 3.06$–4)	1173–1373		123
$MgFe_{2-x}Cr_xO_n$	1173–1373		124,125
$MnO + Fe_2O_3 = MnFe_2O_4$	1064–1373	$-290,004 + 75.6T$	126–128
	1100–1400	$-3,890 - 12T$	127
$NiO + Al_2O_3 = NiAl_2O_4$	973–1473	$1.527 - 1.028 \times 10^{-7}T^{-1}$ $+ 0.927T \log T + 0.5584T$ $-3.597 \times 10^{-5}T^2$	244
$NiO + Al_2O_3 = NiAl_2O_4$	1000–1500		25, 111
$NiO + Cr_2O_3 = NiCr_2O_4$	1000–1500	$-12,930 + 5.17T(\pm 500)$	25
	1300–1500	$-17,550(\pm 570) - 1.07(\pm 410)T$	81, 111
$NiO + Fe_2O_3 = NiFe_2O_4$	1173–1473	$-4,720 - 1.007T(\pm 300)$	25, 112
$NiO + Fe_2O_3 = NiFe_2O_4$	973–1473	$-4,510 - 0.89T$	244
$ZnO + \alpha\text{-}Al_2O_3 = ZnAl_2O_4$	973–1173	$-10,750 + 1.57T (\pm 150)$	246
$ZnO + Cr_2O_3 = ZnCr_2O_4$	973–1173	$-15,000 + 2.05T (\pm 150)$	246

[a] Also see Table 4.1.

Table 4.3. The Standard Free Energies of Formation of Oxy Compounds
Measured by the Solid-Electrolyte Method (Silicates and Spinels not Included)

Reaction (oxygen = gas 1 atm)	Temperature range (°K)	$\Delta G°$ (cal/mol)	Ref.
$CaO(s) + Mo(s) + O_2 = CaMoO_3(s)$	~ 1273	$-144{,}783 + 36.82T$	132
$2CaO(s) + SnO_2(s) = Ca_2SnO_4(s)$	973–1423	$-17{,}040 + 0.85T$	250
$CaO(s) + SnO_2(s) = CaSnO_3(s)$	973–1423	$-17{,}390 + 2.0T$	251
$CaMoO_3(s) + \frac{1}{2}O_2 = CaMoO_4(s)$	~ 1273	$-70{,}350 + 18.19T$	132
$CaO(s) + TiO_{0.5}(s) + 0.75O_2 = CaTiO_3(s)$	1200–1300	$-166{,}340 + 26.63T\,(\pm 500)$	133
$CaO(s) + W(s) + \frac{3}{2}O_2 = CaWO_4(s)$	1090–1373		134
$CoO(s) + 1.235Al_2O_3(s)$ $= CoAl_{2.47}O_{4.70}(s)$	1300–1500	$-12{,}120(\pm 910)$ $+ 3.461(\pm 0.70)T$	135
$Co(s) + TiO_2(s) + 0.5O_2 = CoTiO_3(s)$	1226–1378	$-62{,}320 + 18.69T$	136
$Co(s) + W(s) + 2O_2 = CoWO_4(s)$	1200–1300	$-259{,}432 + 70.54T$	251
		$-260{,}885 + 71.925T$	251
		$-260{,}687 + 71.749T$	251
$CoO(s) + TiO_2(s) = CoTiO_3(s)$	800–1200		11
$Cu_2O(s) + Al_2O_3(s) = Cu_2Al_2O_4(s)$			115, 111
$Cu(s) + Ga_2O_3(s) + \frac{1}{2}O_2 = CuGa_2O_4(s)$	1073–1273		118
$Cu(s) + W(s) + 2O_2 = CuWO_4(s)$	1000–1100	$-248{,}800 + 79.58T$	252
$4Cu(s) + 2Cr_2O_3(s) + O_2 = 4CuCr_2O_2(s)$	1100–1300		253
$3Cu(s) + Fe_3O_4(s) + O_2 = 3CuFeO_2(s)$	970–1300		254
$CuO(s) + GeO_2(s) = CuGeO_3$	923–1023	$-1{,}090(\pm 60) - 2.08$ $(\pm 0.14)T$	116
$2Cu(s) + Nb_2O_5(s) + \frac{1}{2}O_2 = 2CuNbO_3(s)$	1073–1273		118
$Cu_2O(s) + Al_2O_3(s) = 2CuAlO_2(s)$	973–1373	$-5{,}670 + 2.49T\,(\pm 300)$	242
$Cu_2O(s) + 2GeO_2(s) + \frac{1}{2}O_2$ $= 2CuGeO_3(s)$	1073–1273		118
$Cu_2O(s) + 2Nb_2O_5(s) + \frac{1}{2}O_2$ $= 2CuNb_2O_6(s)$	1073–1273		118
$3Cu_2O(s) + 2TiO_2(s) + \frac{3}{2}O_2$ $= 2Cu_3TiO_5(s)$			118
$3Dy_2O_3(s) + W(s) + \frac{3}{2}O_2 = Dy_6WO_{12}(s)$	1200–1400	$-223{,}580 + 54.32T$	255
$FeO(s) + TiO_2(s) = FeTiO_3(s)$	800–1200		11
	1213–1373		137
$2Fe(s) + O_2 + \alpha Al_2O_3 = 2FeO \cdot Al_2O_3$	1023–1809	$-139{,}790 + 32.83T$	243
$2Fe(l) + O_2 + \alpha Al_2O_3 = 2FeO \cdot Al_2O_3$	1809–1973	$-146{,}390 + 36.48T$	243
$\frac{1}{2}Fe_2O_3(s) + \frac{1}{2}Gd_2O_3(s) = GdFeO_3(s)$	1200–1400	$-23{,}800 - 0.61T$	257
$FeO(s) + SO_2(g) = FeS(s) + 1.5O_2(g)$	900–1200	$112{,}500 - 10.2T$	138
$\frac{1}{2}Fe_2O_3(s) + \frac{1}{2}Sm_2O_3(s) = SmFeO_3(s)$	1200–1400	$-22{,}000 - 1.19T$	257
$Fe(s) + W(s) + 2O_2 = FeWO_4(s)$	1200–1300	$-268{,}170 + 68.44T$	258
$\frac{1}{2}Fe_2O_3(s) + \frac{1}{2}Y_2O_3(s) = YFeO_3(s)$	1200–1400	$-19{,}170 - 1.23T$	257
— $\quad$ $LaCoO_3$	1100–1212	$-1{,}284{,}700 + 273.0T$	245
— $\quad$ $LaCuO_2$	1050–1250	$-1{,}010{,}400 + 170.0T$	245
— $\quad$ $LaFeO_3$	900–1225	$-1{,}241{,}000 + 146.9T$	245

————continued

Table 4.3 (*continued*)

Reaction (oxygen = gas 1 atm)	Temperature range (°K)	$\Delta G°$ (cal/mol)	Ref.
$MgO(s) + Mo(s) + O_2(g) = MgMoO_3(s)$	1359–1456	$-135,660(\pm 2000) +$ $+ 36.22\,(\pm 170T)$	19, 20
$MgO(s) + \frac{3}{2}O_2 + Mo(s) = MgMoO_4(s)$	298–1300	$-196,950 - 0.1T\ln T$ $- 2.16 \times 10^{-3}T^2$ $- 10^5 T^{-1} - 0.23$ $\times 10^{-6}T^3 + 62.57T$	139, 221
	1208–1355	$-186,540(\pm 3700)$ $+ 51.55(\pm 3.0)T$	19
$MgO(s) + W(s) + \frac{3}{2}O_2 = MgWO_4(s)$	1220–1370	$-224,370 + 65.67T$	140
$Ni(s) + W(s) + 2O_2 = NiWO_4(s)$	1300–1380	$-264,940 + 72.17T$	259
$MnO(s) + W(s) + \frac{3}{2}O_2 = MnWO_4$	1100–1400	$-214,090 + 56.98T$	259
$3Nd_2O_3(s) + W(s) + \frac{3}{2}O_2 = 3Nd_2WO_6(s)$	1000–1600	$-225,690 + 50.71T$ $\pm [0.77 + 1.66$ $\times 10^{-6}(T - 1323)^2]^{1/2}$	141
$NiO(s) + 1.136Al_2O_3(s)$ $= NiAl_{2.28}O_{4.41}(s)$	1300–1500	$-5,550(\pm 310)$ $- 0.42(\pm 0.23)T$	135
$NiO(s) + TiO_2(s) = NiTiO_3(s)$	800–1200		11
$SrO(s) + W(s) + \frac{3}{2}O_2 = SrWO_4(s)$	1120–1320	$-247,760 + 60.18T(\pm 600)$	142
Sr_3WO_6, Sr_2WO_5 $SrWO_4, Sr_3Al_2O_6$ $SrAl_2O_4$			260
$3Y_2O_3(s) + W(s) + \frac{3}{2}O_2 = 3Y_2WO_6$	1000–1600	$-229,360(\pm 780)$ $+ 54.71(\pm 1.5)T$	141
$2ZnO(s) + TiO_2(rutile) = ZnTiO_4(s)$	930–1100	$-750 - 2.46T\,(\pm 75)$	278
$ZnO(s) + TiO_2(rutile) = ZnTiO_3(s)$	930–1100	$-1600 - 0.199T\,(\pm 50)$	278

The emf of the cell is given by Eq. (4.5) since the activity of AO in solution is related to the two partial pressures by the equation

$$a_{AO} = \left[\frac{p''_{O_2}}{p'_{O_2}}\right]^{1/2} \tag{4.5}$$

$$E = -\frac{RT}{2F}\ln a_{AO} \tag{4.6}$$

The use of the emf method is justified only if the activities of the metal A at both the electrodes and of AO at the AO–A electrode are unity. The mutual solubilities between the oxide and metal phases must be negligible. Further the difference in the free energy of formation between the oxides AO and BO must be sufficiently large (> 20 kcal) even at low concentrations of AO, in order to prevent a possible displacement reaction of the type

$$BO + A = AO + B \tag{4.7}$$

If the two metals can form alloys, then the difference in their free energies of formation should be greater.

The systems investigated have been classified here as (a) binary oxide systems with rock salt structure, (b) binary oxide systems with spinel structure, (c) silicates and glasses.

4.3.2. Binary Oxide Systems with Rock Salt Structure

The earliest measurement of the thermodynamic activities in binary-rock-salt-type solid solutions is that of Engell,[143] who measured "FeO" activities in the systems "FeO"–MgO and "FeO"–MnO using lime zirconia solid electrolyte. Even though "FeO" is nonstoichiometric, the nature of the experiment fixes the "FeO" activity to be that in equilibrium with metallic Fe. High positive deviations from ideality were reported by Engell and were confirmed by other workers for the FeO–MgO[144,145] and FeO–MnO[146,147] systems. The high positive deviation reported for the FeO–MgO systems is in agreement with the results of gas equilibration measurements. However, Schwerttfeger and Muan[148] investigated the FeO–MnO system by the gas equilibration method and found only slight positive deviation. The emf measurements on the systems CoO–MnO[149] and CoO–MgO[150] showed slight positive deviations from ideality for CoO activities, whereas measurements using the equilibration technique showed either ideal behavior[151] or slight positive deviations from ideality[152] for the CoO–MnO system and slight negative deviation[152] for the CoO–MgO system. The recent emf measurements on the system CoO–MgO showed moderately high positive deviations.[273] For the NiO–MnO system, positive deviations from ideality were obtained when both emf technique[153,154,269] and gas equilibration techniques[152,155] were used. The system NiO–MgO has been reported to be ideal based on both emf[15] and gas equilibration[155] measurements. In the case of the CoO–FeO system the emf measurements showed positive deviations[156,272] from ideality, whereas only slight positive deviation from ideality was indicated by the gas equilibration method.[157] The system FeO–(Li$_{0.5}$Fe$_{0.5}$)O has been reported to show negative deviation by Tret'yakov et al.[122] The other systems belonging to this category investigated by the galvanic cell technique are NiO–CoO,[270] CuO–NiO,[271] and CaO–FeO.[274]

Apart from these systems, which are characterized by complete solid solubility, studies have also been carried out on systems where the solid solubility is limited. Among these, nickel oxide doped with lithium oxide has received much attention.[158–161] NiO doped with Fe$_2$O$_3$[161] and Ga$_2$O$_3$[162,163] and ZnO doped with Li$_2$O[164] have also been studied by the emf method. Aronson and Clayton[165] have studied the thermodynamic properties of urania–zirconia solid solutions. They have also investigated the nonstoichiometric urania–thoria solid solutions. Other workers who investigated this system by the emf method are Tanaka *et al.*[166] Roeder and Smeltzer[167] have determined the dissociation pressures of iron–nickel oxides in equilibrium with alloy phases. Kachhawaha, Ganu, and Tare[19] have studied the activities of nickel oxide in the system NiO_4–ZnO using lime zirconia solid electrolyte at 900°C and have also determined the boundaries of the two-phase region in this system.

4.3.3. Spinel Systems

The activities on these systems were measured by techniques similar to those described earlier. Among the spinel systems investigated, the system Fe_3O_4–Mn_3O_4 has received much attention.[54,126,147,168] The other spinel systems studied include Fe_3O–$CoFe_3O_4$,[45,169] Fe_3O_4–$Mg_xFe_{3-x}O_4$,[169,170] $Cu_{1-x}Fe_{2+x}O_4$–Fe_3O_4,[275] and Fe_3O_4–$Ni_{3-x}O_4$.[169,171] The system Fe_3O_4–$Li_{0.5}Fe_{2.5}O_4$[122] has been investigated both at low and high oxygen boundaries.

4.3.4. Silicate Systems and Glasses

The activities in a number of binary silicates have been measured by the solid electrolyte galvanic cell method. If the silicate studied is molten at the temperature of investigation, the use of a solid electrolyte tube or crucible makes it possible to separate the reference electrode from the melt. The selection of a suitable container is a major problem, since all solid electrolytes are not available in the form of tubes. The cell has to be designed in such a way as to hold the silicate melt. The cell design used by Kozuka and Samis[172] for their studies on PbO–SiO_2 melts is shown in Fig. 4.4. The electrolyte tube with the reference electrode is immersed in the molten silicate in this design. Alternatively, the molten silicate itself can be kept inside the solid electrolyte crucible or tube as was done by Matsushita and Goto.[173] A new type of cell designed by Wanibe *et al.*[51] eliminates the slag attack on the electrolyte material and permits the equilibration of the slag with molten silver. The oxygen activity in the molten silver is measured by the solid electrolyte technique. This method can be successfully used for studies on molten slag systems.

The solid electrolytes which have been used for studies on silicates are magnesia, lime-zirconia, yttria-thoria, and lime-thoria. The systems PbO–SiO_2,[174] FeO–SiO_2,[131,175] PbO–Na_2O–SiO_2,[176] and CaO–SiO_2[177] have been investigated using magnesia as the solid electrolyte. The data obtained by using this solid electrolyte must be considered unreliable in view of the fact that the ionic conductivity of magnesia has been shown to vary with oxygen pressure, temperature, and impurity content.[184]

The activities in the binary systems SnO–SiO_2[179] and PbO–SiO_2[172,181,182] have been measured with the help of conventional solid electrolytes. In

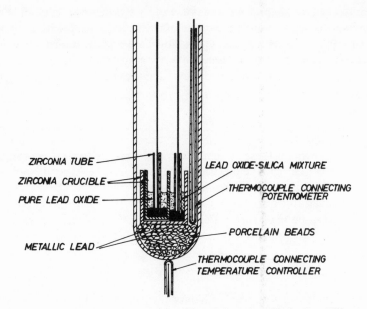

ZIRCONIA TUBE

ZIRCONIA CRUCIBLE

PURE LEAD OXIDE

METALLIC LEAD

LEAD OXIDE-SILICA MIXTURE

THERMOCOUPLE CONNECTING POTENTIOMETER

PORCELAIN BEADS

THERMOCOUPLE CONNECTING TEMPERATURE CONTROLLER

FIG. 4.4. Construction of the cell for molten slags (Reference 172).

addition to these silicates, the activities in the systems $PbO–B_2O_3$,[183] $Na_2O–B_2O_3$,[222] and $PbO–GeO_2$[180] have been measured using lime zirconia as the solid electrolyte.

The application of the emf method for the determination of activities in ternary melts has received relatively little attention so far. The system $CaO–FeO–SiO_2$ has been studied by Erimenko and Filippov[185] and later by Indyk and Bell.[186] It is interesting to note that this system has also been studied using alumina as the solid electrolyte.[187] Recently the system $PbO–Na_2O–SiO_2$ has been studied by Grau and Flengas.[235]

Barmin and Shurjgin[187] have carried out a thermodynamic measurement in a multicomponent silicate system, $CaO–SiO_2–Al_2O_3–Cr_2O_3$, using MgO as the solid electrolyte. The use of magnesia as a solid electrolyte is questionable for reasons given earlier. Lacy and Pask used the lime–zirconia solid electrolyte for measuring the activities of oxides in glasses. The systems studied by them include $NiO–Na_2Si_2O_5$,[188] $Fe_{0.95}O–Na_2Si_2O_5$,[189] and $CoO–Na_2Si_2O_5$.[190]

4.4. Study of the Nonstoichiometry of Metallic Oxides

The range of nonstoichiometry in metal oxides and oxide compounds has been determined by classical methods of chemical analysis, thermogravimetric techniques, and the tensivolumetric method.

Simultaneous measurement of the range of nonstoichiometry and oxygen activity can be carried out by the solid electrolyte galvanic cell method. An elaborate treatment of the various methods of studying the defect equilibria existing in oxide systems has been given by Wagner.[191]

4.4.1. Coulometric Titration

In this method, the composition of a single-phase electrode can be changed at the experimental temperature by permitting and monitoring a current flow in the external circuit. A smaller imposed emf than the equilibrium value will permit a mass transport of oxygen through the electrolyte between the sample and the reference electrode in the spontaneous direction. A larger imposed emf than the equilibrium value will reverse the direction of oxygen transport. Consider a typical cell

$$AO_{1+\delta} \mid \text{oxide solid electrolyte} \mid A, AO_{1+\delta^0} \qquad [4.4]$$

where δ represents the metal deficit or oxygen excess of the electrolyte and δ^0 is the value of δ when the oxide is in equilibrium with pure A. The emf of the cell is given by

$$E = \frac{RT}{2F} \ln a_O \qquad (4.8)$$

where a_O is the oxygen activity of the left-hand electrode. The standard state is the oxide in equilibrium with pure A where oxygen activity is taken to be unity. In the coulometric titration, the nonreversible electrode phase is initially equilibrated at a fixed cell voltage, i.e., at a known oxygen activity. Then a constant current of I A (galvanostatic method) is passed through the cell for a given time of t sec. Then the change $\Delta\delta$ of the excess oxygen is given by

$$\Delta\delta = \frac{It}{2Fn_a} \qquad (4.9)$$

where n_a is the number of gram atoms of metal A in the metal–oxide sample. After the titration, sufficient time is given for the electrode to equilibrate internally and the new cell voltage gives the oxygen activity corresponding to the new composition of the electrode. Oxygen can thus be added coulometrically and the emf of the cell can be determined for each oxygen addition. For the success of the experiment, both mixed-electrode reactions and extraneous electrode–gas reactions must be absent. The cell arrangement used by Tret'yakov and Rapp[200] satisfies this requirement by making the surrounding gas in the electrode chamber part of the system, while excluding oxygen exchange with the gas outside the cell. The treatment presupposes that the transport number of oxygen across the cell is unity in the temperature and partial pressure ranges involved. During the titration, however, oxygen partial pressures far lower than equilibrium values are temporarily induced. In case this is beyond the application range of the solid electrolyte used, the partial electronic conduction of the electrolyte should be taken into consideration. Experiments involving a potentiostatic measurement can also be used for the study of nonstoichiometry. A given initial activity of oxygen is attained throughout the $AO_{1+\delta}$ electrode by applying a predetermined voltage across the cell with the left-hand electrode at negative polarity. For example, if a voltage of 5 mV is applied, then, when the current has dropped to zero, the oxygen activity of the left-hand electrode would be very close to unity. Then the applied voltage is suddenly raised to a new value and the current–time plot is recorded. The current would decay to zero when the oxygen activity corresponding to the new applied voltage has been attained throughout the electrode. If there is a small electronic conduction, then the current would eventually reach a nonzero value and this can be subtracted from the total current in order to yield the ionic current. Thus, correction even for a small amount of electronic conduction in the electrolyte is feasible in this type of measurement. The change $\Delta\delta$ of oxygen excess is given by

$$\Delta\delta = \int_0^t Idt/2Fn_a \qquad (4.9a)$$

An alternate method adopted for the study of the nonstoichiometry of oxides by the emf method is to prepare oxides of known oxygen content and measure the corresponding oxygen activities. The method is tedious and analytical errors could be significant.

4.4.2. Oxides

The nonstoichiometry of a number of oxides has been studied by this method and the systems studied are listed in Table 4.4.

4.5. Some Special Studies of Oxides and Related Systems

There are a few applications which need special reference, since they provide more information than just thermodynamic data. One of them is the study of oxygen activities in ferroelectric materials like $Pb(Zr_{1-x}Ti_x)O_3$[192] [PZT]. These results were used to determine the range of ionic conduction in PZT in the temperature range 450–800°C. Similar work has been carried out by Burt and Krakowski[206] on Nb_2O_5-doped $Pb(Zr_{1-x}Ti_x)O_3$ between 500 and 600°C.

Yet another of the special applications of the oxide electrolyte galvanic cells is the determination of the eutectoid temperature in the Fe–O system.[48] The cell used can be represented as

$$\text{Fe, FeO} \mid \text{ZrO}_2\text{–CaO} \mid \text{Fe, Fe}_3\text{O}_4 \qquad [4.5]$$

Table 4.4. Nonstoichiometric Oxides Studied by the Solid-Electrolyte Method

System	Temperature range (°K)	Reference
Binary systems		
CeO_{2-x}	1020–1273	89, 23
CoO_{1+x}	1200	193, 194
Cu_2O_{1+x}	1223–1323	117
CuO_{1+x}	1223–1323	117
FeO_{1+x}	973–1623	42
	1038–1238	195
	773–1323	47
	1323–1473	196
	873–1273	197
	1178–1308	50
	1273	98
	973–1473	44, 193, 194
Fe_3O_4	1473	194
Fe_2O_{3-x}		121, 198
MoO_{2-x}		99
NbO	1123–1573	220
NbO_{2-x}	1123–1323	73, 99, 199
Nb_2O_{5-x}	1073–1343	74
	1480–1823	70
	1030–1300	19
	1273–1473	82
NiO_{1+x}	1109–1359	79, 194, 200
$PrO_{1.5}$–$PrO_{1.714}$	773–1473	86
$PrO_{1.50-1.81}$	1100–1300	261
Plutonium oxides	973–1413	87
Ta_2O_5		97
$TbO_{1.5-1.74}$	1123–1273	262
TiO_{2-x}	1123–1323	99
	1370–1407	201
Ti_2O_3–$Ti_2O_{3.33}$	1022–1495	236
$TiO_{1.95}$–TiO_2	1123–1323	100
$TiO_{1.78}$	1273	98
$TiO_{1.66}$–TiO_2	1173–1873	41
$TiO_{1.12}$	1273	98
$TiO_{1.32}$	1293	98
UO_{2+x}	1150–1350	103
UO_2–U_3O_8	1073–1473	202
$UO_{2.00}$–$UO_{2.11}$–$UO_{1.24}$–$UO_{2.25}$–$UO_{2.66}$	823–1333	77
$UO_{2.012}$–$UO_{2.188}$	773–1373	104
$UO_{2.125}$–$UO_{2.63}$	1073–1373	203
	823–1333	77
$UO_{2.25}$–$UO_{2.6}$	1123–1373	263

—continued overleaf

Table 4.4. (continued)

System	Temperature range ($^\circ$K)	Reference
$UO_{2.20}-UO_{2.54}$	1073–1373	264, 265
$UO_{2.04}-UO_{2.21}$	1073–1373	101
$VO_{1.26}-V_2O_3-V_3O_5-V_4O_7-V_5O_9-V_6O_{11}-VO_2$	1123–1073	105
$VO_{1.5+y}$	1173–1373	266
$VO-V_2O_3$	1100–1500	204
VO_{2-x}		99
$VO_{1.61}$	980	98
V_nO_{2n-1} $(n = 4 - 9)$	1200–1400	267
$WO_2-WO_{2.72}$	873–1233	125
$WO_{2.702}-WO_{2.976}$	850–1100	108
$WO_2-WO_{2.72}-WO_{2.9}-WO_3$	973–1273	106
$WO_2-WO_{2.722}$		268
Ternary systems		
$Co_xFe_{3-x}O_4$	1473	112, 194
$Cu_{0.984}Fe_{2.016}O_{4+x}$	1223–1323	117
$Cu_{1.011}Fe_{1.989}O_{4+x}$	1223–1323	117
$Cu_{0.551}Fe_{2.449}O_{4+x}$	1223–1323	117
$LiFe_5O_{8-x}$	973–1273	200
$Li_xFe_{3-x}O_{4+y}$	1173–1473	122
$MnFe_2O_{4+x}$		127, 198
$U_yZr_{1-y}O_{2+x}$	1273	167
$U_yTh_{1-y}O_{2+x}$	1150–1350	18, 205
$U_yPu_{1-y}O_{2+x}$		17

Below the eutectoid temperature, the emf of this cell is zero since both the electrodes consist of Fe and Fe_3O_4. The eutectoid temperature was found to be 565°C. Similar measurements have also been carried out by Barbi.[53]

In the case of binary systems with terminal solid solubilities, the oxide electrolyte galvanic cells have been used to measure the activities in the single- and two-phase regions and the data collected used to determine phase boundaries. The systems investigated include CaO–FeO,[274] NiO–ZnO,[280] CoO–ZnO,[281] and CoO–CaO.[279]

5. Thermodynamic Measurements of Oxide Systems with Cells Involving Fluoride Solid Electrolytes

5.1. Introduction

The alkali and alkaline earth fluorides form a versatile class of solid electrolytes. They offer a method of measuring directly the high-temperature thermodynamic properties of oxides of highly electropositive metals like calcium, thorium, and magnesium. The investigation of these systems with oxide solid electrolytes will be difficult on account of the extremely low equilibrium oxygen pressures involved. Though it has long been known that these fluorides are total ionic conductors, they have been getting comparatively little attention so far.

In the recent past, some of the fluoride electrolytes have been employed in galvanic cells for the measurement of the thermodynamic properties of oxide systems. They include CaF_2, SrF_2, BaF_2, and PbF_2. The electrical properties of fluoride electrolytes have been dealt with in Chapter 1. Though magnesium fluoride has been used as a solid electrolyte for thermodynamic measurements, the details of its mode of electrical conduction are not available. However, there is some evidence of total ionic conduction based on the analysis of recent measurements on conductivity.[281]

It has already been pointed out that in the case of a binary solid electrolyte, the nature of the charge-carrying ion is insignificant so long as it is a pure ionic conductor, since the activities of the ionic constituents are related by Gibbs–Duhem equation. Hence binary fluoride electrolytes can be successfully used to measure the thermodynamic properties of metal-oxide systems. The use of the fluoride cells is limited by their low melting points and easy slag-forming tendencies.

5.2. Free Energy of Formation of Oxide Compounds

5.2.1. Principle

An oxygen reversible electrode with fluoride electrolyte can be successfully used for the study of metal–oxide systems. A typical cell involving oxides of metals A and B and calcium fluoride electrolyte can be represented as

$$O_2(g), AO(s), AF_2(s) \mid CaF_2 \mid AF_2(s), ABO_4(s), BO_3(s), O_2(g)$$

$$[4.6]$$

The electrode reaction on the left-hand side will be

$$AO(s) + 2F^- = AF_2(s) + \tfrac{1}{2}O_2(g) + 2e^- \tag{4.10}$$

and that at the right-hand side electrode will be

$$AF_2(s) + \tfrac{1}{2}O_2(g) + 2e^- + BO_3 = ABO_4(s) + 2F^- \tag{4.11}$$

The net cell reaction is

$$AO(s) + BO_3(s) = ABO_4(s) \tag{4.12}$$

The emf of the cell gives the free-energy change of this reaction.

5.2.2. Apparatus and Measurements

The cell assembly used for the study of the oxide compounds with fluoride electrolyte is similar to that used with oxide electrolytes. The electrolyte normally used is in the form of a single crystal. Single crystals have the advantage of high purity and low surface area. For limited periods, polycrystalline material in the form of pressed pellets has been successfully used.[207] However, the electrode components have a tendency to penetrate into the electrolyte and contaminate the polycrystalline pellets. The electrodes in the form of compacts are kept pressed from the bottom by a steel spring. Good electrode compact is important, since the solid electrolyte has no tendency to flow around and on to electrode surface. The cell is enclosed in oxygen atmosphere, free from carbon dioxide and moisture. It has been shown that the presence of moisture in the surrounding atmosphere affects the electrical properties of CaF_2.

The fluoride cells in general are reported to behave sluggishly during the operation of the cell with the oxide systems. The equilibration is sluggish with respect to the reversible electrode reactions and with respect to oxide phase mixtures. To a certain extent this has been overcome by mixing some electrolyte material with the electrode. Benz and Wagner[208] report that the addition of catalysts like PbO, Cr_2O_3 with ions of variable valency improves the cell behavior. The mechanism of action of the catalysts is not clear. Vecher and Vecher,[209] on the other hand, used powdered silver metal along with the oxide–fluoride mixture instead of the catalysts and obtained steady emf values within a short time. The addition of CaF_2 powder to the extent of 50 wt % to the electrode mixture has been found to give satisfactory results[219,279] without the addition of catalysts.

5.2.3. Oxycompounds

The first measurement of the free energy of formation in oxide systems using a fluoride electrolyte is that of Benz and Wagner,[208] who measured the free energy of formation of calcium silicates. These authors established the usefulness of this technique by showing that their results are in good agreement with the previous data available in the literature. Since then many other compounds were studied by the fluoride cell. The oxide systems investigated are listed in Table 4.5. The free energies of formation of many silicates, ferrites, vanadates, niobates, borates, and aluminates apart from oxides were measured by the method and the data reported agreed closely in most cases with those from independent calorimetric measurements. It is seen from Table 4.5 that the most commonly used fluoride electrolyte is calcium fluoride. The most serious limitation of calcium fluoride as an electrolyte at high temperatures is its interaction with the electrode components. Many workers

Table 4.5. Oxide Systems Investigated by the Fluoride Electrolyte Cells

System	Compounds	Electrolyte	Reference
Ca–O	CaO	CaF_2	130
Mg–O	MgO	CaF_2	130
Zn–O	ZnO	CaF_2	130
Al–O	Al_2O_3	CaF_2	130
Ca–Si–O	$CaO \cdot SiO_2$	CaF_2	208
	$3CaO \cdot 2SiO_2$		
	$2CaO \cdot SiO_2$		
Ca–Ti–O	$CaO \cdot TiO_2$	CaF_2	210
	$4CaO \cdot TiO_2$		
Ca–W–O	$CaWO_4$	CaF_2	210
Ca–Fe–O	$CaO \cdot Fe_2O_3$	CaF_2	211
	$2CaO \cdot Fe_2O_3$		
Ca–Cr–O	$CaCrO_4$	CaF_2	212
Ca–V–O	$3CaO \cdot V_2O_5$	CaF_2	207
	$2CaO \cdot V_2O_5$		
	$CaO \cdot V_2O_5$		
Ca–Nb–O	$CaO \cdot Nb_2O_5$	CaF_2	213
	$2CaO \cdot Nb_2O_5$		
Ba–Cr–O	$BaCrO_3$	BaF_2	276
Ba–Zr–O	$BaZrO_3$	BaF_2	277
Mg–Si–O	$MgSiO_3$	MgF_2	224
	Mg_2SiO_4		
Sr–Si–O	$SrSiO_3$	SrF_2	224
	Sr_2SiO_4		
	Sr_3SiO_5		
Ca–B–O	$3CaO \cdot B_2O_3$	CaF_2	214
	$2CaO \cdot B_2O_3$		
	$CaO \cdot B_2O_3$		
Sr–Al–O	$3SrO \cdot 2Al_2O_3$	CaF_2	215
Sr–Ti–O	$SrO \cdot TiO_2$	SrF_2	11
	$4SrO \cdot TiO_2$		
Mg–Al–O	$MgO \cdot Al_2O_3$	MgF_2	11
Mg–Ti–O	$2MgO \cdot TiO_2$	MgF_2	216
	$MgO \cdot TiO_2$		
	$MgO \cdot 2TiO_2$		
Pb–Si–O	$PbO \cdot 2SiO_2$	PbF_2	128
	$PbO \cdot SiO_2$		
	$2PbO \cdot SiO_2$		
	$4PbO \cdot SiO_2$		

found that the measured emf agreed with the thermodynamic values only at the start of the experiment and then dropped gradually. This must be due to the formation of compounds between the fluoride electrolyte and one of the electrode constituents. The instability of the emf observed by Benz and Wagner[208] in their study of the CaO–SiO_2 system could be due to the interaction of CaF_2 with CaO and SiO_2 resulting in the formation of fluorosilicates, $3CaO \cdot 2SiO_2 \cdot CaF_2$ and $11CaO \cdot 4SiO_2 \cdot CaF_2$. The possibility of exchange reaction, solid solubility, and eutectic formation with electrode constituents are other complications which can arise. While determining the free energy of formation of strontium aluminates,[215] CaF_2 has been used as a solid electrolyte in contact with SrF_2 and SrO. In this case, the mutual solubility of SrF_2 and CaF_2 has been overlooked. Moreover, the possibility of an exchange reaction between SrO of the reference electrode and CaF_2 cannot be completely ruled out.

One way to avoid this difficulty is to design a galvanic cell with separated gaseous space which permits no contact between electrode and electrolyte. Such a cell used by Kumorov and Tretyakov[217] is shown in Fig. 4.5.

5.3. Measurement of Activities in Oxide Solid Solutions

Fluoride electrolytes have been used in the measurement of activities in oxide solid solutions. A typical cell used is given below:

$$O_2(g), AO(s), AF_2(s) \mid \text{fluoride electrolyte} \mid AF_2(s), [AO]_{BO}, O_2(g)$$

$$[4.7]$$

The cell gives the activity of the metal oxide AO in its solid solution with BO. The cell reaction is

$$AO \rightarrow [AO]_{BO} \qquad (4.13)$$

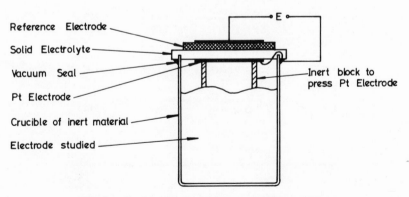

FIG. 4.5. Design of cell with separate gaseous spaces.

The possibility of the reaction $AF_2 + BO = BF_2 + AO$ should be totally absent. The emf gives the partial molar free energy of solution of AO in the solid solution and hence its activity. The oxides should be stable at the cell temperature in an atmosphere of oxygen. The systems investigated include CaO–CdO,[218] CoO–MgO,[218] NiO–MgO,[218] CaO–MnO,[279] and MgO–MnO.[279] The system CaO–CdO was found to be ideal, whereas CoO–MgO showed positive deviations. The activity data obtained by this method agreed well with the gas equilibration results of Hahn and Muan[155] for the NiO–MgO system. High positive deviations were reported in the systems CaO–MnO and MgO–MnO. In the case of systems involving MnO, contamination of the electrolyte when in contact with the electrode led to erratic behavior of the cell.[279] A cell arrangement similar to that shown in Fig. 4.5 was found to be effective in getting reproducible emf's which remained constant. A unique application of this method is the determination of the phase boundaries of the solid solution region in the system CaO–ZrO$_2$ by Pizzini and Morlotti.[219] They used the following cell to measure the CaO activity:

$$Pt \mid O_2 \mid CaO(+CaF_2) \mid CaF_2 \mid ZrO_2\text{–}CaO \mid O_2 \mid Pt \qquad [4.8]$$

For thermodynamic reasons, the reaction between CaF_2 and ZrO_2 to form ZrF_4 could be ruled out. Pure CaO in equilibrium with oxygen at 1 atm pressure served as reference electrode. There was no change in emf when the oxygen pressure was altered from 1 atm to 10^{-4} atm. They found that presaturation of sample with CaF_2 was sufficient to avoid sluggish results. They observed that long-term drift of the measured emf was less than 5 mV over 24 hr in the low-temperature region and even smaller in the high-temperature region.

The phase boundaries in the system ZnO–MgO have been determined by a similar setup with MgF$_2$ solid electrolyte.[279]

6. Concluding Remarks

An attempt has been made to collect and summarize information available on the thermodynamic properties of oxide systems measured with high-temperature galvanic cells incorporating oxide and fluoride solid electrolytes. It is seen that a significant amount of thermodynamic data pertaining to oxide systems has been collected by the solid electrolyte galvanic cell technique. The technique has made possible the accurate study of a number of new oxide systems and nonstoichiometric compounds.

The future interest in the oxide solid electrolytes for work in this area will be concerned with their application at temperatures up to 2000°C, provided the difficulties associated with the operation of the cells at these high temperatures could be overcome. It is solid electrolytes with nonoxygen conductivity, especially the fluorides, that have attracted much interest in the more recent years for studies on oxide systems, but for obvious reasons, these investigations have been limited to low temperatures. The cells incorporating fluoride electrolytes have provided much information on those oxide systems which cannot be conveniently investigated using oxide solid electrolytes, because of the very low equilibrium oxygen pressures involved.

References

1. H. Schmalzried, in *Metallurgical Chemistry*, Symposium Proceedings of the National Physical Laboratory, Middlesex, ed. O. Kubaschewski, Her Majesty's Stationery Office, London (1972), p. 39.
2. S. Pizzini and G. Bianchi, *Chim. Ind.* (*Milan*) **54**, 224 (1972).
3. S. Seetharaman and K. P. Abraham, *J. Sci. Ind. Res.* (*India*) **32**, 641 (1973).
4. K. Kiukkola and C. Wagner, *J. Electrochem. Soc.* **63**, 244 (1959).
5. Ya. I. Gerasimov, I. A. Vasil'Eva, T. P. Chusova, V. A. Geiderikh, and M. A. Timofuva, *Russ. J. Phys. Chem.* **36**, 180 (1962).
6. H. Schmalzried, *Z. Phys. Chem.* **25**, 178 (1960).
7. R. Benz and H. Schmalzried, *Z. Phys. Chem.* **29**, 77 (1961).
8. R. A. Rapp, *Trans. Met. Soc. AIME* **227**, 371 (1963).
9. B. C. H. Steele, in *Electromotive Force Measurements in High Temperature Systems*, ed. C. B. Alcock, Institution of Mining and Metallurgy, London (1968), p. 20.
10. G. G. Charatte and S. N. Flengas, *J. Electrochem. Soc.* **115**, 796 (1968).
11. R. W. Taylor and H. Schmalzried, *J. Phys. Chem.* **68**, 2444 (1964).
12. Discussion on the "EMF Measurements" session, *Thermodynamics*, IAEA, Vienna (1965).
13. K. Goto and Y. Matsushita, *J. Electrochem. Soc. Jpn.* **35**, 1 (1967).
14. R. A. Rapp and D. A. Shores, in *Techniques of Metal Research*, Vol. IV, Part 2, ed. R. A. Rapp, Wiley, New York (1970), p. 162.
15. C. Petot, E. G. Petot, and M. Rigaud, *Can. Met. Quart.* **10**, 203 (1971).
16. K. Goto and W. Plushkell, in *Physics of Electrolytes*, Vol. 2, ed. J. Hladik, Academic Press, New York (1972), p. 563.
17. T. L. Markins and E. J. McIver, in *Pluotonium 1965, Proceedings of the Third International Conference on Plutonium*, eds. A. E. McKay and M. B. Waldron, Chapman and Hall, London (1967), p. 845.
18. H. Tanaka, E. Kimura, A. Yamaguchi, and J. Moriyama, *Nippon Ginzaku Gakkaish* **36**, 633 (1972).
19. J. S. Kachhawaha, M. P. Ganu, and V. B. Tare, *Scripta Met.* **7**, 311 (1973).
20. T. N. Rezukhina and V. A. Levitskii, *IZV. Akad. Nauk. SSSR Neorgan. Materialy* **3**, 138 (1967).
21. G. B. Barbi, *Trans. Faraday Soc.* **62**, 1589 (1960).
22. A. V. Ramana Rao and V. B. Tare, *Scripta Met.* **5**, 807 (1971).
23. F. A. Kuznetsov, V. I. Belyi, T. N. Rezukhina, and Ya. I. Gerasimov, *Dokl. Akad. Nauk SSSR* **139**, 1405 (1961).
24. T. L. Markins, R. J. Bones, and V. J. Wheeler, *Proc. Brit. Ceram. Soc.* **8**, 51 (1967).
25. Yu. D. Tret'yakov and H. Schmalzried, *Ber. Bunsenges. Phys. Chem.* **65**, 396 (1965).
26. W. G. Bugden and J. N. Pratt, *Trans. Inst. Min. Met. Sec. C* **79**, C221 (1970).

27. G. Chattopadhyay, O. M. Sreedharan, and M. S. Chandrasekaraiah, presented at the Indo-Soviet Conference on Solid State Materials, Bangalore, December 1972.
28. T. Matsumura, *Can. J. Phys.* **36**, 1383 (1962).
29. V. A. Levitskii, T. N. Rezukhina, and A. S. Guzei, *Elektrokhimiya* **1**, 237 (1965).
30. Y. Sato, K. Nishimura, I. Sakamoto, T. Yamamura, and Y. Iwano, *Jpn. J. Pwd. Metall.* **2**, 229 (1972).
31. L. R. Bidwell, *J. Electrochem. Soc.* **114**, 30 (1967).
32. C. B. Alcock and T. N. Belford, *Trans. Faraday Soc.* **60**, 822 (1964).
33. G. B. Barbi, *Gazz. Chim. Ital.* **100**, 64 (1970).
34. Y. Matsushita and K. Goto, in *Thermodynamics*, Vol. 1, IAEA, Vienna (1966), p. 108; also *Tetsu-To-Hagane (Overseas)* **4**, 310 (1964).
35. S. C. Schaefer, U.S. Bur. Mines R.I.7549 (1971).
36. S. F. Pal'Guev and A. D. Nevimin, *Trudy Inst. Elektrokhim. Akad. Nauk. SSSR Uralfilial* **1**, 111 (1960).
37. W. A. Dench and O. Kubaschewski, *High Temp. High Pres.* **1**, 357 (1969).
38. J. Osterwald, *Z. Phys. Chem. N.F.* **49**, 138 (1966).
39. A. A. Briggs, W. A. Dench, and W. Slough, *J. Chem. Thermodyn.* **3**, 43 (1971).
40. Yu. D. Tret'yakov, *Izv. Akad. Nauk. SSSR Neorg. Materialy* **2**, 501 (1966).
41. K. Suzuki and K. Sambongi, *Tetsu-To-Hagane* **58**, 1579 (1972).
42. B. E. F. Fender and F. D. Riley, *J. Phys. Chem. Solids* **30**, 793 (1969).
43. B. C. H. Steele, in *Electromotive Force Measurements in High Temperature Systems*, ed. C. B. Alcock, Institution of Mining and Metallurgy, London (1968), p. 16.
44. H. F. Rizzo, R. S. Gordon, and I. B. Cutler, *J. Electrochem. Soc.* **116**, 266 (1969).
45. I. A. Vasileva, S. N. Nudratsova, L. B. Stepina, and A. N. Kornilov, *Russ. J. Phys. Chem.* **43**, 1767 (1969).
46. H. Peters and G. Mann, *Z. Elektrochem.* **63**, 244 (1959).
47. G. B. Barbi, *J. Phys. Chem.* **68**, 2912 (1964).
48. N. Birks, *Nature* **210**, 407 (1966).
49. G. A. Roeder and W. W. Smeltzer, *J. Electrochem. Soc.* **111**, 1074 (1964).
50. M. S. Yakovleva and S. M. Ariya, *Vestn. Leningr. Univ. 18(1b) Ser. Fiz. Khim.* **3**, 130 (1963); *Zh. Fiz. Khim.* **44**, 508 (1970).
51. Y. Wanibe, Y. Yamauchi, K. Kawai, and H. Sakao, *Trans. Iron Steel Inst. Jpn.* **12**, 472 (1972).
52. P. E. C. Bryant and W. W. Smeltzer, *J. Electrochem. Soc.* **116**, 1409 (1969).
53. G. B. Barbi, *J. Phys. Chem.* **68**, 1025 (1964).
54. R. N. Blumenthal and D. H. Whitmore, *J. Am. Ceram. Soc.* **44**, 508 (1961).
55. R. E. Carter, *J. Am. Ceram. Soc.* **43**, 448 (1960).
56. K. A. Kleindinst and D. A. Stevenson, *J. Chem. Thermodyn.* **4**, 565 (1972).
56a. J. V. Smith and D. Chatterji, *J. Am. Ceram. Soc.* **56**, 288 (1973).
57. W. A. Fisher and D. Janke, *Arch. Eisenhuttenwes.* **39**, 89 (1968).
58. G. R. Newns and J. M. Pelmore, *J. Chem. Soc. A* **2**, 360 (1968).
59. D. Chatterji and J. V. Smith, *J. Electrochem. Soc.* **120**, 770 (1973).
60. D. Chatterji and R. W. Vest, *J. Am. Ceram. Soc.* **55**, 575 (1972).
60a. K. A. Klinedinst and D. A. Stevenson, *J. Chem. Thermodyn.* **5**, 21 (1973).
61. H. Kleykamp and J. Paneth, *J. Inorg. Nucl. Chem.*, **65**, 477 (1973).
62. E. S. Ramakrishnan, presented at the Indo-Soviet Conference on Solid-State Materials, Bangalore, December 1972.
63. E. S. Ramakrishnan, *Scripta Metall.* **7**, 305 (1973).
64. C. B. Alcock and S. Zador, *Electrochim. Acta.* **12**, 673 (1967).
65. K. Schwerdtfeger, *Trans. Met. Soc. AIME* **239**, 1276 (1967).

66. J. S. Huebner and M. Sato, *Am. Miner.* **55**, 934 (1970).
67. V. N. Drobyshev and T. N. Rezukhina, *Russ. J. Phys. Chem.* **39**, 75 (1965).
68. V. M. Zhukovskii, T. M. Yaneshkevich, V. P. Lebedkin, V. L. Volkov, and A. D. Nevimin, *Russ. J. Phys. Chem.* **46**, 1542 (1972).
69. G. B. Barbi, *Z. Naturforsch.* **23**, 800 (1968).
70. V. I. Lavreht'ev, Ya. I. Gerasimov, and T. N. Rezukhina, *Dokl. Akad. Nauk SSSR* **136**, 1372 (1961).
71. S. Ignatowicz and M. W. Davies, *J. Less Common Met.* **15**, 100 (1968).
72. V. N. Drobyshev and T. N. Rezukhina, *Trans. Iron Steel Inst. Jpn.* **12**, 472 (1972).
73. M. Hoch, A. S. Iyer, and J. Nelkin, *J. Phys. Chem. Solids* **23**, 1463 (1962).
74. R. N. Blumenthal, J. B. Moser, and D. H. Whitmore, *J. Am. Ceram. Soc.* **48**, 617 (1965).
75. G. B. Barbi, *J. Less Common Met.* to be published.
76. C. Sellars and F. Maak, *Trans. Met. Soc. AIME* **236**, 457 (1966).
77. T. L. Markins, L. E. J. Roberts, and A. Walter, in *Thermodynamics*, IAEA, Vienna (1962), p. 693.
78. B. C. H. Steele and C. B. Alcock, *Trans. Met. Soc. AIME* **233**, 1359 (1965).
79. S. Pizzini and R. Morletti, *J. Electrochem. Soc.* **114**, 1179 (1967).
80. T. L. Markins and M. H. Rand, in *Thermodynamics*, Vol. I, IAEA, Vienna (1966), p. 145.
81. V. A. Levitskii, T. N. Razukhina, and V. G. Dneprova, *Elektrokhimiya* **1**, 933 (1965).
82. C. B. Alcock and G. P. Stavropoulos, *Can. Met. Quart.* **10**, 257 (1971).
83. B. Minushkin, G. Kissel, and V. S. Williams, U.S.A. Report BNL-50176, No. 45 (1969).
84. J. I. Franco and H. Kleykamp, *Ber. Bunsenges. Phys. Chem.* **76**, 691 (1972).
85. H. Kleykamp, *Z. Phys. Chem. N.F.* **71**, 142 (1970).
86. U. Lott, H. Rickert, and C. Keller, *J. Inorg. Nucl. Chem.* **31**, 3427 (1969).
87. T. L. Markins and M. H. Rand, in *Thermodynamics*, Vol. I, IAEA, Vienna (1966), p. 145.
88. J. I. Franco and H. Kleykamp, *Ber. Bunsenges. Phys. Chem.* **75**, 934 (1971).
89. H. Kleykamp, *Z. Phys. Chem. N.F.* **66**, 131 (1969).
90. S. Pizzini and L. Rozzi, *Z. Naturforsch.* **26**, 177 (1971).
91. D. Chatterji and R. W. Vest, *J. Am. Ceram. Soc.* **54**, 73 (1971).
92. F. J. Salzano, H. S. Isaacs, and B. Minushkin, *J. Electrochem. Soc.* **118**, 412 (1971).
93. T. N. Belford and C. B. Alcock, *Trans. Faraday Soc.* **61**, 443 (1965).
94. Z. Kozuka, O. P. Siahaan, and J. Moriyama, *Trans. Jpn. Inst. Met.* **9**, 200 (1968).
95. T. Oishi, T. Hiruna, and J. Moriyama, *Nippon Ginzaku Gakkaishi* **31**, 481 (1972).
96. G. B. Barbi, *Z. Naturforsch.* **25**, 1515 (1970).
97. V. G. Dneprova, T. N. Rezukhina, and Ya. I. Gerasimov, *Russ. J. Phys. Chem.* **42**, 802 (1968).
98. M. S. Yakóvleva and S. M. Ariya, *Russ. J. Phys. Chem.* **37**, 880 (1963).
99. S. Zador, in *Electromotive Force Measurements in High Temperature Systems*, ed. C. B. Alcock, Institution of Mining and Metallurgy, London (1968), p. 145.
100. R. N. Blumenthal and D. H. Whitmore, *J. Electrochem. Soc.* **110**, 92 (1963).
101. D. I. Marchidan and S. Mater, *Rev. Roum. Chim.* **15**, 491 (1970).
102. T. H. Etsell and S. N. Flengas, *Chem. Rev.* **70**, 339 (1970).
103. S. Aronson and J. Belle, *J. Chem. Phys.* **29**, 151 (1958).
104. T. L. Markins and R. J. Bones, Atomic Energy Research Establishment Report AERE R4042 (1963).

105. H. Okinaka, K. Kozuka, and S. Kachi,. *Trans. Jpn Inst. Met.* **12**, 44 (1971).
106. F. E. Rizzo, L. R. Bidwell, and D. F. Frank, *Trans. Met. Soc. AIME* **239**, 1901 (1967).
107. Ya. I. Gerasimov, I. A. Vasil'eva, T. P. Chusova, G. A. Greiderikh, and M. A. Timofuva, *Dokl. Akad. Nauk SSSR* **134**, 1350 (1960).
108. T. N. Rezukhina and Yu. G. Golovanova, *Izv. Akad. Nauk. SSSR Neorg. Mater.* **3**, 867 (1967).
109. R. F. Ksenofontova, I. A. Vask'eva, and Ya. I. Gerasimov, *Dokl. Akad. Nauk SSSR* **143**, 1705 (1962).
110. T. C. Wilder, *Trans. Met. Soc. AIME* **245**, 1370 (1969).
111. H. Schmalzried, *Z. Phys. Chem. N.F.* **25**, 178 (1960).
112. H. Schmalzried and Yu. D. Tret'yakov, *Ber. Bunsenges. Phys. Chem.* **70**, 180 (1966).
113. O. M. Sreedharan and M. S. Chandresekaraiah, *Mater. Res. Bull.* **7**, 1135 (1972).
114. A. A. Slobodyanyuk, Yu. D. Tret'yakov, and A. F. Bessnov, *Russ. J. Phys. Chem.* **45**, 1069 (1971).
115. V. F. Komorov and Yu. D. Tret'yakov, *Russ. J. Phys. Chem.* **45**, 985 (1971).
116. A. A. Slobodyanyuk, Yu. D. Tret'yakov, and A. F. Bessanov, *Russ. J. Phys. Chem.* **46**, 1687 (1972).
117. Yu. D. Tret'yakov, V. F. Komerov, N. A. Prosvirnira, and I. B. Katsensk, *J. Solid State Chem.* **5**, 157 (1972).
118. A. A. Slobodyanyuk, Yu. D. Tret'yakov, and A. F. Bessanov, *Izv. Vyssh. Ucheb. Zaved. Tsvet. Met.* **15**, 18 (1972).
119. T. N. Razukhina, V. A. Levilskii, and P. Ozhigov, *Russ. J. Phys. Chem.* **37**, 358 (1963).
120. T. N. Razukhina, V. A. Levilskii, and B. A. Istomin, *Elektrokhimiya* **1**, 467 (1965).
121. V. F. Komarov, N. N. Obinikov, and Yu. D. Tret'yakov, *Izv. Akad. Nauk. SSSR Neorg. Mater.* **3**, 1064 (1967).
122. Yu. D. Tret'yakov, N. N. Oleynikov, Yu. G. Metlin, and A. P. Erastova, *J. Solid State Chem.* **5**, 191 (1972).   QD 901 . J6818
123. A. S. Guzei, V. I. Lovrent'ev, and T. I. Bulgakova, *Izv. Akad. Nauk SSSR Neorg. Mater.* **3**, 860 (1967).
124. V. I. Lovrent'ev, A. S. Guzej, T. I. Bulgakova, and G. A. Sokolova, *Russ. J. Phys. Chem.* **141**, 1676 (1967).
125. A. S. Guzei, V. I. Lavrenl'ev, T. I. Bulgakova, O. S. Zaitsev, and I. Rosinfel'd, *Izv. Akad. Nauk SSSR Neorg. Mater.* **3**, 909 (1967).
126. I. V. Gordeev, Yu. D. Tret'yakov, and K. G. Khomyakov, *Russ. J. Inorg. Chem.* **9**, 89 (1969).
127. V. I. Roshchupkin and V. I. Lavrent'ev, *Izv. Akad. Nauk SSSR Neorg. Mater.* **3**, 551 (1967).
128. Yu. D. Tret'yakov, *Neorg. Mater.* **1**, 1928 (1965).
129. O. Kubaschewski, *High Temp. High Pres.* **4**, 1 (1972).
130. D. V. Vecher and A. A. Vecher, *Russ. J. Phys. Chem.* **11**, 1565 (1967).
131. V. A. Levitskii and D. D. Ratani, *Izv. Akad. Nauk SSSR Met.* **6**, 65 (1970).
132. V. A. Levitskii, M. Ya. Frenkel, and T. N. Rezukhina, *Elektrokhimiya* **1**, 137 (1965).
133. A. N. Golubenko and T. N. Rezukhina, *Russ. J. Phys. Chem.* **39**, 1587 (1965).
134. T. N. Rezukhina, V. I. Lavrent'yev, V. A. Levitskii, and F. A. Kuznetsov, *Russ. J. Phys. Chem.* **35**, 671 (1961).
135. V. A. Levitskii and T. N. Rezukhina, *Izv. Akad. Nauk SSSR Neorg. Mater.* **2**, 145 (1966).

136. A. N. Golubenko, O. A. Ustinou, and T. N. Rezukhina, *Russ. J. Phys. Chem.* **39**, 616 (1965).

137. V. A. Levitskii, S. G. Popov, and D. D. Ratiani, *Russ. J. Phys. Chem.* **44**, 749 (1970).

138. L. I. Armyanova and S. I. Filippov, *Izv. Vyssh. Ucheb. Zaved. Chem. Met.* **9**, 8 (1972).

139. T. N. Razukhina, V. A. Levitskii, and N. M. Kazimirova, *Russ. J. Phys. Chem.* **35**, 1305 (1961).

140. T. N. Razukhina and V. A. Levitskii, *Russ. J. Phys. Chem.* **37**, 1227 (1963).

141. V. A. Levitskii, V. N. Chentsov, Yu. Ya. Skolis, and Yu. G. Golovanova, *Russ. J. Phys. Chem.* **46**, 151 (1972).

142. V. A. Levitskii and T. N. Rezukhina, *Russ. J. Phys. Chem.* **37**, 599 (1963).

143. H. J. Engell, *Z. Phys. Chem. N.F.* **35**, 192 (1962).

144. I. V. Gordeev, Yu. D. Tret'yakov, and K. G. Khomyakov, Rept. Vestn. Mosk. Univ. No. 6 (1963), p. 59.

145. A. A. Lykasov and V. A. Kozheurov, Fiz. Khim. Osn. Provizvod. State, Mater. Simp. Met. Metalloved., 1968 (Pub. 1971), p. 196.

146. K. K. Prasad, S. Seetharaman, and K. P. Abraham, *Trans. Indian Inst. Met.* **22**, 7 (1969).

147. Yu. D. Tret'yakov, Yu. G. Saksonov, and I. V. Gordeev, *Izv. Akad. Nauk SSSR Neorg. Mater.* **1**, 413 (1965).

148. K. Schwerettfeger and A. Muan, *Trans. Met. Soc. AIME* **239**, 1114 (1967).

149. S. Seetharaman and K. P. Abraham, *Script. Metall.* **3**, 911 (1969).

150. S. Seetharaman and K. P. Abraham, *J. Electrochem. Soc. India* **20**, 54 (1971).

151. E. Aukrvst and A. Muan, *Trans. Met. Soc. AIME* **227**, 1378 (1963).

152. G. P. Popov and S. F. Strokstova, *Russ. J. Phys. Chem.* **46**, 894 (1972).

153. S. Seetharaman and K. P. Abraham, *Trans. Inst. Mining Met. Sec. C.* **77**, C209 (1968).

154. D. J. Cameron and A. E. Unger, *Met. Trans.* **1**, 2615 (1970).

155. W. C. Hahn Jr. and A. Muan, *J. Phys. Chem. Solids* **19**, 338 (1961).

156. S. Seetharaman and K. P. Abraham, *Trans. Indian Inst. Met.* **25**, 16 (1972).

157. J. V. Biggers and A. Muan, *J. Am. Ceram. Soc.* **50**, 230 (1966).

158. S. Pizzini, R. Morolotti, and V. Wagner, *J. Electrochem. Soc.* **116**, 915 (1969).

159. S. Pizzini, R. Morolotti, and V. Wagner, *J. Electrochem. Soc.* **117**, 1529 (1970).

160. J. Deren and G. Rog, *Bull. Acad. Pol. Sci. Ser. Sci. Chim.* **17**, 327 (1969).

161. J. Deren and G. Rog, *Bull. Acad. Pol. Sci. Ser. Sci. Chim.* **15**, 491 (1968).

162. J. Deren and G. Rog, *Akad. Nauk – Oddzial Wkrakowie Praci Komisji Coramiczniz. Ceramika* **16**, 28 (1961).

163. J. Deren, K. Dyrek, J. Pozniczek, M. Rekas, G. Rog, and L. Wenda, *Bull. Acad. Pol. Sci. Ser. Sci. Chim.* **18**, 65 (1970).

164. J. Deren and G. Rog, *Bull. Acad. Pol. Sci. Ser. Sci. Chim.* **18**, 115 (1970).

165. S. Aronson and J. C. Clayton, *J. Chem. Phys.* **35**, 1055 (1961).

166. H. Tanaka, E. Kimura, A. Yamaguchi, and J. Moriyama, *Nippon Ginzaku Gakkaishi* **36**, 633 (1972).

167. G. A. Roeder and W. W. Smeltzer, *J. Electrochem. Soc.* **111**, 1074 (1964).

168. K. Schwerettfeger and A. Muan, *Trans. Met. Soc. AIME* **239**, 1114 (1967).

169. H. Schmalzried and Yu. D. Tret'yakov, *Ber. Bunsenges. Phys. Chem.* **70**, 180 (1966).

170. I. V. Gordeev and Yu. D. Tret'yakov, *Russ. J. Inorg. Chem.* **8**, 943 (1963).

171. I. V. Gordeev and Yu. D. Tret'yakov, *Vestn. Mosk. Univ. Ser. II Khim.* **18**, 32 (1963).

172. Z. Kozuka and C. S. Samis, *Met. Trans.* **1**, 871 (1970).

173. Y. Matsushita and K. Goto, *Trans. Iron Steel Inst. Jpn.* **6**, 131 (1966).
174. O. A. Esin, *Russ. J. Inorg. Chem.* **2**, 237 (1957).
175. O. A. Esin, *Dokl. Akad. Nauk SSSR* **88**, 713 (1953).
176. O. A. Esin, *Zh. Neorg. Khim.* **2**, 87 (1957).
177. K. Sanbongi and Y. Omori, *Sci. Rep. Tohoku Univ.*, 35 (1969).
178. W. C. Hahn, Jr., and A. Muan, *J. Phys. Chem. Solids* **19**, 338 (1961).
179. Z. Kozuka, O. P. Siahaan, and J. Moriyama, *Trans. Jpn. Inst. Met.* **9**, 200 (1968).
180. G. Papst and H. Schmalzried, *Z. Phys. Chem. N.F.* **82**, 206 (1972).
181. Y. Matsushita and K. Goto, *Tetsu-To-Hagane (Overseas)* **4**, 128 (1964); *J. Fac. Eng. Univ. Tokyo* Ser. B. **27**, 217 (1964).
182. G. G. Charatte and S. N. Flengas, *Can. Met. Quart.* **71**, 191 (1968).
183. M. L. Kapoor and M. G. Frohberg, *Can. Met. Quatr.* **7**, 191 (1968).
184. C. B. Alcock, *EMF Measurements in High Temperature System*, Institution of Mining and Metallurgy, London (1968).
185. I. N. Erimenko and S. I. Filippov, *Izv. Vyesh. Ucheb. Zaved. Chem. Met.* **10**, 68 (1967).
186. B. Indyk and H. B. Bell, *J. Iron Steel Inst.* **208**, 1015 (1970).
187. L. N. Barmin and P. M. Shurjgin, *Fiz. Khim. Rasplavlenshlakov* 195, (1971).
188. A. M. Lacy and J. A. Pask, *J. Am. Ceram. Soc.* **53**, 559 (1970).
189. A. M. Lacy and J. A. Pask, *J. Am. Ceram. Soc.* **53**, 676 (1970).
190. A. M. Lacy and J. A. Pask, *J. Am. Ceram. Soc.* **54**, 236 (1971).
191. C. Wagner, in *Progress in Solid State Chemistry*, Vol. 6, ed. H. Reiss and J. O. McCaldin, Pergamon Press, London (1971), p. 1. QC 176 . A1P7
192. A. Ezis, J. C. Burt, and R. A. Krakowski, *J. Am. Ceram. Soc.* **53**, 521 (1970).
193. A. A. Lykasov, Yu. S. Kuznetsov, E. I. Pilkov, I. Shishkov, and V. A. Kozhevrov, *Russ. J. Phys. Chem.* **43**, 1754 (1969).
194. H. G. Sockel and H. Schmalzried, *Ber. Bunsenges. Phys. Chem.* **72**, 745 (1968).
195. F. E. Rizzo and J. V. Smith, *J. Phys. Chem.* **72**, 485 (1968).
196. H. F. Rizzo, R. S. Gordon, and I. B. Cutler, in *Mass Transport in Oxides*, ed. J. B. Wachtman, Jr., U.S. Natl. Bur. Stand. Spec. Publ. 296, Washington, D.C. (1968).
197. H. Asao, K. Ono, A. Yamaguchi, and J. Moriyama, *Mem. Fac. Eng. Kyoto Univ.* **32**, 66 (1970).
198. V. I. Roshchupkin and V. I. Larvrent'ev, *Izv. Akad. Nauk SSSR Neorg. Mater.* **2**, 712 (1966).
199. C. B. Alcock, S. Zador, and B. C. H. Steele, *Proc. Brit. Ceram. Soc.* **8**, 231 (1967).
200. Yu. D. Tret'yakov and R. A. Rapp, *Trans. Met. Soc. AIME* **245**, 1235 (1969).
201. A. N. Golubenko and T. N. Rezukhina, *Izv. Akad. Nauk SSSR Neorg. Mater.* **3**, 101 (1967).
202. K. Kiukkola, *Acta Chem. Scand.* **16**, 327 (1962).
203. D. I. Marchidan and S. Matei, *Rev. Roum. Chim.* **17**, 1053 (1971).
204. M. Hoch and D. Ramakrishnan, *J. Phys. Chem. Solids* **25**, 869 (1964).
205. S. Aronson and J. C. Clayton, *J. Chem. Phys.* **32**, 749 (1960).
206. J. C. Burt and R. A. Krakowski, *J. Am. Ceram. Soc.* **54**, 415 (1971).
207. K. K. Prasad, *Trans. Ind. Inst. Met.* **27**, 259 (1974).
208. R. Benz and C. Wagner, *J. Phys. Chem.* **65**, 1368 (1961).
209. D. V. Vecher and A. A. Vecher, *Zh. Fiz. Khim* **41**, 2916 (1967).
210. T. N. Rezukhina, V. A. Levitskii, and M. Ya. Frenkel, *Izv. Akad. Nauk SSSR Neorg. Mater.* **2**, 325 (1965).
211. T. N. Rezukhina and Ya. Baginsko, *Elektrokhimiya* **3**, 1146 (1967).

212. K. K. Prasad and K. P. Abraham, Proceedings of the Symposium on Materials Science, BARC-NAL, Bangalore (1969).
213. V. G. Druprova, T. N. Rezukhina, and Ya. I. Gerasimov, *Dokl. Akad. Nauk SSSR Fiz. Khim.* **178**, 135 (1968).
214. S. Raghavan and K. P. Abraham, *J. Electrochem. Soc. India* **22**, 149 (1973).
215. V. A. Levitskii, Yu. Ya. Skolis, V. N. Chentsov, and Yo. G. Golovanova, *Russ. J. Phys. Chem.* **46**, 814 (1972).
216. A. K. Shah, K. K. Prasad, and K. P. Abraham, *Trans. Ind. Inst. Met.* **24**, 40 (1970).
217. Yu. D. Tret'yakov and A. R. Kaul, in *Physics of Electrolytes*, ed. J. Hladik, Academic Press, New York (1972), p. 632.
218. K. K. Prasad, Ph.D. thesis, Indian Institute of Science, Bangalore, 1972.
219. S. Pizzini and R. Morlotti, *Trans. Faraday Soc.* **68**, 1601 (1972).
220. W. Nickerson and C. Altstetter, *Script. Metall.* **7**, 229 (1973).
221. T. N. Rezukhina and Z. V. Proshina, *Izv. Akad. Nauk SSSR Neorgan. Mater.* **3**, 138 (1967).
222. S. Sato, T. Yokokawa, H. Kita, and K. Niwa, *J. Electrochem. Soc.* **119**, 1524 (1972).
223. G. M. Mehrotra, M. G. Frohberg, and M. L. Kapoor, *Z. Phys. Chem. N.F.* **99**, 304 (1976).
224. G. Rog, B. Langanke, G. Borchandt, and H. Schmalzried, *J. Chem. Thermodyn.* **6**, 113 (1974).
225. I. A. Vasileva, I. S. Sukhushina, Zh. V. Granovskaya, R. F. Balabaeva, and A. F. Maiorova, *Russ. J. Phys. Chem.* **49**, 1275 (1975).
226. Z. Moser and K. Fitzner, *Rudy Met. Niezelas.* **20**, 510 (1975).
227. D. Janke and W. A. Fischer, *Arch. Eisenhuttenwes.* **46**, 755 (1975).
228. F. N. Mazandarany and R. D. Pehlke, *J. Electrochem. Soc.* **121**, 711 (1974).
229. L. A. Pugliese and G. R. Fitterer, *Met. Trans.* **1**, 1997 (1970).
230. I. Katayama, N. Kemori, and Z. Kozuka, *Nippon Kinzoku Gakkaishi* **39**, 188 (1975).
231. I. Katayama, Z. Kozuka, and Sensaku, *Tech. Rep. Osaka Univ.* **23**, 1121 (1973).
232. G. Palamutcu, Ph.D. thesis, Imperial College, London University (1972).
233. I. Katayama, J. Shibata, and Z. Kozuka, *Nippon Kinzoku Gakkaishi* **39**, 990 (1975).
234. K. T. Jacob, C. B. Alcock, and J. C. Chan, *Acta Metall.* **22**, 545 (1974).
235. A. E. Grau and S. N. Flengas, *J. Electrochem. Soc.* **123**, 352 (1976).
236. R. G. Sommer and E. D. Cater, *J. Electrochem. Soc.* **122**, 1391 (1975).
237. M. Iwase, K. Fujimura, and T. Mon, *Nippon Kinzoku Gakkaishi* **39**, 1118 (1975).
238. H. Charle and J. Osterwald, *Z. Phys. Chem. N.F.* **99**, 199 (1976).
239. E. S. Ramakrishnan, O. M. Sreedharan, and M. S. Chandrasekharaiah, *J. Electrochem. Soc.* **122**, 328 (1975).
240. G. Petot-Ervas, R. Farhi, and C. Petot, *J. Chem. Thermodyn.* **7**, 1131 (1975).
241. S. Seetharaman and L. I. Staffansson, *Scan. J. Met.* **6(3)**, 143 (1977).
241a. T. A. Ramanarayanan and A. K. Bar, *Met. Trans.* **9B**, 485 (1978).
242. K. T. Jacob and C. B. Alcock, *J. Am. Ceram. Soc.* **58**, 192 (1975).
243. J. C. Chan, C. B. Alcock, and K. T. Jacob, *Can. Met. Quart.* **12**, 439 (1973).
244. G. Rog, *Rocz. Chem.* **50**, 147 (1976).
245. O. M. Sreedharan, M. D. Karkhanavala, and M. S. Chandrasekharaiah, *Conf. Int. Thermodyn. Chim. (C.R.) 4th*, **3**, 115 (1975).
246. K. T. Jacob, *Thermochim. Acta* **15**, 79 (1976).
247. S. Shiomi, N. Sano, and Y. Matsushita, *Tetsu-To-Hagane* **61**, 177 (1975).
248. L. I. Armyanova and S. I. Filippov, *Kinet. Zakonomern. Sovmestnogo Vosstanov Okislor Zheleza Drugikh Mater.* (1973), p. 56; *Chem. Abstr.* **85**, 38677U (1976).
249. V. A. Lentok, Yu. G. Golovanova, S. G. Popov, and V. N. Chentsov, *Russ. J. Phys. Chem.* **49**, 971 (1975).

250. K. T. Jacob and J. C. Chan, *J. Electrochem. Soc.* **121**, 534 (1974).
251. T. N. Rezukhina and T. A. Kashina, *J. Chem. Thermodyn.* **8**, 513 (1976).
252. T. A. Kashina and T. N. Rezukhina, *Russ. J. Phys. Chem.* **49**, 755 (1975).
253. E. Rosen, *Chem. Scr.* **8**, 43 (1975).
254. H. Paulsson, E. Rosen, and R. Tegman, *Chem. Scr.* **8**, 193 (1975).
255. V. A. Lentski, V. N. Chentsov, Yu. Khekimov, and Ya. I. Gerasimov, *Russ. J. Phys. Chem.* **49**, 347 (1975).
256. K. T. Jacob and C. B. Alcock, *Met. Trans.* **68**, 215 (1975).
257. S. Yamanchi, K. Fueki, T. Mukaibo, and C. Nakayama, *Bull. Chem. Soc. Jpn.* **48**, 1039 (1975).
258. T. N. Rezukhina and T. A. Kashina, *J. Chem. Thermodyn.* **8**, 519 (1976).
259. T. N. Rezukhina and T. A. Kashina, *Zh. Fiz. Khim.* **48**, 2894 (1974).
260. V. A. Levitskii and Yu. Ya. Skolis, *J. Chem. Thermodyn.* **6**, 1181 (1974).
261. I. A. Vasil'eva, Ya. I. Gerasimov, and A. F. Maiorova, *Dokl. Akad. Nauk SSSR* **226**, 369 (1976).
262. I. A. Vasil'eva, Ya. I. Gerasimov, A. F. Maiorova, and I. V. Pirvova, *Dokl. Akad. Nauk SSSR* **221**, 865 (1975).
263. D. I. Marchidan and S. Tanasescu, *Rev. Roum. Chim.* **19**, 1435 (1974).
264. D. I. Marchidan and S. Tanasescu, *Rev. Roum. Chim.* **20**, 1365 (1975).
265. D. I. Marchidan and S. Tanasescu, *Conf. Int. Thermodyn. Chim.* [*C.R.*] *4th* **3**, 189 (1975).
266. I. A. Vasil'eva and I. S. Sukhushina, *J. Chem. Thermodyn.* **7**, 5 (1975).
267. I. A. Vasil'eva, I. S. Sukhushina, and R. I. Balabaeva, *J. Chem. Thermodyn.* **7**, 319 (1975).
268. I. A. Vasil'eva, L. P. Ogorodova, and L. I. Stepanets, *Vestn. Mosk. Univ. Khim.* **17**, 47 (1976).
269. S. Labus and G. Rog, *Rocz. Chem.* **49**, 339 (1975).
270. K. Torkor and R. Schneider, *J. Solid State Chem.* **18**, 89 (1976).
271. M. G. Rog, J. Deren, P. Grange, and H. Charcosset, *Bull. Soc. Chim. France*, No. 3–4, 471 (1975).
272. I. A. Maksutov and A. A. Lykasov, *Izv. Vysch. Uchib. Zaved. Chem. Met.* **2**, 18 (1974).
273. M. Rigand, G. Giovannetti, and M. Hone, *J. Chem. Thermodyn.* **6**, 993 (1974).
274. V. B. Tare and B. Deo, *Int. Conf. Thermodyn. Chim.* [*C.R.*] *4th* **3**, 104 (1975).
275. M. Fredriksson and E. Rosen, *Chem. Scr.* **9**, 118 (1976).
276. B. Deo and V. B. Tare, *Mater. Res. Bull.* **11**, 469 (1976).
277. B. Deo, J. S. Kachhawaha, and V. B. Tare, *Mater. Res. Bull.* **11**, 653 (1976).
278. K. T. Jacob and C. B. Alcock, *High Temp. High Pres.* **7**, 433 (1975).
279. S. Raghavan, Ph.D. thesis, Indian Institute of Science, Bangalore, 1976.
280. J. S. Kachhawaha, M. P. Ganu, and V. B. Tare, *Scr. Metall.* **7**, 311 (1973).
281. V. V. Gorlach and V. M. Lisitsyn, *Fiz. Tverd. Tela* **16**, 1988 (1974).

Kinetic Studies

V. B. Tare, A. V. Ramana Rao, and
T. A. Ramanarayanan

1. Introduction

Solid electrolyte cells may be conveniently used to study the kinetics of reactions occurring at the electrode–electrolyte interface, at the interface formed by electrode components, and transport phenomena in the electrode and in the electrolyte. Kinetic studies using solid electrolytes generally involve the measurement of either the current or the voltage across an electrochemical cell as a function of time. Since currents and voltages can be measured more precisely than volume, weight, composition, etc., the accuracy involved in electrochemical kinetic measurements is much higher. However, the technique is limited to specific systems.

2. Polarization Studies

The current–voltage characteristics of the electrode–electrolyte interface often show nonohmic behavior. This behavior is due to the slowness of one or more of the several kinetic steps occurring at an interface or in the bulk phase. This phenomenon is termed "polarization."

V. B. Tare · Department of Metallurgical Engineering, Institute of Technology, Banaras Hindu University, Varanasi 221005, India. *A. V. Ramana Rao* · Department of Metallurgical Engineering, Regional Engineering College, Warangal 506004, India. *T. A. Ramanarayanan* · Department of Metallurgical Engineering, Indian Institute of Technology, Kanpur 208016, India. Dr. Ramanarayanan's present address: Corporate Research Laboratories, Exxon Research and Engineering Company, Linden, New Jersey 07036.

2.1. Polarization Studies Involving Gaseous Electrodes

In this category, attention has been mostly given to solid oxide electrolytes. Extensive studies have been carried out on the properties of electrodes such as oxide electrolyte/Pt, $O_2(g)$ and oxide electrolyte/Pt, $CO(g)$, $CO_2(g)$. Increasing interest in fuel cells has led to a study of current-carrying electrodes of this type. The overall reaction involved is

$$\tfrac{1}{2}O_2(g) + V_O^{\cdot\cdot}(\text{electrolyte}) + 2e'(\text{metal}) = O_O \qquad (5.1)$$

The rate of this reaction, of course, can be expected to increase when the effective three-phase contact area is increased. Porous metal or oxide have been used as supporting electrodes. The gas probably migrates through the electrode by permeation and surface diffusion. It is unlikely that the reaction is restricted to three-phase contact zones. Reaction probably occurs across the entire electrolyte-supporting electrode area. A reaction at the gas–electrolyte interface, such as

$$\tfrac{1}{2}O_2(g) + V_O^{\cdot\cdot}(\text{electrolyte}) + 2e'(\text{electrolyte}) = O_O \qquad (5.2)$$

is also possible when the electrolyte has some electronic conductivity. The contribution from this reaction would depend upon the range of oxygen pressures and temperatures involved in the particular study. Raleigh[1] has presented an excellent review of the work in this area. The results depend on whether strongly anodic, strongly cathodic, or intermediate currents are used. In some of the early steady-state dc current–voltage studies[2,3] with the Pt, O_2/ 0.85ZrO$_2$–0.15CaO system at 800–1100°C, only ohmic polarization was observed at current densities up to 0.1 mA/cm^2. In a study using air for the gas phase, Neuimin et al.[2] observed 40-mV nonohmic polarization at 20 mA/cm^2 at 900°C and 100 mA/cm^2 at 1000–1100°C. These results indicate fairly rapid kinetics. Slower kinetics were observed by Neuimin[2] when a $CO-CO_2$ electrode was used. At 10-mA/cm^2 anodic current density, overvoltages of the order of 330 mV were observed. Anodic polarization data have been obtained by Binder and co-workers[4] for H_2-H_2O and $CO-CO_2$ gas mixtures with Pt as supporting electrode and 0.85ZrO$_2$–0.15CaO as the electrolyte. For H_2-H_2O gas mixtures, negligible polarization was observed. For a CO/CO_2 ratio of 9, polarization was small, but considerable for ratios of 1 and 0.1. Möbius and Rohland[5] found negligible polarization with an O_2-Ag electrode, but severe polarization with a $CO-CO_2-Ag$ electrode. However, with $U_3O_8-CO-CO_2$ electrode on calcia-stabilized zirconia, the observed polarization was very small.

It must be emphasized that the observed kinetics depend strongly on the type of supporting electrode used. Kröger and co-workers[6,7] have studied Pt–O_2 electrodes on 0.85ZrO$_2$–0.15CaO using various types of platinum

(porous Pt, Pt foil, etc.). They studied the direct current–voltage characteristics of symmetrical cells of the type

$$\text{Pt, } O_2(\text{I}) \mid Zr_{0.85}Ca_{0.15}O_{1.85} \mid \text{Pt, } O_2(\text{II}) \qquad [5.1]$$

with $P_{O_2}(\text{I}) = P_{O_2}(\text{II})$. The current–voltage characteristics were nonohmic in
all cases. The authors have summarized the rate-limiting steps for the various
types of electrodes as follows.

(a) For foils at low applied voltages, diffusion through Pt is rate limiting,
while at high voltages, electronic current through the calcia-stabilized
zirconia electrolyte with a $P_{O_2}^{-1/2}$ dependence influence the rate.

(b) For sputtered films, diffusion through Pt determines the rate only at
intermediate voltages, while at low voltages P_{O_2}-independent ionic conductance and at high voltages electronic conductance with a $P_{O_2}^{-1/4}$ dependence
determine the rate.

(c) For unfluxed Pt paste, diffusion through Pt with $P_{O_2}^{1/2}$ dependence
determines the rate at low voltages, while at high voltages the rate is determined by the reaction

$$O(\text{ads}) + V_O = O_O \qquad (5.3)$$

(d) For fluxed Pt paste, diffusion through Pt at low voltages, and electronic conduction with a $P_{O_2}^{-1/4}$ dependence at high voltages determine the
rate.

Pizzini[8] has recently assessed some aspects of the kinetics of ion transfer
across interfaces. Considering an O_2–metal electrode and oxide electrolyte,
the following steps could be rate-controlling for the cathodic reaction:

(a) The migration of oxygen vacancies in the electrolyte.

(b) Oxygen diffusion in the gas phase or oxygen diffusion in the pores
of the electrode.

(c) Oxygen chemisorption and dissociation at the metal surface.

(d) Oxygen diffusion at the metal surface or at the grain boundaries of
metal.

(e) Dissolution and diffusion of oxygen into the metal phase.

(f) Oxygen chemisorption and dissociation at the electrolyte surface
followed by diffusion.

(g) Dissolution and diffusion of oxygen in the electrolyte (high-voltage
range).

(h) Electrochemical reaction involving electron transfer across the electrode–electrolyte interface.

In the case of cathodic oxygen reduction reaction, with metallic oxides
like $PrCoO_3$ and Pr_2NiO_4 as the porous supports, the overall reaction rate
seems to depend on the kinetics of the molecular or Knudsen pore diffusion.

The use of Pt as the porous metallic support seems to hinder surface adsorption and diffusion processes, which are fast in the case of metallic oxides.

A versatile technique has been devised by Bauerle[9] for use in the moderate current regime. He measured the ac impedance of symmetric cells consisting of porous Pt electrodes on $ZrO_2-Y_2O_3$ as a function of frequency over a wide range of oxygen pressures at 400–800°C. Complex admittance figures could be constructed from the data and these figures in turn could be used to derive an equivalent circuit for the electrode. The technique supplies a considerable amount of information.

2.2. Polarization Studies Involving Liquid Electrodes

Polarization characteristics have been studied for several liquid-metal electrodes with dissolved oxygen on a stabilized zirconia electrolyte. Goto et al.[10] studied the polarization characteristics of the cell:

$$\text{air, } \underline{O} \text{ (in liq. Ag)} \mid \text{oxide solid electrolyte} \mid \underline{O} \text{(in liq. Ag), air} \qquad [5.2]$$

Cell [5.2] is symmetrical so that under open-circuit conditions the voltage across the cell should be zero. When a constant dc current is passed through the cell in either direction, oxygen ions would be transferred through the electrolyte from one electrode–electrolyte interface to the other. The overall process includes the following steps:

(a) Transfer of O^{2-} ions through the electrolyte to the electrolyte/liquid-silver interface.

(b) The electrochemical reaction at the electrode–electrolyte interface, which may be further subdivided into several consecutive steps.

(c) Transport of oxygen through liquid silver.

When oxygen ions are transferred from left to right in cell [5.2], the chemical potential of oxygen decreases at the left-hand electrode–electrolyte interface and increases at the right-hand electrode–electrolyte interface. If the current is now cut off, the galvanic cell will have a definite open-circuit emf resulting from the difference in the chemical potential of oxygen at the right-hand and left-hand interfaces. The emf decays to zero with time.

If the interfacial reaction is fast as compared with diffusion in the liquid-metal phase, an oxygen concentration gradient will be established in the liquid-metal phase. The shape of the emf decay curve will then be established by the unsteady-state diffusion of oxygen in the liquid metal. The emf decay curves can then be used to measure oxygen diffusion coefficients in the liquid metal. If the interfacial reaction is slow as compared with diffusion, the oxygen concentration will be nearly uniform in the liquid metal. In practice, diffusion in the metal phase is found to be the slower step and this gives rise to the behavior shown in Fig. 5.1.

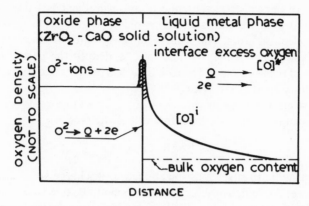

FIG. 5.1. Oxygen density profile during oxygen transfer from oxide phase to liquid phase (Reference 10).

Similar results were obtained by Etsell and Flengas[11] with the cell

$$\text{Ar, } O_2, \underline{O}(\text{in liq. Ag}) \mid ZrO_2 + CaO \mid O_2 \qquad [5.3]$$

The oxygen content of silver could be changed by changing the Ar/O_2 ratio. In a symmetrical cell with 1 atm of oxygen (no silver) at both electrodes, only resistance polarization was observed even at a current density of 200 mA/cm² at 1000°C. In cell [5.3], with the left-hand electrode as the cathode, diffusion overpotential, arising from the depletion of oxygen at the silver–electrolyte interface, was readily observed at temperatures between 1000 and 1200°C.

2.3. Polarization Studies Involving Solid Electrodes

Raleigh[1,12] has given a detailed discussion on solid electrodes. These may be divided into several categories.

2.3.1. Inert Electrodes

By an inert electrode is meant any good electronic conductor which has no ion in common with the electrolyte and is chemically inert towards it. An example is the AgBr/Pt or C system. Such electrodes are best known for their use in "polarization cells" to study low-level electronic conduction in predominantly ionic electrolytes. Consider, for instance, the cell

$$\text{Cu} \mid \text{CuBr} \mid \text{graphite} \qquad [5.4]$$

CuBr is a predominantly ionic conductor. If a potential below the decomposition potential of CuBr is applied across cell [5.4] with the Cu electrode at negative polarity, there results a depletion of copper near the CuBr/graphite interface. Thus a concentration gradient of Cu is set up across the electrolyte. At steady state, the concentration gradient is balanced by the potential gradient and conduction occurs only by virtue of the flow of excess electrons or electron holes. The steady-state electronic currents as a function of the applied potential may be analyzed to deduce the partial electronic conductivity of the electrolyte. The technique has been used to study electronic conduction in several electrolytes.[13-21]

An inert electrode–electrolyte interface over a finite range of potential behaves as a capacitance interface. In a cell of the type

$$\text{Ag} \mid \text{AgBr} \mid \text{Pt or C} \qquad\qquad [5.5]$$

potentials cathodic to silver must be applied to discharge silver at the right-hand interface. At sufficiently high anodic potentials, one would expect Br_2 discharge, halide formation, or anodization of the inert electrode. In an intermediate range of potentials, the interface would behave essentially like a capacitor in series with the electrolyte. Most of the previous work in this area attributed interface charging to a diffuse ionic space-charge layer only.[22] However, more recent work has shown that several alternative charging mechanisms could be operative.[12]

2.3.2. Parent Metal Electrodes

In the case of parent metal electrodes, the situation is different. Consider, for example, the cell

$$\text{Ag} \mid \text{AgI} \mid \text{Ag} \qquad\qquad [5.6]$$

which has been investigated by Raleigh.[12] When a current is passed through a cell, Ag from the electrode dissolves into the electrolyte as silver ions at the anode. This could lead to the generation of voids at the anodic interface. At the cathode, on the other hand, crowding occurs as the metal ions arriving at the interface must be accommodated. Conditions are favorable for dendritic growth at the cathode. The above effects, however, can be avoided by having current densities which are lower than the exchange current density. Under such conditions, one can identify nearly equal cathodic and anodic overvoltages. The terrace-kink-ledge[23] model can be applied to the metal electrode and the overvoltages can be attributed to atom migration to and from a limited number of electroactive sites on the metal.

Studies on polarization at the interface formed by components of the electrode also have been carried out. Rickert and O'Brian[24] studied transfer across the Ag/Ag_2S interface. Contreras and Rickert[25,25(a)] conducted studies on the Ag/Ag_2Te and Ag/Ag_2Se interfaces. It was concluded in the latter study that equilibrium for the electrons is established at the Ag/Ag_2Se and Ag/Ag_2Te phase boundaries during the transport of silver. The observed polarization was attributed to the transfer of silver ions. In these studies, when pressure was applied to strengthen the Ag/Ag_2Se or Ag/Ag_2Te contact, a decrease in overvoltage was observed.

Raleigh[1] has formed the following general conclusions about parent metal interfaces. When a dc current which is smaller than the exchange current is used, the overvoltage at a carefully formed interface is determined primarily by the concentration of electroactive sites on the metal surface. In the case of $Ag/AgBr$, the rate-limiting step appears to be the migration of Ag^+ ions in the electrolyte to or from such sites. At current densities higher than the exchange current density, a steady-state anodic mismatch is generated. An applied load tends to lessen the degree of mismatch. The overvoltage at the anode is thus dependent on the degree of mechanical mismatch and the intensity of the applied load. At cathode, contact is always maintained, but dendritic deposition may occur.

2.3.3. Parent Solid-Solution Electrodes

Next we consider the case of parent metal solid-solution electrodes, which are those that contain a component of the electrolyte in solid solution. $CuBr/Cu–Au$ and $ZrO_2(CaO)/Cu(+dissolved\ O)$ are good examples. If the solid-solution electrode is used as the one where the solute ion is discharged, then the supply of solute at this electrode will be limited by diffusion into the electrode. If current passage is in the reverse direction, then the removal of solute from the electrode is limited by the diffusion of solute out of the electrode. In either case, one has a means of determining the interdiffusivity in the solid-solution electrode. This will be treated in detail in Section 3.1.

2.3.4. Multiphase Electrodes

We consider next multiphase solid electrodes. In this category, metal–metal-oxide electrodes are of particular interest. They are used to provide buffered reference electrodes of fixed oxygen activity in galvanic cells using oxide solid electrolytes. Raleigh[1] has pointed out that even in open-circuit measurements, such electrodes must be considered as current sources since a high enough exchange current is needed to assure electrode equilibrium and buffer against contaminants from the gaseous atmosphere. Usually diffusional processes in the electrode limit the reversibility of such electrodes. In the case of some refractory alloy–metal-oxide electrodes[26,27] and alloy–inter-metallic-metal-oxide three-phase electrodes,[28] long equilibration times have been found to be necessary.

Metal–metal-oxide electrodes have often been used as current sources in coulometric titration experiments.[29–33] In such uses it is helpful to know the degree of reversibility of the various metal–metal-oxide electrodes. Worrell and Iskoe[34] have studied the polarization behavior of cells of the type

$$\text{M, MO} \mid \text{ZrO}_2 \text{(CaO)} \mid \text{M, MO} \qquad [5.7]$$

where M, MO is Cu, Cu_2O; Fe, FeO; or Ni, NiO. Small currents of less than 100 μA were passed across cell [5.7] at temperatures between 800 and 1000°C. Simultaneously the voltage across the cell was measured with a high-impedance electronic voltmeter. Steady voltages were obtained within minutes when the currents were less than 100 μA. When the $i\Omega$ drop contribution was subtracted from the steady-state voltages, steady-state overvoltages were obtained; thus

$$E_{st} = \eta + i\Omega \qquad (5.4)$$

Here E_{st} is the measured steady-state voltage and η is the overvoltage. The overvoltage could be related to the diffusivity of oxygen in the metal particles through the relation

$$\eta = i \frac{RT}{F} \frac{\Delta X}{2FAC_O D_O} \qquad (5.5)$$

where R is the gas constant, F is the Faraday constant, ΔX is the diffusion distance in the metal particle, A is the contact area between the electrode and the electrolyte, C_O is the saturation solubility of oxygen in the metal, and D_O is the diffusivity of oxygen in the metal. The overvoltages calculated using Eq. (5.5) were in good agreement with the experimental values. Thus the overvoltages are caused by oxygen diffusion in the metal particles of the metal–metal-oxide electrode. The overvoltages were found to be maximum for the Ni, NiO electrode and minimum for the Cu, Cu_2O electrode, with the Fe, FeO electrode coming in between. Figure 2.4 of Chapter 2 shows some results obtained at 900°C. Thus in applications where the metal–metal-oxide electrode is used as a current source, the Cu, Cu_2O electrode seems to be superior to the Fe, FeO and the Ni, NiO electrodes.

2.4. Choice of Electrodes

The analysis of the polarization characteristics of an electrode/electrolyte system can be used for a variety of applications such as the determination of electronic and ionic conductivities discussed in detail in Chapter 1 and the determination of diffusion coefficients in the electrode materials discussed later in this chapter. Apart from this, it is necessary to have nonpolarizable electrodes for fuel cell applications, for thermodynamic measurements, and for kinetic measurements. In fact lack of proper nonpolarizable electrodes seriously limits the use of solid electrolytes.

It is clear from the various electrode systems discussed in earlier sections that the gaseous electrodes are the best nonpolarizable electrodes for calcia-stabilized zirconia electrolytes. Currents of the order of 200 mA/cm² can be passed through these electrodes at 1000°C without appreciable polarization.[11] As far as the use of liquid or multiphase electrodes are concerned, the polarizability possibly depends upon diffusional processes within the electrode. Among metal/metal-oxide electrodes, the Cu/Cu_2O electrode is found to be the least polarizable.

3. Diffusion Measurements

Solid electrolytes may be conveniently used to measure diffusion coefficients of oxygen in liquid and solid metals and chemical diffusivities in nonstoichiometric compounds and alloys. Techniques to determine diffusivities have been discussed in several review articles.[35,36] In general three methods are commonly employed to measure diffusion coefficients: potentiostatic, galvanostatic, and potentiometric techniques.

3.1. Potentiostatic Techniques

In the potentiostatic technique, a constant voltage is applied across an electrochemical cell and the current in the external circuit measured as a function of time.

Raleigh and Crowe[37] used the cell Ag | AgBr | Au to determine the diffusion coefficient of Ag in Au at 400°C. An emf was applied so that Ag was transported from the left-hand electrode to Au. By applying successive voltage steps and measuring the corresponding changes in current with time, the diffusion coefficient of Ag at various Au–Ag compositions could be obtained. The Cottrell equation, namely,

$$i = \frac{2FA(C - C^0)D^{1/2}}{(\pi t)^{1/2}} \tag{5.6}$$

was used to evaluate the diffusion coefficient. In this expression, i is the diffusion current, t is the time, C is the concentration of the electroactive species established at the electrode surface by the imposed potential, C^0 is the bulk concentration, and D is the interdiffusion coefficient.

It is important to note that in Eq. (5.6), the current, i, must be due to diffusion only. At higher applied voltages, the measured current in the external circuit contains a contribution from the electronic leakage current in the electrolyte used. Therefore a correction must be made for this effect. In AgBr, which was the electrolyte used by Raleigh and Crowe,[37] the electronic carriers are n-type. In such a case, the electronic leakage current reaches a saturation value on the application of higher potentials.[38]

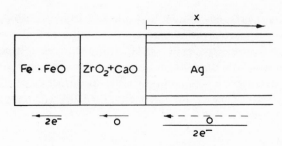

FIG. 5.2. Arrangement of cell for measurement of oxygen diffusion in the linear geometry (Reference 31).

Goldman and Wagner[39] used the cell Cu | CuCl | Au to determine the diffusion coefficient of Cu in single crystalline gold in the temperature range 355–400°C. A potentiostatic technique, similar to that employed by Raleigh and Crowe, was used. However, CuCl is a *p*-type electrolyte in which the electronic leakage current at higher applied potentials does not reach a saturation value.[38] Thus a time-dependent electronic contribution appears and a correction must be made for this. In addition, Goldman and Wagner had to apply a correction for a current contribution arising from the inherent double-layer capacitance at the CuCl–Au interface.

Figure 5.2 shows the arrangement used by Rickert and Steiner[31] to measure the diffusivity of oxygen in solid silver. To begin with, the silver sample has a certain oxygen content, C^0. At $t > 0$, the oxygen content at the metal/electrolyte interface is brought to a very small value by applying a constant potential across the cell

$$\text{Fe, FeO} \mid \text{ZrO}_2(\text{Y}_2\text{O}_3) \mid \text{Ag}(\underline{\text{O}}) \qquad [5.8]$$

The diffusion of oxygen out of the metal is followed by recording the current in the external circuit as a function of time. From the solution to Fick's second law, using the relevant boundary conditions for linear geometry, the current can be related to the oxygen diffusivity through

$$i = 2FAC^0 \left(\frac{D_\text{O}}{\pi t} \right)^{1/2} \qquad (5.7)$$

Thus a plot of i vs. $t^{-1/2}$ should be a straight line from the slope of which the diffusion coefficient may be calculated if C^0 is known.

Rickert and El Miligy[40] determined the diffusivity of oxygen in liquid silver and liquid copper using a potentiostatic technique involving cylindrical geometry. They used the cell

$$\text{Pt, air} \mid \text{ZrO}_2(\text{Y}_2\text{O}_3) \mid \text{liquid Ag or Cu} + \underline{\text{O}} \qquad [5.9]$$

Using a long-time solution to Fick's second law, it was possible to calculate the diffusivity directly.

Rapp and co-workers[41-44] determined the diffusivity of oxygen in liquid silver,[41] liquid copper,[41] liquid tin,[42] solid silver,[42] solid nickel,[42] liquid lead,[43] and liquid iron[44] using a technique similar to that of Rickert and El Miligy, but with a modified interpretation. They used a cell of the type

$$\text{Pt, air} \mid \text{ZrO}_2 \ (3\text{-}4 \text{ wt \% CaO}) \mid \text{solid or liquid metal} + \underline{O} \qquad [5.10]$$
$$\mu'_{O_2} \qquad\qquad\qquad\qquad\qquad\qquad\qquad \mu''_{O_2}$$

A cylindrical geometry was used in the experiments. The experimental arrangement of cell [5.10] is shown in Fig. 5.3. Initially, the liquid or solid metal was equilibrated at an oxygen concentration, $C_0(1)$ by applying a pre-selected voltage, $E_{appl}(1)$, across the cell with the metal at negative polarity. At steady state, there is a constant electronic current in the external circuit and the ionic current has dropped to zero. The applied voltage is now quickly changed to a new value, $E_{appl}(2)$, so that almost instantaneously a concentration $C_0(2)$ is established at the electrode–electrolyte interface. The resulting diffusion of oxygen into or out of the metal is followed from the variation of the current in the external circuit as a function of time. At steady state, a constant electronic current prevails. This is subtracted from the total current

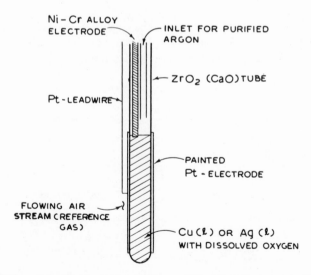

FIG. 5.3. Cell used by Rapp *et al.* to measure diffusivity of oxygen in liquid metals (Reference 41).

to obtain the ionic current–time plot. For the boundary conditions

$$C_0 = C_0(1), \qquad 0 \leqslant r \leqslant a, \qquad t = 0$$

and

$$C_0 = C_0(2), \qquad r = a, \qquad t > 0$$

the long-time solution to Fick's second law can be written in the form

$$i_{\text{ion}} = 8\pi hFD_0[C_0(1) - C_0(2)] \exp\left[-\frac{(2.045)^2 D_0 t}{a^2}\right] \tag{5.8}$$

where a is the radius of the solid electrolyte tube and h is the height of the metal column. From the slope of the log i vs. t plot, the diffusion coefficient may be calculated using Eq. (5.8). Typical current–time plots and log i vs. t plots are shown in Figs. 2.2 and 2.3 of Chapter 2, respectively.

Pastorek and Rapp[33] have used the cell

$$\text{FeO, Fe}_3\text{O}_4 \,|\, \text{ZrO}_2\,(15 \text{ mol \% CaO}) \,|\, \text{Cu} \,|\, \text{ZrO}_2\,(15 \text{ mol \% CaO})|\, \text{FeO, Fe}_3\text{O}_4$$
$$[5.11]$$

to determine the diffusivity of oxygen in solid copper potentiostatically. The circuit used is shown in Fig. 5.4. The copper specimen was in the form of a cylindrical disk. Initially the copper was brought to a known oxygen activity throughout by imposing a voltage, $E_{\text{appl}}(1)$, across both cells of the double cell [5.11]. The emf was then changed to $E_{\text{appl}}(2)$ and the diffusion of oxygen into copper was followed by recording the current in the external circuit as a function of time. The current–time plot was combined with the relevant solution to Fick's second law in order to determine the diffusivity. At long

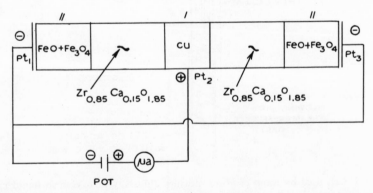

FIG. 5.4. Experimental setup for determining the diffusivity of oxygen in solid copper potentiostatically (Reference 33).

times, the cell current could be related to the diffusivity through an equation of the form

$$\ln i = \ln B - \frac{\pi^2 D_0 t}{l^2} \qquad (5.9)$$

where i is the current, B is a constant (at a given temperature), and l is the thickness of the copper specimen. Thus D_0 may be calculated from the long-time slope of the $\ln i$ vs. t plot. Recently Reddy and Rapp[70] used fluoride double cells similar to cell [5.11] in construction to determine the diffusivity of fluorine in solid Cu and solid Ni.

Ramana Rao and Tare[45] used a cell of the type

$$Cu, Cu_2O \mid Cu \mid ZrO_2 (CaO) \mid Cu, Cu_2O \qquad [5.12]$$

to determine the oxygen diffusivity in solid copper (Fig. 5.5). The oxygen concentration in the copper at the copper–electrolyte interface was fixed by applying a voltage across the cell, while that at the other interface is fixed by the Cu–Cu$_2$O equilibrium. At steady state, there is a constant current through

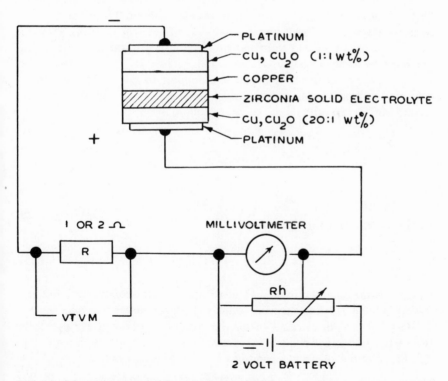

FIG. 5.5. Schematic setup used for measurement of diffusion coefficient of oxygen in solid copper using a steady-state technique (Reference 45).

the cell, corresponding to a fixed oxygen concentration gradient in the copper disk, $\Delta C_0/\Delta X$. This current may be related to the diffusivity of oxygen in copper through Fick's first law; thus

$$i = 2FAD_0\left[\frac{\Delta C_0}{\Delta X}\right] \tag{5.10}$$

Thus prior knowledge of the solubility of oxygen in copper is necessary in order to calculate the diffusivity.

Table 5.1 lists some diffusivity values for oxygen in liquid and solid metals obtained using the potentiostatic electrochemical technique.

The potentiostatic technique has also been used to determine chemical diffusivities in nonstoichiometric compounds. Chu et al.[46] have used this technique to determine the chemical diffusivity of iron in wüstite. The cell used by them may be represented as

$$P_{O_2}, \text{Pt}_1 \mid \text{ZrO}_2 \text{ (10 mol \% Y}_2\text{O}_3) \left| \underset{x=0}{\text{Fe}_{1-\delta(x)}\text{O}} \right| \underset{x=1}{\text{Pt}_2, \text{N}_2} \qquad [5.13]$$

One side of the solid electrolyte was in contact with air. The other side consisted of the wüstite being investigated. The cell is schematically shown in Fig. 5.6. A uniform Fe/O ratio was first established throughout the wüstite specimen by applying a preselected voltage across cell [5.13]. The applied voltage was then changed in steps corresponding to $E = 5$–20 mV, and the current was observed as a function of time. From the solution to Fick's second law for the initial and boundary conditions,

$$C(x = 0, t) = C' \qquad \text{at } t > 0$$

$$C(x, t = 0) = C'' \qquad \text{at } 0 < x \leqslant l$$

$$\left(\frac{\partial C}{\partial x}\right)_{x=l} = 0 \qquad \text{at } t > 0$$

it follows that for short times $(t \ll l^2/\tilde{D})$,

$$i = 2FA\left(\frac{\tilde{D}}{\pi t}\right)^{1/2} (C'' - C') \tag{5.11}$$

where A is the cross-sectional area of wüstite, C is the concentration of iron in wüstite, and $\tilde{D}$ is the chemical diffusivity of iron in wüstite. According to Eq. (5.11), $\tilde{D}$ may be calculated from the slope of a plot of i vs. $t^{-1/2}$ provided $(C'' - C')$ is known. Such a plot is shown in Fig. 5.7. For long times, $t > l^2/4\tilde{D}$, the solution to Fick's second law leads to the expression

$$\log i = \log\left[\frac{4FA\tilde{D}(C'' - C')}{l}\right] - \frac{1.071\tilde{D}t}{l^2} \tag{5.12}$$

Table 5.1. *Diffusivity Values of Oxygen in Liquid and Solid Metals*

Investigator	Metal	Temperature range (°C)	Diffusivity (cm²/sec)
		Potentiostatic technique	
Rickert and Steiner[31]	solid silver	800	1.8×10^{-5}
Rickert and El Miligy[40]	liquid silver	990–1220	$26.3 \times 10^{-4} \exp(-7,300/RT)$
Rickert and El Miligy[40]	liquid copper	1100–1250	$1.22 \times 10^{-2} \exp(-14,400/RT)$
Oberg et al.[41]	liquid silver	1000–1200	$2.8 \times 10^{-3} \exp(-8300/RT)$
Oberg et al.[41]	liquid copper	1100–1350	$6.9 \times 10^{-3} \exp(-12,900/RT)$
Ramanarayanan and Rapp[42]	liquid tin	750–950	$9.9 \times 10^{-4} \exp(-6300/RT)$
Ramanarayanan and Rapp[42]	solid silver	750–950	$4.9 \times 10^{-3} \exp(-11,600/RT)$
Ramanarayanan and Rapp[42]	solid nickel	1393	1.3×10^{-6}
Szwarc et al.[43]	liquid lead	740–1080	$1.44 \times 10^{-3} \exp(-6200/RT)$
Oberg et al.[44]	liquid iron	1620 ± 5	1.5×10^{-4}
Pastorek and Rapp[33]	solid copper	800–1030	$1.7 \times 10^{-2} \exp(-16,000/RT)$
Ramana Rao and Tare[45]	solid copper	800–1000	$5.08 \times 10^{-3} \exp(-16,430^a/RT)$
		Galvanostatic technique	
Pastorek and Rapp[33]	solid copper	950–1040	$1.7 \times 10^{-2} \exp-(15,600/RT)$
Ramanarayanan and Worrell[47]	solid copper	800–1000	$2.4 \times 10^{-2} \exp(-18,600/RT)$
Kawakami and Goto[47a]	liquid iron	1550	$(1.9 \pm 0.7) \times 10^{-4}$
		Potentiometric technique	
Masson and Whiteway[48]	liquid silver	970–1200	$5.15 \times 10^{-3} \exp(-9900/RT)$
Sano et al.[49]	liquid silver	1000–1200	$3 \times 10^{-3} \exp(-8700/RT)$
Sano et al.[50]	liquid lead	800–1100	$9.65 \times 10^{-5} \exp(-4800/RT)$
Otsuka and Kozuka[51]	liquid silver	1000–1150	$32.8 \times 10^{-4} \exp(-9100/RT)$
		Combined potentiostatic and potentiometric	
Otsuka and Kozuka[52]	liquid lead	900–1100	$14.8 \times 10^{-4} \exp(-4660/RT)$
Otsuka and Kozuka[52a]	liquid copper	1125–1300	$5.7 \times 10^{-3} \exp(-11,900/RT)$

[a] A correction has been made for a systematic error in the original calculations.

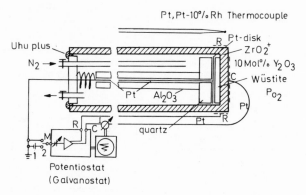

FIG. 5.6. Schematic diagram of cell arrangement to measure diffusivity of iron in wüstite (Reference 46). 1, Foreign reference input voltage; 2, mercury relay switch; M, working electrode; R, reference electrode; C, counter electrode.

From the slope of the straight line in the log i vs. t plot, the chemical diffusivity can be determined without a knowledge of the concentration difference, $C'' - C'$. Such a plot is shown in Fig. 5.8. $\tilde{D} = 3.2 \times 10^{-6}$ cm² sec⁻¹ for $\delta = 0.106$ at 1000°C.

A similar technique was used by Ramanarayanan and Jose[71] to determine the chemical diffusivity of Cu in nonstoichiometric cuprous sulfide at 400°C.

3.2. Galvanostatic Techniques

In the galvanostatic method, a constant current is passed through an electrochemical cell and the voltage across the cell measured as a function of time. Pastorek and Rapp[33] used cell [5.11] also to determine the diffusivity of oxygen in solid copper galvanostatically. The circuit used is shown in Fig. 5.9. Initially the copper specimen was equilibrated throughout at the oxygen

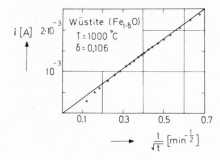

FIG. 5.7. i vs. $1/t^{1/2}$ plot at 1000°C for short times in the potentiostatic diffusion experiment with wüstite (Reference 46).

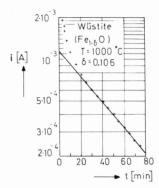

FIG. 5.8. Log i vs. t plot at long times in the potentiostatic diffusion experiment with wüstite (Reference 46).

activity corresponding to FeO–Fe_3O_4 coexistence with the switch in position 1. After equilibration, the switch was moved to position 2 so that a constant predetermined current passed through the cell on the left while measuring the voltage across the cell as a function of time, from which the diffusivity was evaluated.

Ramanarayanan and Worrell[47] used a galvanostatic method to determine the solubility–diffusivity product of oxygen in solid copper. They used the cell

$$Cu, Cu_2O \left| \underset{\Delta x}{Cu} \right| ZrO_2 (CaO) \left| \underset{\Delta x}{Cu} \right| Cu, Cu_2O \qquad [5.14]$$

The electrodes were thin copper foils saturated with oxygen. The copper foil electrodes on both sides of the electrolyte had the same thickness. When a small current (less than 50 μA) is passed through the cell from left to right, there is a depletion of oxygen at the left-hand electrode and accumulation (supersaturation) at the right-hand electrode. Within about 10 min, a steady-state voltage is attained. On subtracting the $i\Omega$ drop across the electrolyte

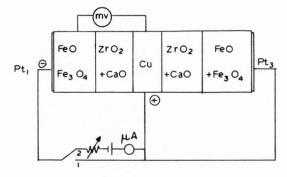

FIG. 5.9. Galvanostatic determination of diffusivity of oxygen in solid copper (Reference 33).

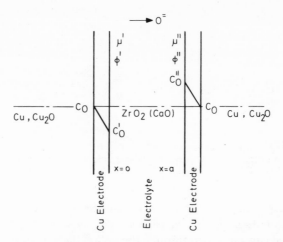

FIG. 5.10. Concentration gradients in oxygen-saturated copper electrodes in the galvanostatic determination of the oxygen diffusivity in solid copper (Reference 47).

from the steady-state voltage, a steady-state overvoltage is obtained. As shown in Fig. 5.10, at steady state there is a fixed oxygen concentration gradient in the copper foil electrodes on both sides. Equation (5.5) may be used to determine the solubility–diffusivity product of oxygen from the steady-state overvoltage, η, and the cell current, i. ΔX is the thickness of the copper foil electrode.

Table 5.1 also lists the diffusivity values of oxygen in solid copper and molten iron determined galvanostatically.

Kawakami and Goto[47a] determined the diffusivity of oxygen in molten iron by assembling the following galvanic cell:

$$\text{Mo, MoO}_2 \mid \text{ZrO}_2 + (\text{CaO}) \mid \text{molten iron with } \underline{\text{O}} \qquad [5.14a]$$

When a constant current is supplied to the cell there is an oxygen overpotential at the iron electrode only, the overpotential at Mo–MoO$_2$ being negligible. Solving the appropriate diffusion equation with constant oxygen flux at the interface the following equation can be obtained:

$$\exp\left(\frac{2F\eta}{RT}\right) = 1 + \frac{i}{\pi^{1/2}FC_0D^{1/2}}\, t^{1/2} \qquad (5.12a)$$

In Eq. (5.12a) η is the oxygen overpotential, i is the current density, and C_0 is the bulk oxygen concentration.

C_0 is obtained from the reversible emf of the cell at zero current. Thus from the plot of $\exp(2F\eta/RT)$ against $t^{1/2}$, the diffusivity was calculated to be $(1.9 \pm 0.7) \times 10^{-4}\ \text{cm}^2/\text{sec}$ at 1550°C.

3.3. Potentiometric Techniques

In the potentiometric technique, the open-circuit voltage across a cell is studied as a function of time and this dependence is analyzed to obtain the diffusion coefficient.

Masson and Whiteway[48] used this technique to measure the diffusivity of oxygen in liquid silver. Sano and co-workers[49,50] employed the method to measure the diffusivity of oxygen in liquid silver and liquid lead. The arrangement used by Sano *et al.* is shown in Fig. 5.11. Gaseous oxygen is absorbed into the liquid-metal column of length L contained within an alumina tube with a solid electrolyte disk $(0.85 ZrO_2 + 0.15 CaO)$ sealed to the bottom with glass (12 wt % Na_2O–20 wt % Al_2O_3–68 wt % SiO_2). Initially, the oxygen content of liquid silver was brought down to practically zero by passing dry hydrogen. At $t > 0$, the concentration at the surface of silver was brought to C_0 by introducing a proper gas mixture. The change in the open-circuit cell voltage was followed as a function of time. A typical plot is shown in Fig. 5.12. For the initial condition,

$$t = 0, \qquad C = 0$$

and boundary conditions,

$$x = 0, \qquad C = C_0$$

$$x = L, \qquad \frac{\partial C}{\partial x} = 0$$

the solution to Fick's second law leads to the expression

$$\frac{C}{C_0} = \frac{4}{\pi} \sum_{n=0}^{\infty} \frac{(-1)^n}{2n + 1} \exp\left[-\frac{D(2n + 1)^2 \pi^2 t}{4 l^2} \right] \qquad (5.13)$$

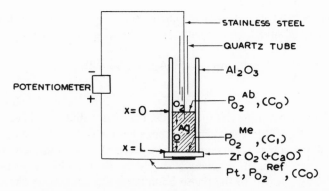

FIG. 5.11. Schematic cell arrangement used by Sano *et al.* (Reference 50).

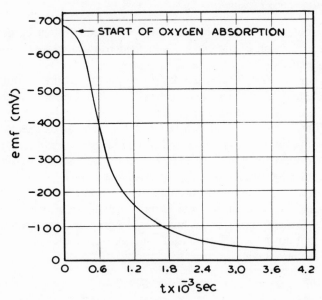

FIG. 5.12. Variation of cell emf with time in the potentiometric determination of oxygen diffusion in liquid silver (Reference 50).

The open-circuit cell emf is given by

$$E = \frac{RT}{2F} \ln \frac{C}{C_0} \qquad (5.14)$$

Upon combining Eqs. (5.13) and (5.14), the diffusivity of oxygen in the liquid metal was determined with the aid of the dimensionless parameter, Dt/L^2.

Otsuka et al.[51] used a similar technique to determine the diffusivity of oxygen in liquid silver. Table 5.1 lists some of the oxygen diffusivity values determined potentiometrically.

Chu et al.[46] have used a potentiometric technique to determine the chemical diffusivity of silver in silver sulfide at 200°C. For this, the galvanic cell

$$\text{Pt, Ag} \left| \text{AgI} \right| \underset{\substack{x=0 \\ | \\ \text{Pt}}}{\text{Ag}_{2+\delta(x)}\text{S}} \left| \underset{x=L}{} \text{S(l)} \right. \qquad [5.15]$$

was used in which, on the right-hand side, the silver sulfide was in equilibrium with liquid sulfur. The experimental arrangement of cell [5.15] is shown in Fig. 5.13. Initially a constant current or a constant emf was imposed to achieve a stationary concentration gradient of silver in silver sulfide from the phase boundary $\text{Ag}_{2+\delta}\text{S}/\text{AgI}$ to the phase boundary, $\text{Ag}_{2+\delta}\text{S}/\text{S(l)}$. In these experiments, the emf of cell [5.15] was only 10–20 mV away from the value in equi-

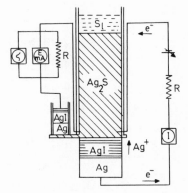

FIG. 5.13. Arrangement for the determination of the chemical diffusivity of silver in silver sulfide (Reference 46).

librium with sulfur. The circuit was now opened and the cell emf recorded as a function of time. The solution to Fick's second law with the initial and boundary conditions,

$$C = C'' - (C'' - C') \frac{x}{L} \qquad \text{at } t = 0$$

$$C = C' \qquad \text{at } x = L, \, t \geqslant 0$$

and

$$\left(\frac{\partial C}{\partial x}\right)_{x=0} = 0 \qquad \text{at } t > 0$$

leads to

$$C = C'' - \frac{2(C'' - C')}{L} \left(\frac{\tilde{D}t}{\pi}\right)^{1/2} \qquad (5.15)$$

for $t \ll 4L^2/\pi^2 \tilde{D}$.

Prior knowledge of the solubility of silver in silver sulfide from coulometric titration curves was necessary in order to calculate the diffusion coefficient using Eq. (5.15). The chemical diffusivity was found to be equal to 0.47 cm^2 sec^{-1} at 200°C for silver sulfide in equilibrium with sulfur.

3.4. Combined Potentiostatic and Potentiometric Techniques

Several investigators have used a combination of potentiostatic and potentiometric techniques to determine diffusivities. Otsuka and Kozuka used the method to determine the diffusivity of oxygen in liquid lead[52] and liquid copper.[52a] The cell used by them in their experiments with liquid lead may be schematically represented as

$$\text{Ni–NiO} \left| \text{ZrO}_2(+\text{CaO}) \right| \underset{x=0}{\text{O in liq. Pb}} \left| \underset{x=L}{\text{ZrO}_2(+\text{CaO})} \right| \text{FeO, Fe}_3\text{O}_4 \qquad [5.15a]$$

$$\underbrace{\qquad\qquad \text{potentiometer} \qquad\qquad} \qquad \underbrace{\qquad\qquad \text{potentiostat} \qquad\qquad}$$

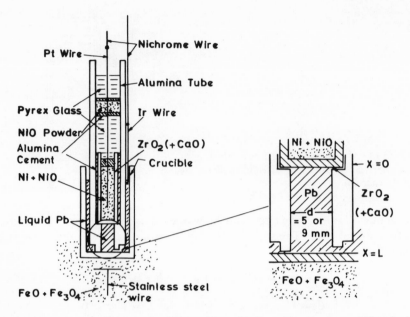

FIG. 5.14. Schematic diagram of cell assembly and cell used by Otsuka and Kozuka in the combined potentiostatic and potentiometric technique (Reference 52).

The experimental cell arrangement is shown in Fig. 5.14. Initially, a suitable oxygen concentration is attained throughout the liquid lead by applying a suitable voltage across the right-hand cell of the double cell [5.15a] with the help of the potentiostat. When a uniform concentration has been attained, the oxygen concentration at $x = L$ is quickly changed to a new value by changing the applied voltage. The variation in the oxygen concentration at $x = 0$ is measured as a function of time by recording the open-circuit voltage of the left-hand cell using a potentiometer.

For the initial and boundary conditions,

$$C = C_0, \quad 0 \leqslant x \leqslant L, t = 0$$
$$C = C_1, \quad x = L, \qquad t > 0$$
$$\frac{\partial C}{\partial x} = 0, \qquad x = 0, \qquad t \geqslant 0$$

Fick's second law leads to the solution

$$\frac{C - C_0}{C_1 - C_0} = 1 - \frac{4}{\pi} \sum_{n=0}^{\infty} \frac{(-1)^n}{2n + 1} \exp\left[\frac{-D(2n + 1)^2\pi^2 t}{4L^2}\right] \cos\left[\frac{(2n + 1)\pi x}{2L}\right]$$

$$(5.15a)$$

If the voltage applied by the potentiostat is large enough, then $C_1 \approx 0$ and one obtains

$$\frac{C}{C_0} = \frac{4}{\pi} \sum_{n=0}^{\infty} \frac{(-1)^n}{2n+1} \exp\left[-\frac{D(2n+1)^2\pi^2t}{4L^2}\right] \qquad (5.15b)$$

The difference between the initial emf ($t = 0$) and the emf at any time measured by the potentiometer is given by

$$\Delta E = \frac{RT}{2F} \ln \frac{C}{C_0} \qquad (5.15c)$$

where C is the oxygen concentration at $x = 0$. At large values of t, only the first term in the summation in Eq. (5.15b) makes any significant contribution. Thus, upon combining Eqs. (5.15b) and (5.15c),

$$\Delta E = \frac{RT}{2F} \ln \frac{4}{\pi} - \frac{RT}{2F} \frac{\pi^2 Dt}{4L^2} \qquad (5.15d)$$

D can be determined from the slope of the ΔE vs. t plot. A typical plot is shown in Fig. 5.15.

A combination of potentiostatic and potentiometric techniques was used by Hartmann, Rickert, and Schendler[52b] to determine the chemical diffusivity of Ag in $Ag_{2+\delta}S$ as a function of silver activity at 200 and 300°C. The experimental cell arrangement used by them is shown in Fig. 5.16. A concentration, C_0, of Ag is established initially throughout the $Ag_{2+\delta}S$ sample by means of

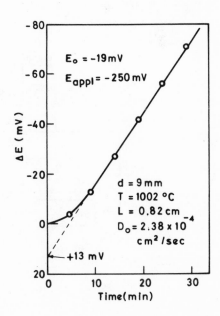

$E_0 = -19\,mV$

$E_{appl} = -250\,mV$

$d = 9\,mm$
$T = 1002\,°C$
$L = 0.82\,cm$
$D_0 = 2.38 \times 10^{-4}\,cm^2/sec$

$\longleftarrow +13\,mV$

FIG. 5.15. ΔE plotted against time (Reference 52).

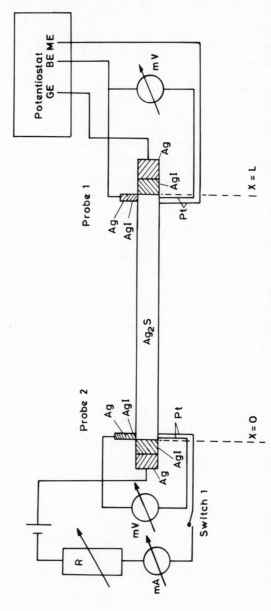

FIG. 5.16. Arrangement used by Hartmann, Rickert, and Schendler to study the chemical diffusivity of silver in silver sulfide using a combined potentiostatic and potentiometric technique (Reference 52b).

the potentiostat. Switch 1 is closed and a suitable current is passed so that Ag enters the silver sulfide specimen at $x = 0$. Finally, a steady-state situation is reached when the concentration at $x = 0$ no longer changes. Let this silver concentration be C_1. Probes 1 and 2 are used to determine the silver concentrations at $x = L$ and $x = 0$, respectively. If switch 1 is now opened, silver diffuses out of silver sulfide until the silver concentration returns to C_0 throughout the specimen. The diffusion process may be followed by recording the voltage of probe 2 as a function of time. For the initial and boundary conditions

$$C = C_1 - (C_1 - C_0)\frac{x}{L} = C_1 - \Delta C \frac{x}{L}, \qquad t = 0$$

$$\frac{dC}{dx} = 0, \qquad\qquad\qquad \text{for } x = 0 \text{ and } t > 0$$

$$C = C_1 - \Delta C, \qquad\qquad \text{for } x = L, t \geqslant 0$$

The solution to Fick's second law leads to the expression

$$C(0, t) = C_1 - \frac{2\Delta C}{L}\left(\frac{Dt}{\pi}\right)^{1/2} \qquad \text{for } t \ll \frac{4L^2}{\pi^2 D} \qquad (5.15e)$$

where D is the chemical diffusivity of silver in silver sulfide. The voltage o probe 2 vs. time plot may be analyzed to determine D using Eq. (5.15e). At 200°C, D was found to be 0.08 cm²/sec for $\delta = 2 \times 10^{-3}$ and 0.37 cm²/sec for $\delta = 1.2 \times 10^{-4}$. At 300°C, D was equal to be 0.08 cm²/sec for $\delta = 2.5 \times 10^{-3}$ and 0.26 cm²/sec for $\delta = 2.9 \times 10^{-4}$.

3.5. Evaluation of Experimental Techniques

Of the various techniques used above, the potentiostatic technique is the most popular and has been used by many investigators. However, a simple potentiostatic technique can be successfully used only when transport in the electrode (in which the diffusivity is being measured) is slow enough so that transport in the electrolyte would not interfere with the measurements. Recently, Jose[52c] measured the chemical diffusivity of copper in cuprous sulfide at 400°C potentiostatically using the cell Cu | CuBr | $Cu_{2-\delta}S$. The electrodes and the electrolytes were in the form of cylindrical pellets. It was found that in order to obtain correct diffusivity values, the ratio of electrode length ($Cu_{2-\delta}S$) to electrolyte length should be greater than 12. At smaller ratios, transport within the copper bromide electrolyte interfered with the measurements.

The galvanostatic technique has been used only to a very limited extent. Pastorek and Rapp,[33] in their study of the oxygen diffusivity in solid copper, found that galvanostatic experiments did not give data which were as reproducible as the potentiostatic experiments. Considering the limited number of studies where the galvanostatic method has been employed, it would appear that investigators have had problems with this technique.

The potentiometric technique is quite good and has been used by several investigators. It is particularly useful when one is concerned about transport in the electrolyte interfering with the diffusion measurements. Thus Chu, Rickert, and Weppner[46] used the method to determine the very high chemical diffusivity values of silver in silver sulfide. A combination of potentiostatic and potentiometric techniques is more versatile and can be used when one wants to measure the diffusivity as a function of the deviation from stoichiometry in the case of nonstoichiometric compounds.

4. Kinetics of Phase-Boundary and Diffusion-Controlled Reactions

4.1. Phase-Boundary Reactions

4.1.1. Potentiostatic Techniques

Kobayashi and Wagner[53] made one of the early attempts to measure the rate of surface reactions using solid electrolytes. They investigated the reduction of silver sulfide by hydrogen using the cell

$$\text{Ag(s)} \mid \text{AgI(s)} \mid \text{Ag}_2\text{S(s)} \mid \text{Pt} \qquad\qquad [5.16]$$

An applied emf across the cell was used to fix the activity of silver in silver sulfide by means of the relation

$$\mu_{\text{Ag}} - \mu_{\text{Ag}}^{\circ} = -EF \qquad\qquad (5.16)$$

where μ_{Ag}° is the chemical potential of silver in pure silver and μ_{Ag} is the chemical potential of silver in silver sulfide. When hydrogen is passed over the cell, sulfur from silver sulfide is removed by the reaction

$$\text{S(in Ag}_2\text{S)} + \text{H}_2\text{(g)} = \text{H}_2\text{S(g)} \qquad\qquad (5.17)$$

Since μ_{Ag} in silver sulfide is fixed by the applied voltage, an equivalent amount of silver is removed from the silver sulfide electrochemically. The removal of silver can be measured as a current in the external circuit. Thus the rate of H_2S formation can be readily obtained by measuring the current across the cell. By varying the applied emf, the rate of formation of H_2S can be studied as a function of the silver activity in silver sulfide. The silver activity and the sulfur activity in silver sulfide are related through the Gibbs–Duhem equation. Kobayashi and Wagner found that the rate was nearly proportional to the sulfur activity in silver sulfide.

Roy and Schmalzried[54] have also investigated the $Ag_2S-H_2-H_2S$ system. The silver activity in Ag_2S was fixed electrochemically. The experimental arrangement was such that the reactive gases had ready access to the silver sulfide but were excluded from interaction with the reference electrode. Also, the exposure of the Pt lead connected with the silver sulfide sample to sulfur vapor was avoided. Mass transport in the gas phase was avoided by conducting the experiments at sufficiently high flow rates. The rate of reduction of S in silver sulfide by hydrogen was found to be proportional to the partial pressure of hydrogen, but independent of the sulfur activity in silver sulfide.

According to Bechtold,[55] the size of the Pt electrode has an effect on the rate of the reduction reaction. Wagner[56] has suggested that this is possibly the explanation for the different rate laws observed by Kobayashi and Wagner, and by Roy and Schmalzried.

Belton and Harvey[57] used a steady-state electrochemical technique to study the effect of tellurium on the rate of absorption of oxygen by liquid lead from water-vapor–hydrogen–argon mixtures. The cell which they used is shown in Fig. 5.17. At a given applied voltage, the current was measured at various compositions of the gas in contact with the upper surface of the metal. The results have been explained by assuming the formation of surface films of TeO and $Te(OH)_2$.

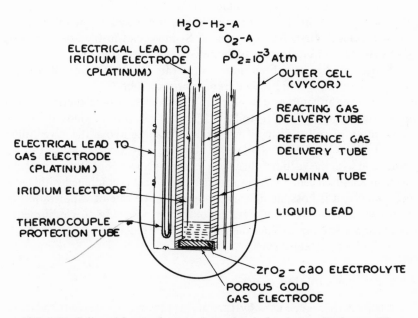

FIG. 5.17. Cell used for steady-state electrochemical technique (Reference 57).

4.1.2. Galvanostatic Techniques

Rickert[58] used the galvanostatic technique to study the vaporization of silver sulfide into vacuum in the temperature range 200–400°C. The Pt/Ag/AgI/Ag$_2$S/Pt electrochemical cell was basic to this study. The emf across this cell is related to the activity of sulfur in silver sulfide through the relation

$$a_S = \exp[2(E - E^*)F/RT] \tag{5.18}$$

where E^* is the emf of the cell when silver sulfide is in equilibrium with sulfur. The cell was placed in vacuum and silver was removed from silver sulfide by imposing a current. Concentration gradients within silver sulfide are evened out very quickly because of the extremely high chemical diffusion coefficient of silver in silver sulfide. At steady state, an equivalent amount of sulfur vaporizes from the silver sulfide and the current is a measure of the rate of vaporization. The vaporization rate of sulfur is related to the current density through

$$\dot{v} = I/F \tag{5.19}$$

where $\dot{v}$ is the vaporization rate of sulfur in eq cm^{-2} sec^{-1} and I is the current density in A cm^{-2}. The attainment of a steady state is determined by measuring the activity of silver in Ag$_2$S by means of an Ag/AgI probe. At steady state, the activity must be steady. Upon varying the current, a new steady state is attained. Thus the evaporation rate can be measured as a function of the activity of sulfur in silver sulfide. Rickert found that the vaporization rates were proportional to the square of the sulfur activity in silver sulfide. It was concluded that the formation of S$_2$ molecules in the adsorbed layer from adsorbed sulfur atoms was the rate-determining step.

Similar studies have been done on the rate of vaporization of selenium from silver selenide[59] and iodine from cuprous iodide.[60]

Sulfur vapor contains a number of molecular species in thermodynamic equilibrium, ranging from S$_2$ to S$_8$.[61] If, in the cell used to study the rate of vaporization of sulfur from silver sulfide, the space above the silver sulfide electrode is enclosed by a glass vessel allowing only a small orifice to make connection with the vacuum, then such an arrangement constitutes an electrochemical Knudsen cell.[62] When such an arrangement is combined with a mass spectrometer, it is possible not only to measure the rate of vaporization but also to determine the partial pressures of the various sulfur species present in equilibrium in the vapor. Thus the presence of the species S$_2$, S$_3$, S$_4$, S$_5$, S$_6$, S$_7$, and S$_8$ has been established.[58]

4.1.3. Potentiometric Techniques

Figure 5.18 is a cell used by Tare and Schmalzried[63] to study the oxidation rate of iron by CO$_2$. It consists of a calcia-stabilized zirconia disk on one

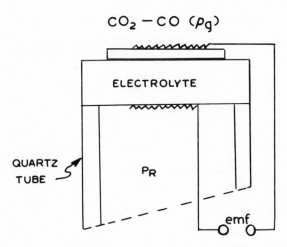

FIG. 5.18. Schematic design of the cell used by Tare and Schmalzried (Reference 63).

side of which a metal foil of known thickness is pressed. A gas mixture of CO and CO_2 is passed over the metal foil. p_g is the oxygen potential in the CO–CO_2 gas phase, p_e is the oxygen potential at the electrolyte metal interface, and p_r is the oxygen potential in the reference gas. The cell emf is given by the relation

$$E = \frac{RT}{4F} \ln \frac{p_e}{p_r} \qquad (5.20)$$

If the metal phase is in equilibrium with the surrounding gas phase, then $p_e = p_g$. If now p_g is suddenly changed to p'_g and conditions are such that diffusion through the metal foil electrode is fast, then the reaction occurring at the electrode–gas interface would determine the emf of the galvanic cell. Thus the rate can be followed by following the emf of the cell as a function of time.

Figure 5.19 is a typical plot of the emf as a function of time. Prior to heating to the experimental temperature, the CO/CO_2 ratio was held slightly on the reducing side. When the desired temperature was attained, the gas ratio was suddenly changed so that the oxygen potential was above that of the Fe/FeO equilibrium, but below that of the FeO/Fe_3O_4 equilibrium. Under these conditions, only FeO is the stable oxide. Thus during the initial period, A, oxygen is adsorbed on the surface and diffuses into the foil until wüstite is nucleated at the low point of the curve. Wüstite continues to form until the entire foil is transformed to oxide (period B) after which the emf starts to drop slowly to the value corresponding to the CO/CO_2 ratio. Part C indicates the oxidation of nonstoichiometric FeO to the composition of FeO coexisting in equilibrium with the prevailing CO–CO_2 gas mixture. From the time taken

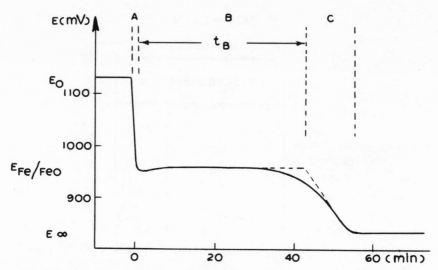

FIG. 5.19. Trends in cell voltage during oxidation of iron and wüstite (Reference 63).

to complete the oxidation of iron to wüstite (period B), k, the rate constant, can be calculated.

Fitterer and Cassler[64] extended this work using foils of various thicknesses. The cell used by them is shown in Fig. 5.20 and the results are shown in Fig. 5.21. They did not observe a plateau as in Tare and Schmalzried's experiments. When the gas ratio was changed, the emf dropped until wüstite was nucleated; then the voltage rose to point C, which corresponds to the Fe–FeO equilibrium. The time taken to go from A to C was taken as the time for complete oxidation of iron to wüstite. The reaction occurring at the surface of the foil was found to be rate limiting up to a thickness of 9.6×10^{-3} in. Beyond this thickness, the parabolic rate indicative of diffusion control, was observed.

Schmalzried and Wagner[65] have investigated the reduction of silver sulfide by hydrogen. For silver to nucleate when the reduction occurs there must be a supersaturation, i.e., the activity of silver must be greater than unity. After the silver is nucleated, the supersaturation is expected to decrease. To test this, Schmalzried and Wagner used the $Pt/Ag/AgI/Ag_2S/Pt$ electrochemical cell. At 400°C, an emf of 160 mV was applied across the cell with the Ag at negative polarity. This corresponded to an Ag/S ratio of nearly 2 in the silver sulfide. The circuit was now opened and a gentle stream of hydrogen allowed to pass over the silver sulfide. Simultaneously, the cell voltage was measured as a function of time. The silver activity in silver sulfide is related to the cell voltage through

$$a_{Ag} = \exp(-EF/RT) \qquad (5.21)$$

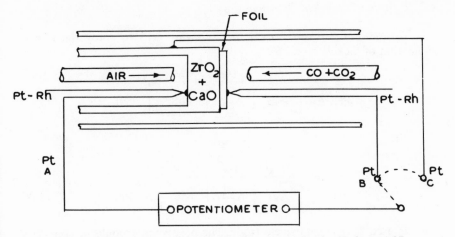

FIG. 5.20. Experimental setup used by Fitterer and Cassler (Reference 64).

The cell voltage rapidly decreased and reached a minimum value between -1 and -2 mV (i.e., a silver activity between 1.017 and 1.034) corresponding to the onset of nucleation. The voltage then rose and reached a quasistationary value between 0 and -0.3 mV corresponding to a_{Ag} between 1 and 1.005.

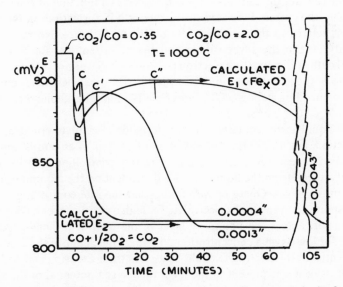

FIG. 5.21. Typical curves of voltage vs. time in oxidation of iron to wüstite (Reference 64).

4.2. Diffusion-Controlled Reactions

Mrowec and Rickert[66] used the following cells to study the rate of sulfidation of nickel:

$$
\overset{2e^-}{\longrightarrow} \quad \overset{2Ag^+}{\longleftarrow} \quad \overset{2Ag^+}{\longleftarrow} \quad \overset{\longleftarrow 2e^-}{\quad} Ni^{2+}
$$

$$
\text{Pt1, Ag} \; \Big| \; \text{AgI} \; \Big| \; \text{Ag}_2\text{S} \; \Big| \; \text{NiS} \; \Big| \; \text{Ni} \qquad\qquad [5.17]
$$

$$
2e^- \Big| \text{Pt2}
$$
$$
\downarrow
$$

$$
\overset{\longrightarrow}{\underset{2e^-}{}} \quad \overset{2Ag^+}{\longleftarrow} \quad \overset{2Ag^+}{\longleftarrow} \quad \overset{Ni^{2+}}{\longleftarrow} \quad \overset{\longrightarrow}{\underset{2e^-}{}}
$$

$$
\text{Pt1, Ag} \; \Big| \; \text{AgI} \; \Big| \; \text{Ag}_2\text{S} \; \Big| \; \text{NiS} \; \Big| \; \text{Ni, Pt3} \qquad [5.18]
$$

Ag was used as the negative electrode. The positive pole could be connected to Pt2 or Pt3. When silver ions are withdrawn through silver sulfide, an equivalent amount of sulfur reacts with Ni to form NiS. Thus the current measured in the external circuit gives the rate of formation of NiS on Ni. In cell [5.17] both electrons and nickel ions pass through NiS, whereas cell [5.18] involves the passage of nickel ions only. Two different electrochemical techniques were used. In the galvanostatic method a constant current was passed through the cell, thus maintaining a constant rate of NiS formation, and the cell voltage (which gives the sulfur activity on the NiS surface) was measured as a function of time. In the potentiostatic method, a predetermined voltage was applied across the cell and the cell current was followed as a function of time. In both techniques, the current is related to the rate of NiS formation on Ni and the voltage is related to the sulfur activity on the NiS surface. No reaction was observed between the silver sulfide and the nickel sulfide. Both cells gave identical data, indicating that electron transfer in NiS was not rate controlling. Diffusion of nickel ions via doubly ionized nickel vacancies was found to be the rate-controlling step. At $400°C$ and $E = 140$ mV a parabolic rate constant equal to 5×10^{-9} cm^2 sec^{-1} was calculated.

The liquid metal contained in a solid oxide electrolyte crucible, heated by induction, has a uniform concentration of oxygen. If an emf is applied so that oxygen is removed from the melt, the rate-controlling step for the removal of oxygen from the liquid metal is dependent on the amount of oxygen present in the metal. Oberg et al.[67] found that the rate-controlling step was the transport of oxygen in the electrolyte for high oxygen concentration ranges and transport across the boundary layer in the melt at the melt/electrolyte interface for low oxygen concentrations. Voltages much above the decomposition potential of the electrolyte could be used in these experiments. As long as sufficient oxygen was present in the melt, the oxygen potential at the electrolyte/melt interface did not reach a value low enough to decompose the stabilized zirconia electrolyte. Mass transfer coefficients of oxygen in the deoxidation of induction stirred copper melts were obtained in the above experiments.

If a reducing gas such as CO or H_2 is bubbled through a liquid metal containing oxygen and a solid-electrolyte oxygen probe is used to measure the oxygen content of the melt, then information regarding bubble–metal interaction can be obtained. Geiger and Nanda [68] measured the rate of deoxidation of copper, copper–tin, and copper–silver alloys by bubbling CO and using stabilized zirconia as a solid electrolyte to monitor the oxygen content of the metal bath. The analysis of the data showed that liquid-phase mass transfer was the rate-limiting step. The appropriate mass transfer coefficient could thus be calculated. Bandopadhyay and Ray [69] used a modified technique to obtain the diffusion coefficient and mass transfer coefficient of oxygen in liquid lead.

5. Concluding Remarks

In the above review, attention has been mostly restricted to kinetic studies using oxide electrolytes, silver halide electrolytes, and copper halide electrolytes. Recently several fast-ion-conducting electrolytes based on the β-alumina structure and compounds such as $RbAg_4I_5$, $Ag_6I_4WO_4$, Ag_3SI, Ag_2HgI_4, etc. have been developed. Kinetic investigations involving these electrolytes may become prominent especially in relation to energy conversion applications.

References

1. D. O. Raleigh, in *Electroanal. Chem. – A Series of Advances VI*, ed. A. J. Bard, Marcel Dekker, New York (1973), p. 87.
2. A. D. Neuimin, S. V. Karpachev, and S. F. Palguev, *Proc. Acad. Sci. USSR Phys. Chem. Sect.* **141**, 875 (1961).
3. J. Weissbart and R. Ruka, *J. Electrochem. Soc.* **109**, 723 (1962).
4. H. Binder, A. Köhling, H. Krupp, K. Richter, and G. Sanstede, *Electrochim. Acta* **8**, 781 (1963).
5. H. H. Möbius and B. Rohland, *Z. Chem. (Leipzig)* **4**, 158 (1966).
6. H. Yanagida, R. J. Brook, and F. A. Kröger, *J. Electrochem. Soc.* **117**, 593 (1970).
7. R. J. Brook, W. L. Pelzmann, and F. A. Kröger, *J. Electrochem. Soc.* **118**, 185 (1971).
8. S. Pizzini, in *Fast Ion Transport in Solids*, ed. W. van Gool, North-Holland, Amsterdam (1973), p. 461.
9. J. E. Bauerle, *J. Phys. Chem. Solids* **30**, 2657 (1969).
10. K. Goto, M. Someno, M. Sano, and K. Nagata, *Met. Trans.* **1**, 23 (1970).
11. T. H. Etsell and S. N. Flengas, *Met. Trans.* **2**, 2829 (1971).
12. D. O. Raleigh, in *Fast Ion Transport in Solids*, ed. W. van Gool, North-Holland, Amsterdam (1973), p. 476.

13. C. Wagner, *Proc. Int. Comm. Electrochem. Thermodyn. Kinet.* (*C.I.T.C.E.*) **7**, 361 (1957).
14. M. H. Hebb, *J. Chem. Phys.* **20**, 185 (1952).
15. C. Wagner, *Z. Electrochem.* **60**, 4 (1956); **63**, 1027 (1959).
16. B. Ilschner, *J. Chem. Phys.* **28**, 1109 (1958).
17. D. O. Raleigh, *J. Phys. Chem. Solids* **26**, 329 (1965).
18. J. B. Wagner and C. Wagner, *J. Chem. Phys.* **26**, 1597 (1957).
19. J. B. Wagner and C. Wagner, *J. Electrochem. Soc.* **104**, 509 (1957).
20. A. Morkel and H. Schmalzried, *J. Chem. Phys.* **36**, 3101 (1962).
21. J. W. Patterson, E. C. Bogren, and R. A. Rapp, *J. Electrochem. Soc.* **114**, 752 (1967).
22. D. O. Raleigh, *J. Phys. Chem. Solids* **29**, 261 (1968).
23. B. E. Conway and J. O'M. Bockris, *Electrochim. Acta* **3**, 340 (1961).
24. H. Rickert and C. D. O'Brian, *Z. Phys. Chem. N.F.* **31**, 71 (1962).
25. L. Contreras and H. Rickert, in *Fast Ion Transport in Solids*, ed. W. van Gool, North-Holland, Amsterdam (1973), p. 523.
25a. L. Contreras, Ph.D. thesis, University of Dortmund, Dortmund, West Germany, 1975.
26. S. C. Singhal and W. L. Worrell, in *Proceedings of the International Symposium on Metallurgical Chemistry*, ed. O. Kubaschowski, Her Majesty's Stationery Office, London (1972), p. 65.
27. S. C. Singhal and W. L. Worrell, *Met. Trans.* **4**, 895, 1125 (1973).
28. P. J. Meschter, Ph.D. dissertation, University of Pennsylvania, 1974.
29. C. B. Alcock and T. N. Belford, *Trans. Faraday Soc.* **60**, 822 (1964).
30. T. N. Belford and C. B. Alcock, *Trans. Faraday Soc.* **61**, 443 (1965).
31. H. Rickert and R. Steiner, *Z. Phys. Chem. N.F.* **49**, 127 (1966).
32. H. F. Rizzo, R. S. Gordon, and I. B. Cutler, *J. Electrochem. Soc.* **116**, 266 (1969).
33. R. J. Pastorek and R. A. Rapp, *Trans. Met. Soc. AIME* **245**, 1711 (1969).
34. W. L. Worrell and J. L. Iskoe, in *Fast Ion Transport in Solids*, ed. W. van Gool, North-Holland, Amsterdam (1973), p. 513.
35. R. A. Rapp and D. A. Shores, in *Physicochemical Measurements in Metals Research*, Part 2, ed. R. A. Rapp, Interscience, New York (1970), p. 123.
36. B. C. H. Steele, in *Heterogeneous Kinetics at Elevated Temperatures*, eds. G. R. Belton and W. L. Worrell, Plenum Press, New York (1970), p. 135.
37. D. O. Raleigh and H. R. Crowe, *J. Phys. Chem.* **70**, 689 (1966).
38. D. O. Raleigh, in *Progress in Solid State Chemistry*, Vol. 3, ed. H. Reiss, Pergamon Press, New York (1967), p. 83.
39. J. Goldman and J. B. Wagner, *J. Electrochem. Soc.* **121**, 1318 (1974).
40. H. Rickert and A. A. El Miligy, *Z. Metallkd.* **59**, 635 (1968).
41. K. E. Oberg, L. M. Friedman, W. M. Boorstein, and R. A. Rapp, *Met. Trans.* **4**, 61 (1973).
42. T. A. Ramanarayanan and R. A. Rapp, *Met. Trans.* **3**, 3239 (1972).
43. R. Szwarc, K. E. Oberg, and R. A. Rapp, *High Temp. Sci.* **4**, 347 (1972).
44. K. E. Oberg, L. M. Friedman, R. Szwarc, W. M. Boorstein, and R. A. Rapp, *J. Iron Steel Inst.* **210**, 359 (1972).
45. A. V. Ramana Rao and V. B. Tare, *Z. Metallkd.* **63**, 70 (1972).
46. W. F. Chu, H. Rickert, and W. Weppner, in *Fast Ion Transport in Solids*, ed. W. van Gool, North-Holland, Amsterdam (1973), p. 181.
47. T. A. Ramanarayanan and W. L. Worrell, *Met. Trans.* **5**, 1773 (1974).
47a. M. Kawakami and K. S. Goto, *Trans. Iron Steel Inst. Jpn.* **16**, 204 (1976).
48. C. R. Masson and S. G. Whiteway, *Can. Met. Quart.* **6**, 199 (1967).
49. N. Sano, S. Honma, and Y. Matsushita, *Met. Trans.* **1**, 301 (1970).

50. N. Sano, S. Honma, and Y. Matsushita, *Met. Trans.* **2**, 1494 (1971).
51. S. Otsuka and Z. Kozuka, *J. Jpn. Inst. Met.* **37**, 364 (1973).
52. S. Otsuka and Z. Kozuka, *Met. Trans.* **6B**, 389 (1975).
52a. S. Otsuka and Z. Kozuka, *Met. Trans.* **7B**, 147 (1976).
52b. B. Hartmann, H. Rickert, and W. Schendler, *Electrochim. Acta* **21**, 319 (1976).
52c. P. D. Jose, M.Tech. thesis, Indian Institute of Technology, Kanpur, India, August 1976.
53. H. Kobayashi and C. Wagner, *J. Chem. Phys.* **26**, 1609 (1957).
54. P. Roy and H. Schmalzried, *Ber. Bunsenges. Phys. Chem.* **67**, 539 (1963).
55. E. Bechthold, *Ber. Bunsenges. Phys. Chem.* **69**, 328 (1965).
56. C. Wagner, in *Advances in Catalysis*, Vol. 21, Academic Press, New York (1970), p. 323.
57. G. R. Belton and F. J. Harvey, in *Heterogeneous Kinetics at Elevated Temperatures*, eds. G. R. Belton and W. L. Worrell, Plenum Press, New York (1970), p. 433.
58. H. Rickert, in *Electromotive Force Measurements in High Temperature Systems*, ed. C. B. Alcock, Institution of Mining and Metallurgy, London (1968), p. 59.
59. R. J. Ratchford and H. Rickert, *Z. Elektrochem.* **66**, 497 (1962).
60. S. Mrowec and H. Rickert, *Z. Elektrochem.* **66**, 14 (1962).
61. D. Detry, J. Drowart, P. Goldfinger, H. Keller, and H. Rickert, *Z. Phys. Chem.* **55**, 314 (1967).
62. N. Birks and H. Rickert, *Ber. Bunsenges Phys. Chem.* **67**, 97 (1963).
63. V. B. Tare and H. Schmalzried, *Trans. Met. Soc. AIME* **236**, 444 (1966).
64. G. R. Fitterer and C. D. Cassler, in *Heterogeneous Kinetics at Elevated Temperatures*, eds. G. R. Belton and W. L. Worrell, Plenum Press, New York (1970), p. 165.
65. H. Schmalzried and C. Wagner, *Trans. Met. Soc. AIME* **222**, 539 (1963).
66. S. Mrowec and H. Rickert, *Z. Phys. Chem.* **28**, 422 (1961).
67. K. E. Oberg, L. M. Friedman, W. M. Boorstein, and R. A. Rapp, *Met. Trans.* **4**, 75 (1973).
68. G. H. Geiger and C. R. Nanda, *Met. Trans.* **2**, 1101 (1971).
69. G. Bandopadhyay and H. S. Ray, *Met. Trans.* **2**, 3055 (1971).
70. S. N. S. Reddy and R. A. Rapp, *Met. Trans.* **9B**, 559 (1978).
71. T. A. Ramanarayanan and P. D. Jose, *J. Electrochem. Soc.* **125**, 1684 (1978).

6

Technological Applications of Solid Electrolytes

K. P. Jagannathan, S. K. Tiku, H. S. Ray,
A. Ghosh, and E. C. Subbarao

1. General

1.1 Introduction

The previous chapters have adequately discussed and illustrated the present status of solid electrolytes from the point of view of fundamental research and information. The scope is indeed very wide. Potential uses of solid electrolytes of various kinds in commercial applications are also numerous. However, in actual practice such applications remain rather limited due to problems associated with commercially available materials as well as actual technological difficulties in incorporating such materials in contemplated devices.

Common solid electrolytes, as discussed earlier, are basically of three kinds: simple or complex halides, simple or complex oxides, or oxide–solid solutions. Of these, the oxide–solid solutions (e.g., ZrO_2–CaO, ZrO_2–MgO, ZrO_2–Y_2O_3, ThO_2–Y_2O_3) are by far the most commonly used in solid-electrolyte work, followed by complex oxides such as β-alumina and complex halides (e.g., $RbAg_4I_5$). Commercial applications of solid electrolytes also are,

K. P. Jagannathan, H. S. Ray, A. Ghosh, and E. C. Subbarao • Department of Metallurgical Engineering, Indian Institute of Technology, Kanpur 208016, India. *S. K. Tiku* • Interdisciplinary Programme in Materials Science, Indian Institute of Technology, Kanpur 208016, India. Dr. Tiku's present address: Department of Materials Science, University of Southern California, Los Angeles, California 90007. Dr. Jagannathan's present address: Research and Development, Hindustan Steel Ltd., Ranchi 834002, India.

so far, restricted mostly to these. Many of these applications, however, can be extended to other kinds of solid electrolytes at least in principle.

The fundamental principles governing the commercial applications of a solid electrolyte may be briefly summarized as follows.

(a) The material provides an impervious barrier to gases and liquids. The material, however, allows one or more kinds of ions to migrate through its lattice when a tendency for such migration exists. This tendency is induced by a potential gradient generated either through an applied voltage or through a chemical potential gradient of the migrating ions.

(b) A solid electrolyte allows the measurement of the difference of chemical potentials of the migrating species on either side of it in terms of electromotive force in a properly constituted cell.

(c) Solid electrolytes are stable compounds not easily corroded by high-temperature environment.

The purpose of the present chapter is to review the main commercial applications of solid electrolytes. Special emphasis is laid on discussing the underlying principles and future possibilities.

1.2. Classification of Technological Applications

The main technological applications of solid electrolytes fall into the following categories.

(*a*) *Open-Circuit Applications.* These involve equilibrium measurements. In these applications the solid electrolyte is incorporated in suitable emf cells which serve, e.g., as meters for oxygen in gases or in metals.

(*b*) *Closed-Circuit Applications.* Here the solid electrolyte is a medium of mass transfer, under induced external voltage. Application of external voltage necessarily implies a closed circuit.

(*c*) *Energy Conversion.* In electrochemical energy conversion, a fuel is electrochemically oxidized to give rise to direct electric power. Here the solid electrolyte provides the electrolytic medium necessary both for ionic migration as well as development of electromotive force.

(*d*) *Solid State Batteries.* Ionic transport is utilized for energy storage.

(*e*) *Solid State Ionics.* This includes a range of devices which utilize the ionic conductivity of solids.

All these applications are discussed in detail in the subsequent sections.

2. Oxygen Meters

2.1. Introduction

By virtue of natural prevalence and chemical reactivity, oxygen is the most important reacting species taking part in innumerable reactions. A large number of industrial processes are based on reactions directly involving oxygen. Combustion of fuels, corrosion of metals and alloys, extraction and refining of metals, etc. may be cited as examples. In all reactions underlying these processes, the concentration of oxygen in various phases is an important parameter which often needs to be measured, monitored, and controlled.

Over the years, numerous techniques have been employed for oxygen analysis.[1] The commonly used Orsat apparatus can determine the oxygen concentration in a gas mixture at high concentration ranges only. For lower ranges, various instrumental methods such as Hersch electrochemical cell (less than 1000 ppm oxygen), spectrometry, gas chromatography, etc. have been employed. Other methods of oxygen estimation in gases as well as liquids include redox titration, polarography, electrode potential measurement, solid-electrode voltametry, etc. Fluometric methods can measure oxygen concentration in some inorganic substances. Oxygen dissolved in metals can be analyzed either by deoxidation in hydrogen or by employing the vacuum fusion technique. In addition to all these, there is also the radio-isotopic method as one of general utility.

The discovery of the oxygen-ion-conducting solid electrolyte and the ability of an electrochemical cell based on it to measure oxygen potential has led to a breakthrough in the field of continuous oxygen concentration measurement. The advantages of the solid-electrolyte oxygen meters (or sensors) are many. Some outstanding advantages are the following.

(a) Emf response of a solid electrolyte cell is usually specific, rapid, and continuous.

(b) Emf measurement can be made directly on the system to be investigated.

(c) A single sensor can often cover an exceedingly wide range of oxygen potential and temperature.

(d) Sensor output (the cell voltage) is amenable to a very precise measurement.

(e) The meter draws very little current and, therefore, hardly disturbs the system.

(f) Because the output voltage is amenable to fast and continuous measurement, it is ideally suited as a transducer for recording and feedback control.

(g) It is versatile and, in principle, applicable for oxygen analysis in any material since it measures oxygen potential specifically.

There are, of course, some inherent limitations also. For example, no device using zirconia–calcia or thoria–yttria can be used below 500°C or so because of high resistance and irreversibility. Secondly, oxide solid electrolytes have very poor thermal shock resistance and great care is needed in order to prevent thermal cracking. Therefore, the solid-electrolyte device must either by protected from rapid temperature fluctuations or be designed as a consumable article.

Measurement of oxygen concentration (or p_{O_2}) in gases and determination of oxygen in liquid metals and alloys are the two most important commercial applications of solid-electrolyte cell as oxygen meter.

2.2. Determination of Oxygen in Gases

2.2.1. Principle of Operation of Oxygen Meter

The oxygen meter is an oxygen concentration cell of the following type:

$$\text{reference gas, Pt} \mid \text{solid oxide electrolyte} \mid \text{Pt, sample gas}$$
$$p^{\circ}_{O_2}, \mu^{\circ}_{O_2} \qquad\qquad (O^{2-}) \qquad\qquad p_{O_2}, \mu_{O_2} \qquad [6.1]$$

The emf of the cell follows the well-known equation

$$E = -(\mu_{O_2} - \mu^{\circ}_{O_2})/4F = RT/4F(\ln p_{O_2}/p^{\circ}_{O_2}) \qquad (6.1)$$

where F is the Faraday constant, μ is chemical potential, and p is partial pressure. R and T are, respectively, the universal gas constant and the absolute temperature. Superscript $^{\circ}$ refers to the reference gas. If the oxygen in gases behaves nonideally, the pressure terms must be replaced by fugacities. This, however, is seldom necessary. From the determination of E, the unknown pressure p_{O_2} may be calculated when the partial pressure of oxygen in the reference gas, $p^{\circ}_{O_2}$, is known. Concentration of oxygen in moles per unit volume may be known from the equation

$$p = (n/v)RT \qquad (6.2)$$

where n is the number of moles of oxygen and v the total volume of gas.

2.2.2. Some Basic Considerations

Since both electrodes are gaseous, there is need to completely separate the two electrodes. This may be achieved through either of the two arrangements shown in Fig. 6.1. In the emf cell [6.1], the gas atmospheres need electronic conductors for charge transfer reactions at the electrodes. Platinum has been the only metal employed for this purpose. It is worth considering as to what should be the general characteristics of the electrode material.

The conductor surface experiences the electrode potential, which is measured with the help of the conductor. In order to generate the correct emf corresponding to the oxygen potentials in the gas phases, the surface of the conductor must rapidly come into equilibrium with respect to oxygen in the gas phase. A reactive metal would be unsuitable since it would undergo

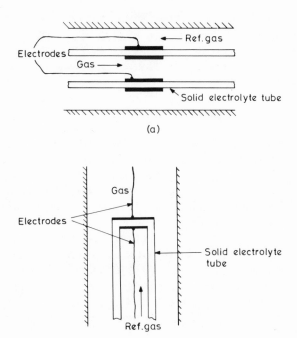

FIG. 6.1. Cell arrangements for gas probe. (a) Open-end tube type; (b) closed-end tube type.

oxidation of a normally irreversible nature. Besides, this would introduce a metal/oxide couple which may interfere with the electrode potential. A liquid metal would be unsuitable as electrode material as this would normally have a large oxygen absorption capacity making rapid equilibration with gas phase difficult. Moreover, there would be experimental difficulties such as the availability of a suitable lead etc.

Platinum is suitable because of its low reactivity and high melting point. Its capacity to allow quick oxygen adsorption on its surface makes possible rapid attainment of equilibrium with respect to the gas phase. It may be possible to substitute other noble metals such as Rh, Ir, Ru, etc., for platinum but the former are more expensive.

Although a platinum electrode does not exhibit good reversibility at room temperature,[2] its reversibility has been found to be satisfactory at high temperatures in molten carbonates[3] and molten silicates.[4-6] These findings have been further corroborated by polarization studies at high temperature with solid electrolytes. Etsell and Flengas[7] demonstrated good reversibility of Pt electrodes with $CaO–ZrO_2$ electrolyte at 700–1100°C, except in $CO–CO_2$ atmosphere. Since mass transfer would then be the rate-controlling step in the overall electrode process, the actual form of the electrode (porous paste or foil) becomes somewhat unimportant, as shown by experiment. Indications are that oxygen is exchanged at the common interface of gas/metal/solid electrolyte and this is the reason why any appreciable current cannot be drawn.[6a]

2.2.3. Leakage of Oxygen through Solid Electrolyte Tubes

One problem with the solid-electrolyte method is leakage of oxygen through even the best commercially available electrolyte tubes due to either pores or electronic conduction[7] and the increase of leakage, if present, with temperature. The leakage can be tolerated if the buffering capacity of the gas is large. But this is not so in all cases (e.g., in Ar–O_2 mixtures with low values of p_{O_2}) and considerable error may result.

Let us consider the effect of a leak rate of 10^{-5} cm^3/sec of O_2. It can be shown that in order to cut down the error to less than 1%, the flow rates of the Ar–O_2 mixture should be at least 0.1, 10, and 10^3 cm^3/sec. for p_{O_2} values of 10^{-2}, 10^{-4}, and 10^{-6} atm, respectively. Therefore, although a reasonably accurate measurement is possible down to a p_{O_2} value of nearly 10^{-4} atm, difficulties may be encountered in the measurement of lower values unless very high flow rates are used.

If, in place of the Ar–O_2 mixture, a CO–CO_2 mixture be the sample gas, then the buffering capacity of the gas is considerably enhanced as a result of the reaction

$$CO + \tfrac{1}{2}O_2 = CO_2 \qquad (6.3)$$

$$K = p_{CO_2}/(p_{CO}p_{O_2}^{1/2})$$

or

$$p_{O_2}^{1/2} = (p_{CO_2}/p_{CO})(1/K) \qquad (6.4)$$

Let Δp_{O_2} be the absolute error in p_{O_2} as a consequence of the leak mentioned above. Then, from Eq. (6.4)

$$\tfrac{1}{2}p_{O_2}^{-1/2}\Delta p_{O_2} = (1/K)\Delta(p_{CO_2}/p_{CO}) \qquad (6.5)$$

or

$$\Delta(p_{O_2}) = (2/K)p_{O_2}^{1/2}\Delta(p_{CO_2}/p_{CO})$$

$$= 2(p_{CO}/p_{CO_2})p_{O_2}\Delta(p_{CO_2}/p_{CO}) \qquad (6.6)$$

Assuming that $p_{CO}/p_{CO_2} = 1/1$ (which corresponds to p_{O_2} of about 10^{-14} atm at 1200°K),

$$\Delta p_{O_2} = 2p_{O_2}\Delta(p_{CO_2}/p_{CO}) \qquad (6.7)$$

If all the leaking oxygen reacts with CO to form CO_2 then

$$\Delta(p_{CO_2}/p_{CO}) = 2\dot{v}_{O_2}/(0.5\dot{v} - 2\dot{v}_{O_2}) \qquad (6.8)$$

where $\dot{v}$ is the flow rate of $CO + CO_2$ in cm^3/sec (STP). For $\Delta p_{O_2}/p_{O_2} \leqslant 0.01$ and with the above-mentioned leakage rate of oxygen ($\dot{v}_{O_2}$),

$$\dot{v} \geqslant 1.6 \times 10^{-2} \, cm^3/sec \, (STP)$$

Even if p_{CO}/p_{CO_2} is only 10^{-2} then $\dot{v} \geqslant 0.75 \, cm^3/sec$ (STP).

These calculations readily demonstrate that the effect of some leakage through the electrolyte can be successfully tackled with a gas of high buffering capacity such as a CO/CO_2 mixture at a reasonable flow rate. The concentration polarization would tend to adversely affect the results and make the situation worse. Sample calculations can again be presented to substantiate this contention. Thus one may conclude that considerable precaution should be necessary for $Ar-O_2$, N_2-O_2, or $He-O_2$ mixtures with low p_{O_2}.

Clegg[8] determined the leak rate in $Ar-O_2$ mixture by measurement of p_{O_2} as a function of volumetric flow rate of argon and the following relationship was established:

$$p_{O2} = (\dot{v}_{O_2}/\dot{v}_{Ar}) + p^\circ_{O_2} \tag{6.9}$$

As Fig. 6.2b shows, p_{O_2} was found to vary linearly with $1/\dot{v}_{Ar}$. The slope gave $\dot{v}_{O_2}$, which was found to be $1.15 \times 10^{-5} \, cm^3/sec$ (STP). This leakage was attributed to the porous nature of the electrolyte. Etsell and Flengas[9] also found a considerable dependence of emf reading on flow rate at low flow rates in $Ar-O_2$ mixture dilute in oxygen. They attributed this to the permeability of the solid-electrolyte tube and/or electronic conductivity in the material. Figure 6.3 presents the results of their investigation. Permeability measurements of calcia-stabilized zirconia tubes between 1400–1600°C revealed the tubes to be permeable to oxygen only and not to inert gases.[10] More evidence of this nature is appearing now[104,177] to indicate that it is primarily the electronic conduction that is responsible for such oxygen permeation. Again, leakage of oxygen through other high-temperature materials

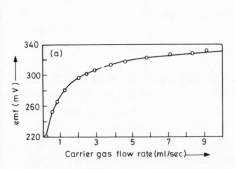

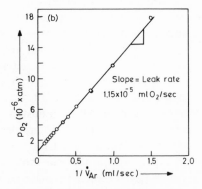

FIG. 6.2. Influence of flow rate on emf measurements by gas probe (Reference 8).

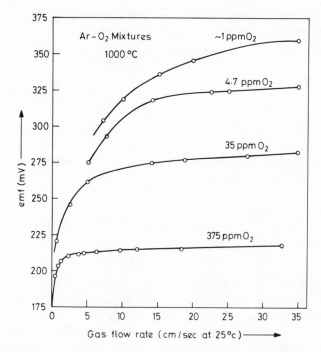

FIG. 6.3. Influence of flow rate at different gas compositions on gas probe measurements (Reference 9).

used in the construction to the cell, and joints also can upset the results and it is important to ascertain these effects.

2.2.4. Ionic and Electronic Conduction

Any oxide solid electrolyte which is a predominantly ionic conductor at appropriate temperature and oxygen potential ranges may be employed for measurements. However, it seems from the available information that only calcia-stabilized zirconia (and, to a limited extent, yttria-doped thoria) has so far been tried. Calcia-stabilized zirconia tubes are readily available commercially and are, therefore, widely used. However, their poor thermal shock resistance causes failure of the material under temperature fluctuations which cannot be avoided in industrial furnaces and sometimes in laboratory experiments as well. This problem may be overcome by maintaining the cell inside another small furnace of its own at a constant temperature or by improving the thermal shock resistance by using a lower CaO content (2–4 wt % CaO) when the material is partially stabilized.[11] However, the resistance of such an electrolyte is much higher.[12] Also, the material is more prone to electronic conductivity. Presence of electronic conductivity leads to local current in the solid electrolyte and causes transfer of oxygen from the high oxygen potential side to low oxygen potential side. Such an inference was drawn by Steele and Alcock[13] and Jagannathan[14] in connection with investigations on condensed systems. This leads to leakage of oxygen through the tube with associated problems as discussed in Section 2.2.3.

2.2.5. Operating Temperature Range

There is an operating temperature range for the solid-electrolyte oxygen meter. At low temperature, the resistivity of the electrolyte is very high and irreversibility sets in the electrode reaction. Moreover, the contact between platinum electrode (foil, sheet, or gauze) and the electrolyte becomes poor. The minimum operating temperature of the cell is held to be around 600°C by various investigators.

At high temperatures, leakage rates are greatly enhanced, the electronic conductivity of the electrolyte increases, and volatilization of platinum as oxides begins. The construction and maintenance of the assembly also becomes more difficult. The highest temperature reported so far is around 1100–1150°C.[9,15,16] Etsell and Flengas,[9] in their experiments, first raised the temperature to 1000°C so that platinum was softened and a good contact was established. They recommended a temperature of 1000°C as the optimum one, whereas Spacil[11] recommended 850°C.

2.2.6. Reference Electrode

Air and pure oxygen have been the most widely used reference gases because of the convenience as well as good reversibility of the reference electrode. Condensed electrodes such as Ni/NiO mixture have been thought of for sample gases of very low oxygen partial pressure in order to minimize the difference in p_{O_2} on the two sides of the electrolyte. However, there are problems of maintenance and stability of these references. Moreover, it is evident from the preceding discussions that a large difference in p_{O_2} between reference and sample tends to adversely affect only the inert gas–oxygen mixtures of low p_{O_2}, and not sample gases with reasonably large buffering capacity. Again, the errors in the former case may be caused by diffusion and leakage through other parts of the meter. If for some reason it becomes very desirable to use a reference of low p_{O_2}, it is best to either employ a gas of low, but known, value of p_{O_2} as reference or alternatively, use a double-tube arrangement[12] with an intermediate gas in between air (or oxygen) and the sample.

Figure 6.4 shows the cell assembly of Etsell and Flengas,[9] which is one of the more sophisticated ones.

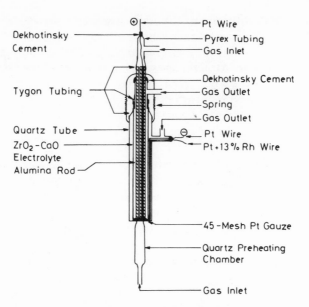

FIG. 6.4. Cell employed by Etsell and Flengas (Reference 9).

2.3. Measuring and Recording

2.3.1. Some General Considerations

One of the characteristics of oxygen sensors is that the cell emf changes appreciably on the passage of a small current. While the electrode kinetics play a major role in this, there are also other contributing factors which should be noted.

(a) The resistance of a ZrO_2–15-mol %-CaO disk of 1-cm² area and 3-mm thickness (which roughly corresponds to many experimental situations) is approximately 300 and 10 ohms at 700 and 1000°C, respectively. Therefore, the iR drop can cause some error especially at lower temperatures.

(b) The gas flow rate may not be enough to flush out the oxygen at low p_{O_2} values. A sample calculation shows that if the current drawn during experiment is 1 μA, and an Ar–O_2 mixture is flowing across one of the electrodes at 5 cm³/sec, then the error in p_{O_2} due to flow of current would be about 5% at p_{O_2} value of 10^{-6} atm. The error would be higher if p_{O_2} were smaller.

2.3.2. Measuring and Recording of emf

In trying to decide on the measuring and recording instruments, the following points are to be borne in mind.

(a) The current flow through the solid electrolyte during measurements must not be appreciable because it leads to error due to oxygen transfer, iR drops as well as polarization. A conventional potentiometer using a Kelvin–Varley bridge and a mechanical galvanometer would tend to draw too much current during balancing and hence a series resistor should be employed when the bridge is unbalanced. Vacuum tube instruments with high resistance (greater than a megohm) are being used extensively. Shielding of the circuit from stray emf's is very much necessary in these high-resistance circuits and various precautions are to be taken.[12] Many varieties of measuring instruments have been employed such as potentiometers,[16–18] radiometer, universal pH meter (impedance $= 10^9$ ohms), electrometer,[19] vibrating reed electrometer,[12] etc. In all these instruments it is difficult to get accuracy of better than 1% of the full-scale value. This is considered to be adequate for most applications.

(b) Recording of emf can be easily accomplished, in principle, by connecting the cell to a suitable recorder provided the impedance of the measuring circuit is high. The response of the cell to any change in gas composition is very fast and the new steady value is reached within a few seconds.[8,9] Therefore, a recorder would be able to reveal fluctuations in gas composition reasonably well. Both analog and digital recording are practiced.[12]

2.4. Estimation of Gas Composition from p_{O_2} Measurement

The oxygen meter determines oxygen potential (μ_{O_2}) or fugacity (f_{O_2}) and, *inter alia* for perfect gases, determines p_{O_2}. If the gas is a mixture of oxygen and an inert diluent (Ar, CO_2, etc.), then

$$p_{O_2} = (\% \ O_2/100)P \tag{6.10}$$

where P is the total pressure of the sample gas.

If the gas contains, besides oxygen and inert gas, any species reactive to oxygen such as hydrogen, then an equilibrium similar to the following prevails:

$$2H_2(g) + O_2(g) = 2H_2O(g) \qquad K_H \tag{6.11}$$

and

$$p_{O_2} = (1/K_H)(p_{H_2O}/p_{H_2})^2 \tag{6.12}$$

Similarly, if the reactive gas is CO, then

$$p_{O_2} = (1/K_c)(p_{CO_2}/p_{CO})^2 \tag{6.13}$$

Now,

$$K_H \approx 2 \times 10^{11} \qquad \text{at } 1500°K \tag{6.14a}$$

and

$$K_c \approx 3 \times 10^{10} \qquad \text{at } 1500°K \tag{6.14b}$$

If pure H_2O at 1 atm is maintained at a temperature of 1500°K, then it would dissociate partly into H_2 and O_2, such that

$$p_{H_2} = 2p_{O_2} \tag{6.15}$$

Assuming that p_{H_2O} is nearly 1 atm, the values of p_{H_2} and p_{O_2} are approximately 3×10^{-6} and 1.5×10^{-6} atm, respectively. Since 1500°K is about the maximum temperature for the solid electrolyte device being considered and since pure H_2O yields the highest p_{O_2} in the H_2O/H_2 system, the oxygen concentration obtained from dissociation of H_2O in other circumstances would be less than the above figure. Similarly, for CO_2/CO mixture, the highest oxygen concentration through dissociation of CO_2 is about 4×10^{-6}. These calculations, therefore, lead to the following conclusions:

(a) In $H_2O-H_2-O_2$ mixture, containing free H_2 or CO_2-CO-O_2 mixture containing free CO, the partial pressure of oxygen is negligible. The entire mixture may be considered, in effect, to be composed of only H_2O and H_2 or CO and CO_2.

(b) If the partial pressure of oxygen is greater than about 10^{-6} atm then the sample gas would contain a negligible quantity of H_2 or CO and the system behaves as an inert gas–oxygen mixture.

Again, in the first situation (when free H_2 or CO is present), measurement of oxygen potential allows determination of gas composition with the help of Eq. (6.12) and (6.13). Another characteristic is that at a fixed composition (fixed H_2O/H_2 or CO_2/CO ratio), the p_{O_2} measured by the sensor would depend on the measurement temperature, because K_H and K_c are functions of temperature. The implication of this is that if the measurement temperature is different from the temperature of operation, then a correction is required as shown below for H_2O/H_2:

$$\log p_{O_2} \text{ (measured)} = (-\log K_H)_{T_m} + 2 \log (p_{H_2O}/p_{H_2}) \tag{6.16}$$

and

$$\log p_{O_2} \text{ (operation)} = (-\log K_H)_{T_o} + 2 \log (p_{H_2O}/p_{H_2}) \tag{6.17}$$

where T_m is measurement temperature (°K) and T_o is operational temperature (°K).

Combining Eqs. (6.16) and (6.17),

$$\log p_{O_2} \text{ (operation)} = \log p_{O_2} \text{ (measured)} - (\log K_H)_{T_o} + (\log K_H)_{T_m} \tag{6.18}$$

In industrial practice, leaving aside N_2, which is only a diluent, presence of H_2, H_2O, CO, CO_2, and O_2 are to be considered. As discussed previously, if the oxygen concentration is above 10^{-6}, free H_2 and CO can be ignored. This is the situation with complete combustion with theoretical or excess oxygen. For deficit oxygen, concentration of free oxygen is to be ignored. The chemical analysis of such a gas may be attempted with the help of the oxygen meter if some additional information is available. Suppose the ratio of hydrogen/carbon in the fuel is known, we have

$$n_{H_2}^{\circ} = n_{H_2} + n_{H_2O}$$
$$n_C^{\circ} = n_{CO} + n_{CO_2} \tag{6.19}$$

where $n_{H_2}^{\circ}$ and n_C° are the numbers of moles of hydrogen and carbon of fuel under consideration and are fixed arbitrarily maintaining $n_{H_2}^{\circ}/n_C^{\circ} = r$ in the fuel.

Now the meter gives p_{CO_2}/p_{CO} and p_{H_2O}/p_{H_2} through Eqs. (6.12) and (6.13). Let

$$n_{CO_2}/n_{CO} = a$$
$$n_{H_2O}/n_{H_2} = b \tag{6.20}$$

If one further assumes that there is no nitrogen, then the four unknowns, viz., n_{H_2}, n_{CO}, n_{H_2O}, and n_{CO_2}, can be solved with the help of Eqs. (6.19) and (6.20), and percentage composition of the flue gas can thus be calculated. If air is the only source of nitrogen, then

$$n_{N_2}/n_{O_2} = n_{N_2}/(\tfrac{1}{2}n_{CO} + \tfrac{1}{2}n_{H_2O} + n_{CO_2}) = 0.79/0.21 \tag{6.21}$$

In that case all the five unknowns can be determined. Oxygen meters were employed by Ghosh and co-workers[12a] to determine the gas compositions in mixtures of CO and CO_2. The general conclusion is that the equilibrium

$$2CO(g) + O_2(g) \rightleftharpoons 2CO_2(g) \tag{6.21a}$$

is not always established. Hence special care is required to ensure it. This is because the reaction rates are slow in this system and catalysts such as Ag, Pt, and Ni are to be present to help in the equilibration. The performance of the meter was checked against a mixture of hydrogen and water vapor of known composition. Excellent agreement was observed.

2.5. Industrial Applications

Two important approaches are possible regarding the actual use of oxygen meters. The sensor may be introduced into the reaction tube and furnace and direct measurements may be made. Alternatively, the sensor can be located as a separate instrument with its own heating arrangement in which case thermal, mechanical, and chemical damage to the probe are minimized. Here the sample gas is to be fed into this device by some method. Both these approaches have their advantages and disadvantages and a judicious selection has to be made.

In industrial applications, if the sensor tip is inserted into the furnace chamber, it would be subjected to severe environmental conditions and thermal and mechanical shocks. Also, dust and corrosive gas species can chemically damage the sensor by reaction and/or slag formation. These problems may be avoided by having a separate heater for the probe and drawing the sample gas into it,[11] after removing the dust particles by filtration. with proper care, cells with air reference have been used for 18 months continuously.

Rudd[15] employed a silicon carbide protective sheath around the solid electrolyte tube and inserted the meter into the furnace chamber. SiC, being a good thermal conductor, protected the tube from thermal and mechanical shocks. Gas sampling and attendant problems were avoided. Handman *et al.*[15a] utilized partially stabilized zirconia electrolyte containing 3.5% by weight of CaO for better thermal shock resistance. They also employed a clay–graphite shield for further protection against thermal shock, etc. So far this kind of meter is being put to use to primarily monitor and control the furnace conditions rather than to make accurate measurements. Applications have been in combustion control,[20,21] basic oxygen furnace,[22] annealing of metal products,[11,15] continuous annealing line for tin plates, and analysis of effluent gases in blast furnaces. The generation and control of atmospheres for obtaining a specific carbon potential for carburization, etc. appears to be another area of application for the meter. The heart of this kind of application is the justified assumption that gas phase equilibrium prevails and carbon and oxygen potentials are related.

The sensor is very well suited to the determination of the stoichiometric combustion point (SCP) because there is a drastic change of p_{O_2} from reducing to oxidizing atmosphere at that point as the air/fuel ratio is varied through the SCP, as illustrated in Fig. 6.5.[11]

Several commercial oxygen meters for analysis of oxygen in gases have been marketed by, e.g., Westinghouse Electric Corporation, U.S.A., SIMAC Ltd., U.K., Leeds and Northrup, U.S.A., Electro-Nite, Belgium.

2.6. Oxygen Determination in Liquid Metals

2.6.1. Introduction

Dissolved oxygen and entrapped oxides decide the properties of the refined metals to quite an extent. An accurate and rapid determination of these is vital in the control of the refining operation. The major aim in all the modern refining operations is the enhanced production rate and, where oxygen is present in the molten metals, the key to the control of the operation lies in the accurate and rapid determination of oxygen. Until recently the techniques employed in this connection were vacuum/inert gas fusion or neutron activa-

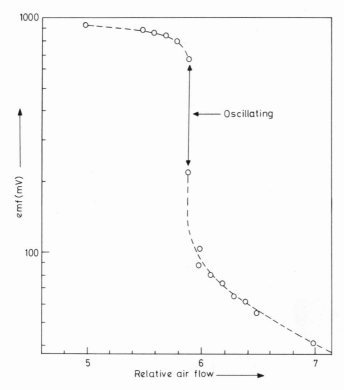

FIG. 6.5. Measured emf sensor output near stoichiometric combustion point (Reference 11).

tion methods. Both these suffer from inherent disadvantages, viz., the procedures are slow and give only the total oxygen.[23]

With the advent of the solid oxide electrolyte cells, particularly calcia-stabilized zirconia and yttria-doped thoria, there has been a revolution in the determination of oxygen in high-temperature systems. The advantages in using oxygen probes for the determination of oxygen in gases (Section 2.1 above) are also applicable when such are used for the determination of oxygen in molten metals. Some of the additional advantages in this case are that (a) the cell measures only the dissolved oxygen and (b) the use of the cell avoids the need for sampling hot metals.

Numerous investigations aimed at perfecting an oxygen probe have resulted in the commercial production of probes for the iron and steel industry. Probes have also been used in investigations involving the determination of oxygen in molten copper, lead, tin, silver, and sodium.[23]

In the pioneering work of Kiukkola and Wagner,[24] the cell was constituted by solid electrodes and a solid electrolyte. Extension of application to the measurement of oxygen in liquid metals posed some new problems, but did not pose any major challenge. However, it was when efforts were made to measure oxygen in molten steel that serious difficulties were encountered owing to the greatly enhanced corrosion, leakage, electronic conductivity, lack of suitable high-temperature materials, etc. The application to molten steel meant use under extreme conditions.

It is worthwhile to note that the oxygen probe cannot be employed in conjunction with reactive metals such as Al, Mg, etc. because the dissolved oxygen content in these cases is negligible. Also these metals may attack the solid electrolyte. Barring these reactive metals, other common metals can be divided into two categories—metals melting below 1200°C (Cu, Pb, Sn, Ag, etc.) and metallic materials melting above 1200°C, such as iron and steel. Assuming that equilibrium prevails amongst dissolved oxygen and other solutes, it is possible to determine the concentration of other solutes from oxygen probe data.

Therefore, the discussion in this review is divided into two parts, viz. (a) measurement of oxygen content in molten metals below 1200°C and (b) measurement of oxygen content in molten steel.

2.6.2. Principle

The galvanic cell in the measurement of oxygen content in molten metals can be set up as follows:

$$
\begin{array}{c|c|c}
\text{reference} & \text{solid oxide} & \text{oxygen in} \\
\text{electrode} & \text{electrolyte} & \text{molten metal} \\
(p^{\circ}_{O_2}) & (O^{2-}) & (p_{O_2})
\end{array}
\qquad [6.2]
$$

The emf for the cell [6.2] is given by Eq. (6.1) if $t_{O^{2-}} \approx 1$. The equilibrium relationship between oxygen partial pressure and oxygen content is brought out through the following equations:

$$h_O = K_h p_{O_2}^{1/2} \qquad (6.22)$$

where h_O is the Henrian activity of oxygen and K_h is a constant, whose value is known from the literature for common metals.

Again,

$$[h_O] = (\text{wt } \% \text{ O}) f_O \qquad (6.23)$$

and

$$\log f_O = e_O^A (\text{wt } \% \text{ A}) + e_O^B (\text{wt } \% \text{ B}) + \cdots \qquad (6.24)$$

where e_O^A, e_O^B, ... are the interaction coefficients on the Henrian scale for the added elements[25,26] and f_O is the activity coefficient of oxygen on the Henrian scale.

Thus the wt % of oxygen can be determined by a solid oxide electrolyte probe with the help of Eqs. (6.1) and (6.22)–(6.24). A proper diagnosis of errors in the measurement with oxygen probes is relevant both from the point of view of design of such probes and also the eventual utility of the data collected from them. Errors in emf readings arise from the following: (a) thermoelectric emf due to the use of dissimilar leads on the two electrodes and/ or due to temperature difference across the cell, (b) errors resulting from the electrolyte, e.g., electronic conductivity, porosity, (c) errors caused by wrong temperature reading, and (d) errors caused by irreversibility of the electrodes. All these factors are discussed at the appropriate places in Sections 2.6.3 – 2.7.1.

2.6.3. Determination of Oxygen in Molten Metals below 1200°C

Oxygen content has been determined by the solid-electrolyte method in molten copper,[27-34] copper alloys,[35] tin,[36,37] tin alloys,[38-40] lead,[41,42] silver,[28] lead–silver alloys,[43] and sodium.[115] Only calcia-stabilized zirconia tubes have been employed in all these instances. CaO was the most common dopant, the concentration of which ranged from 3–4 wt %[27,28,37] to $7\frac{1}{2}$ wt %.[28] Reasons for using lower CaO content were to improve thermal shock resistance as well as to increase the resistance with a view to cut down the short-circuiting effect due to electronic conduction.[28] Yttria-doped thoria in conjunction with yttria-stabilized zirconia was also used to minimize the latter effect.[28] Some investigators have also employed MgO-stabilized zirconia for measurement in molten copper[30,32] and copper alloys.[44] The major advantages in the use of tubes are their commercial availability and separation of the two electrode compartments. However, extra care is required to protect the tubes from cracking due to thermal shocks.

Oxygen or air have been widely used as reference electrodes.[28-32] Ni/NiO was also employed in a few studies.[27,28] Diaz and Richardson[28] found the emf to decrease with time with a Ni/NiO reference at 1100°C. This was attributed to the electronic conductivity of the electrolyte. However, increasing electrolyte resistance had hardly any effect. A Ni/NiO electrode was used also by Wilder,[27] who did not report any difficulty.

Metallic lead wires when immersed in liquid metals get attacked. Therefore the lead wires were dipped in the melts for a few minutes and emf measured. Table 6.1 gives briefly the various leads that were employed.

All the above arrangements gave satisfactory results. Diaz and Richardson[28] suggested that continuous immersion would result in less scatter and more accuracy, and they employed Cermet (Metamic 612 of Morgan, Inc.) as the lead material. This enabled them to continue their experiments for long duration. Choudary and Ghosh[46] in connection with their studies on Cu–Ag alloys dipped inconel instantaneously, Cermet (Metamic 612) continuously,

Table 6.1. *Oxygen Determination in Liquid Metals*
(Nonferrous)

Investigators	System	Temperature range (°C)	Lead
Belford and Alcock[36,41]	Sn, Pb	510–750	Ir
Osterwald[29]	Cu	1100–1300	Cr_2O_3
Puschkell and Engell[30]	Cu	1150	Mo
Rickert and Wagner[32]	Cu	1100–1420	Pt
Baker and West[45]	Ag	1000–1500	Rh
Wilder[27]	Cu	1100–1200	Ta

and also employed Cu_2O-coated Pt. All of them gave satisfactory results. However, with Cermet and Cu_2O-coated Pt the data had less scatter. The Cermet was found to be good for tin alloys also.[47,48] Some of the typical cell arrangements are shown in Figs. 6.6 and 6.7.

A commercial oxygen probe employing calcia-stabilized zirconia for the measurement of oxygen in molten copper has been developed and patented as OXYCELL,[33] which is claimed to give better control in cast copper wires with attendant higher productivity. Thermal shock resistance has been considerably improved by using very thin tubes and through adjustment of the composition of the electrolyte. Continuous immersion in molten copper even for 250 hr has been achieved.[34] Analytical instruments are now available for on-line measurement of oxygen in liquid sodium for nuclear applications.

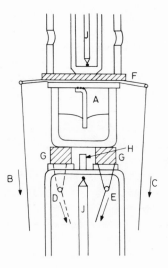

FIG. 6.6. Cell assembly for determination of oxygen activity by coulometric method (Reference 41). A, Electrolyte crucible containing Pb, Ir probe; B and C, leads to electrometer and coulometer circuits, respectively, via silica ballards; D and E, leads to earth and coulometric circuits, respectively, via silica ballards; F, Pt foil; G, semicircular metal oxide pellets; H, Alumina spacer; J, Thermocouples.

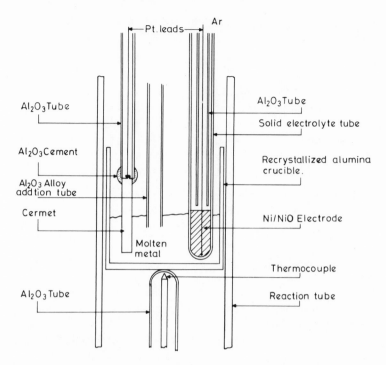

FIG. 6.7. Cell assembly for determination of oxygen in molten metals (Reference 40).

2.7. Determination of Oxygen in Liquid Steel

It has already been mentioned that serious difficulties were encountered in trying to extend the application of solid electrolytes to the determination of oxygen in molten steel. Sustained efforts were made mainly because of the commercial interest involved in such measurements. Some of the noteworthy investigations over the last decade which led to the perfection of a probe for measurements in liquid steel are those by Fischer *et al.* in Germany,[49] Fitterer,[50-52] and research scientists belonging to U.S. Steel Corp.,[53-57] Matsushita, Goto, Kawai, and others in Japan,[58,59] Swinkels and co-workers in Australia,[60,61] and Yavoiskii and others in the U.S.S.R.[62-64] Brahma Deo and Tare[23] have reviewed these efforts.

Fischer and co-workers[49] as well as Matsushita and Goto[58] and Yavoiskii *et al.*[62,63] made attempts to employ solid-electrolyte tubes in a manner similar to the studies on low-temperature melts with a view to continuously measuring the activity of oxygen in molten steel. However, of late, solid-electrolyte tablets fused to silica or alumina tubes have become popular. These are not suitable for use over any extended period and are designed as disposable after one use. The reasons for popularity of such disposable probes are as follows.

(a) Normally in steelmaking the need to determine oxygen arises only once or twice in a heat.

(b) A solid-electrolyte tablet fused to silica or alumina is much cheaper than a solid electrolyte tube and hence is expendable. Also, if properly designed, resistance to thermal shock can be increased to a large extent.

(c) Expendable immersion thermocouples have been in use and an oxygen probe coupled with them would enable measurement of temperature and oxygen content of the steel bath simultaneously within a few seconds.

A successful oxygen probe should achieve the ends listed in Section 2.6.1 and minimize the errors outlined in Section 2.6.2. Some of the difficulties and sources of errors can now be examined in greater detail.

2.7.1. Solid-Electrolyte Material

Various high-temperature oxides were investigated as to their suitability as solid-electrolyte material at steelmaking temperature. Baker and West[45] tested a variety of materials, viz., ZrO_2–CaO, ZrO_2–MgO, Al_2O_3, MgO, ThO_2–Y_2O_3, etc. Electronic conductivity was observed for all of them beyond a certain temperature and p_{O_2} combination and was significant at the steelmaking temperatures. Fruehan *et al*,[54] observed that CaO–ZrO_2 electrolyte can be used up to 10 ppm and yttria-doped thoria up to 1 ppm oxygen content in molten steel. This aspect has been reviewed in Chapter 1. The reason why materials like MgO,[58,65] mullite,[49,116] and Al_2O_3[62,63] have found a measure of acceptance is because at such a high temperature, these exhibit appreciable ionic conductivity. Fischer and co-workers[49] found mullite to yield theoretically expected emf even down to 0.001% oxygen in iron. Although MgO also yielded a linear relationship between emf and $\log[\%O]$,[65] it was found to exhibit irreversibility. In these materials, electronic conductivity is present to a large extent and Etsell and Flengas[66] contend that results obtained from such studies could at best be only qualitative. The present status is that calcia-stabilized zirconia has been found to be most acceptable. Richards *et al.*[61] emphasize that even in calcia-stabilized zirconia, electronic conductivity cannot be prevented under steelmaking conditions. Swinkels has proposed a rapid method for the experimental determination of electronic conductivity limit,[67] using the approach originally proposed by Schmalzried.[68]

According to Schmalzried, when $t_{O^{2-}} \approx 1$,

$$E = (RT/F) \ln[p_{\ominus}^{1/4} + p_{O_2}^{1/4}(2)/p_{\ominus}^{1/4} + p_{O_2}^{1/4}(1)] \qquad (6.25)$$

where (1) and (2) refer to the two electrodes, and $p_{\ominus}$ is the oxygen partial pressure at which the electronic and ionic conductivities are equal. Measurements of $p_\ominus$ for stabilized zirconia by various investigators show a wide range of values even differing by several orders of magnitude (Fig. 6.8).[67] This suggests that different investigators perhaps studied materials differing in chemical and/or physical properties.

If $p_\ominus \ll p_1$ or p_2, then a knowledge of $p_\ominus$ is not required. However, Fig. 6.8 indicates that this may not be the case at 1600°C. Noting that $p_\ominus$ may vary widely from material to material, it is advisable to determine $p_\ominus$ for each batch of electrolyte. Swinkels[67] proposed a simple method of determination of $p_\ominus$. The sample is placed between a liquid metal (e.g., silver) and a Pt/air electrode. Current is passed to remove oxygen from the liquid metal until the oxygen activity in the liquid metal approaches zero. Then $p_{O_2}(2) \gg p_\ominus$ and $p_{O_2}(1) \ll p_\ominus$. Under these conditions,

$$E = (RT/4F) \ln[p_{O_2}(2)/p_\ominus] \tag{6.26}$$

and can be measured by momentarily stopping the current employed in the transfer of oxygen. Once $p_\ominus$ is known in this manner Eq. (6.25) can be employed to calculate oxygen content in steel.

Another significant factor that should be controlled is the electrolyte porosity, which would cause oxygen permeation from one electrode to the other resulting in erroneous emf. Further, entrapped metal in pores would lead to enhanced electronic conductivity.[23,49,51,61] Scaife et al.[68a] have conducted further investigations into properties of ZrO_2–CaO, ZrO_2–Y_2O_3, and ZrO_2–Sc_2O_3 electrolytes. They contend that for accuracy of $\pm 5\%$, the properties of the stabilized zirconia electrolytes should be known so that

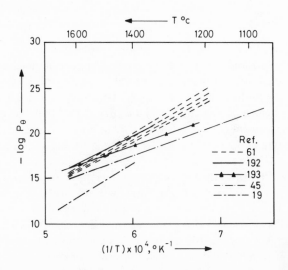

FIG. 6.8. Electronic conductivity limits for calcia-stabilized zirconia electrolytes.

appropriate allowances may be made in calculating the oxygen activity. Experimentally determined Seebeck coefficients agreed with those theoretically estimated at 1600°C. It was demonstrated that corrections for permeability and thermal emf could be made very low by proper choice of reference electrode and probe geometry. However, corrections to data are required to take into account the influence of electronic conductivity of the electrolyte.

2.7.2. Reference Electrode

Various reference electrodes have been employed such as oxygen,[50] air,[49] CO_2,[51] a gas mixture arising out of $(COOH)_2 \cdot 2H_2O$ decomposition,[61] graphite,[58] $Fe-Fe_xO$,[50] $Mo-MoO_2$,[49] $Cr-Cr_2O_3$,[54-57,60,61] etc. It is preferable that oxygen potential of the reference electrode be close to that prevailing in molten steel ($p_{O_2} = 10^{-10}-10^{-13}$ atm) for Fe-C-O melts at 1600°C because it cuts down the electronic conductivity and leakage of oxygen. Gas references have two disadvantages: (a) portability of the device is lost and (b) gas flow rate is critical since a slow flow changes gas composition and a rapid flow cools down the reference electrode and leads to appreciable thermoelectric emf,[60] since the thermoelectric coefficient of $CaO-ZrO_2$ is about 0.5 mV/°C. Fitterer[51] claims to have obtained good emf traces with CO_2 reference. But his results are at best qualitative only.

Schwerdtfeger,[53] Fruehan et al.,[54-57] and Richards et al.[60,61] have found $Cr-Cr_2O_3$ to be good ($p_{O_2} \approx 10^{-13}$ atm at 1600°C) for use as a reference electrode.

2.7.3. Probe Design

The criteria to be employed for a good probe design have already been mentioned. Accordingly, probes developed combined thermocouple and emf probes in a single unit, in which calcia-stabilized zirconia electrolyte tablets were fused onto a silica or alumina tube. The need for accurate temperature measurement very near the emf probe lies in the fact that 5°C error in temperature measurement can lead to 4% error in oxygen concentration. Sealing of the electrolyte tablet to a silica tube has been perfected. Sealing according to Fruehan et al.,[54] was carried out as follows. The tablet was placed inside a matching silica tube and the tube was made to collapse around the electrolyte tablet by heating in an oxyhydrogen flame without directly exposing the electrolyte to the flame. The tube was then cooled to cherry-red heat and quenched in water. Earlier, Fitterer[51] also perfected the silica–electrolyte seal and found not even a single failure in 100 trials. Probes with electrolyte tablets fused to silica or alumina tubes had improved thermal shock resistance, contributed less to thermal emf, and had rapid emf response.[51,54-57,59-61] Although silica

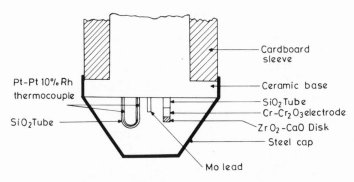

FIG. 6.9. Consumable oxygen probe with Cr–Cr$_2$O$_3$ as the reference electrode (adapted from Reference 55).

has better thermal shock resistance than Al$_2$O$_3$, it forms a low-melting Pt–Si alloy under reducing conditions. Hence probes have also been developed with an Al$_2$O$_3$ tube and a calcia-stabilized zirconia tablet fused together.[60,61] Den Hartog and Slangen[68b] found a reaction between the zirconia electrolyte and steel containing a high concentration of oxygen. Electron probe micro-analysis revealed penetration of FeO into the electrolyte. Again, when the oxygen concentration in the steel was low, the quartz tube reacted with it, presumably leading to dissolution of SiO$_2$ in steel. These effects caused erroneous readings. Vigorous shaking of the probe during use reduced the error.

Various leads for immersion in liquid steel have been employed such as refractory-coated mild steel,[61] iron rod,[68] carbon electrode,[50] and molyb-denum rod.[54,55] Figures 6.9 and 6.10 show some typical probes. It can be seen from these figures that the electrolyte is protected from slag attack by a steel cap and the lance used for connection purposes is protected by a cardboard sleeve. The size of the probe should be as small as can be manufactured.

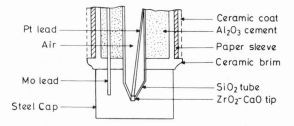

FIG. 6.10. Consumable oxygen probe with air as the reference electrode (Reference 59).

2.7.4. Performance and Applications

As a result of concerted efforts, the probe has attained a high degree of perfection. Swinkels *et al.* have discussed the various sources of errors and how they have been minimized.[60] As shown in Fig. 6.11, study by the device yielded, for $\underline{C} + \underline{O} = CO(g)$ equilibrium, the following result[61]

$$[a_O]_{probe} = 0.98[a_O]_{equilibrium}^{1.01} \qquad (6.27)$$

where a_O is oxygen activity, which is in good agreement with the previous workers. Large-scale plant trials showed a reliability of 85%–95% and a high confidence level of the measured values, that is, within 10% error. The meter responded within 5 sec when the emf and temperature stabilized.[51,55,60] In laboratory trials with Fe–O, Fe–Cr–O, Fe–Al–O, Fe–C–O, and other melts corresponding to steelmaking and deoxidation conditions, satisfactory and reliable results were obtained.[23,56,60,61] A major problem faced in these trials was with standard samples used for the calibration purposes. Results based on vacuum fusion analysis tended to be inaccurate due to oxide inclusions, more so when trials were undertaken under plant conditions. From this point of view, since earlier Fe–C–O equilibrium studies were good, those samples which were free from inclusions were found to be good specimens for calibration.[23]

Measurements in plants for various steelmaking practices with this device have yielded valuable and interesting information, viz., carbon–oxygen

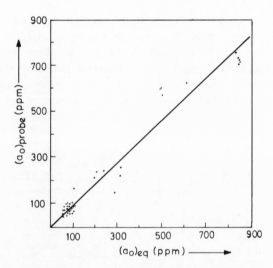

FIG. 6.11. Oxygen activity as measured by the probe vs. oxygen activity from the C–O equilibrium in an Fe–C–O melt (Reference 61).

equilibrium is far more closely approached than has so far been determined.[60] A commercial probe has been developed by CNRM and Hogoven and marketed by Electro-Nite in Belgium.[105] Relationship between carbon and oxygen in a large basic oxygen vessel has been established. Again they found that C–O equilibrium is achieved closely. Therefore the oxygen probe may be used for instantaneous carbon determination. The use of these probes in the steel industry and their possible applications for steelmaking control have been discussed by Shanahan.[68c]

3. Oxygen Transfer to and from Materials

3.1. Introduction

In recent years several kinetic measurements involving oxygen transfer through solid electrolytes have been reported in the literature. The principles of some of these applications have been reviewed by Rickert,[69,70] Goto and Matsushita,[71] Rapp and Shores,[72] and Tare and Ghude.[73] The two main principles are the following.

(a) The emf of a solid electrolyte galvanic cell gives thermodynamic information concerning chemical potentials or activities at the electrodes provided conduction is completely ionic. Therefore the course of a reaction at an electrode may be followed by recording the emf as a function of time if the instantaneous value is assumed to be steady at all stages of reaction.

(b) The current through a cell such as

$$p'_{O_2}, Pt/ZrO_2 + CaO/Pt, p''_{O_2} \qquad [6.3]$$

is a measure of the rate at which oxygen is passed from one side of the cell to the other. Since only oxygen ions can pass through the electrolyte when the electrical circuit through the electron-conducting leads is closed, it also follows that oxygen transfer across the solid electrolyte may be controlled by controlling the current provided the current does not exceed the limiting current for the transport processes.

While the first principle is useful from the point of view of fundamental studies regarding kinetics of certain reactions, the second principle not only indicates useful fundamental studies but also the potentially significant industrial application of a solid-electrolyte cell as a device for controlled oxygen transfer to and from gaseous or liquid metallic media. The oxygen flux (J) from one electrode to the other is related to the current density (i) as follows:

$$i = 2FJ \qquad (6.28)$$

where F is the Faraday constant. A similar equation holds good for a large number of nonoxide solid electrolytes as well. For example, consider the following:

$$Pt/Ag/AgI/Ag_2S/Pt \quad (200-450°C) \qquad [6.4]$$

Silver ion is the current-carrying species in AgI. In the kinetic applications, the electric current passing through the cell is a measure of the rate of addition or removal of silver from the silver sulfide. This principle has been used for measuring the rate of silver loss from the sulfide phase.[106] A review of coulometry using nonoxide solid electrolytes is available elsewhere.[107] Some observations on coulometric titrations involving oxygen have been made by Tare and Ghude,[73] Rickert and Steiner,[74] and Rapp.[75]

3.2. Oxygen Pump and Probe

Heyne[76] employed calcia-stabilized zirconia set up in order to remove oxygen from a closed system. Yuan and Kröger[77] made the first attempt to remove oxygen from nitrogen gas by employing a calcia-stabilized zirconia tube. Their setup is shown in Fig. 6.12a. When electrode I is made cathodic and electrode II anodic, and a current is passed, oxygen can be transferred from the nitrogen stream into the air. This is the pumping action. To what extent oxygen was being removed was monitored by measuring the emf between points I and IV. Therefore the electrode I was common to both pump and probe. This is the so-called "three-electrode arrangement." Agarwal et al.[78] preferred a "four-electrode arrangement" with complete independence of the pump circuit and measuring circuit, as shown in Fig. 6.12b. They have demonstrated this to be a better arrangement. As shown in Fig. 6.13, shorting of the inner electrodes gave a larger value of open-circuit voltage ($E_{\text{o.c.}}$) in the measuring circuit. Similar behavior was observed also by Choudhary et al.[79] This seems to be one of the reasons why Yuan and Kröger[77] reported attainment of an incredibly low p_{O_2} in the inert gas stream (10^{-38} atm at 530°C). Completely separate pump and probe were also employed by Kleitz et al.[117]

Because of pumping of oxygen, a concentration gradient would be set up in and around the porous platinum paste. If this concentration gradient persists when the gas arrives at the measuring electrodes, the value of p_{O_2} calculated from open-circuit voltage would not represent the average value of p_{O_2} in the exit gas. For their conditions, Yuan and Kröger[77] made diffusion calculations and ascertained that good mixing could be presumed. Agarwal et al.[78] experimentally confirmed this in their setup too. Therefore it appears that gas mixing and homogenization are fairly rapid and a probe situated somewhat downstream gives the average p_{O_2} of the exit gas stream.

Figure 6.13 shows the current–voltage relationship in the pump, viz.,

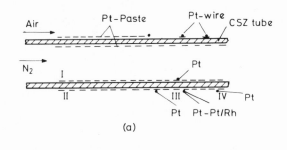

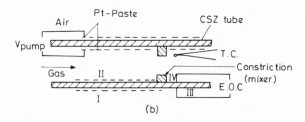

FIG. 6.12. Pump designs. (a) Three-electrode pump assembly (Reference 77); (b) four-electrode pump assembly (Reference 78).

I_{pump} vs. V_{pump}. It also shows the extent of oxygen transfer. This is the generally observed trend. The rapid increase of current beyond 1200 mV is accompanied by the onset of electronic conduction in the calcia-stabilized zirconia electrolyte. This would explain why the extent of oxygen removal was not as large in ZrO_2 + 3–4 wt % CaO as it was in ZrO_2 + $7\frac{1}{2}$ wt % CaO, because the former exhibits more electronic conduction. Yuan and Kröger[77] employed V_{pump} up to 9 V. They found an irreversible change at $V_{pump} > 4$ V. This irreversibility resulted from an increase of current with time at fixed V_{pump} above 4 V. This was explained by decomposition of ZrO_2 and CaO, which is likely since the decomposition voltages of the above are 2.52 and 2.84 V, respectively, at 530°C. In addition, polarization and consequent changes in oxygen/metal ratio in the electrolyte cannot be ruled out.

On the basis of the above arguments, at high value of V_{pump}, Yuan and Kröger[77] postulated the existence of ionically conductive zirconia in the outer and central sections of the tube, electronically and ionically conducting zirconia below the inner surface, and zirconium (and calcium) metal at the inner surface. These complications preclude use of a large value of V_{pump}.

Introduction of buffering gases such as CO_2 or H_2 or any other reactive gas is expected to alter the characteristics of the pump as well as attainable p_{O_2} in the exit gas considerably, and it has been found to be so.[77] This results from factors such as thermodynamic equilibria in gas mixture, kinetics of electrode reactions, etc.

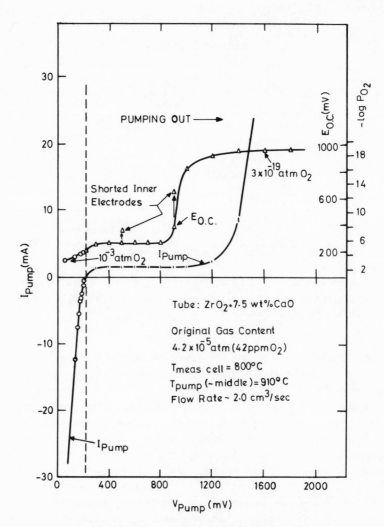

FIG. 6.13. Pump characteristics (Reference 78).

Alcock and Zador[194] also employed a "four-electrode" setup. The pump and probe were similar and mounted at the two ends of the high-temperature chamber. They deoxidized nitrogen gas containing controlled amounts of free oxygen and water or carbon dioxide at applied voltages ranging from 0.5–3.0 V. The current–voltage relationships observed were similar to those of Yuan and Kröger[77] and Agarwal et al.[78] The electrolyzing current-applied voltage relationships at low applied voltage indicated that the circuit resistance was mainly determined by the initial free-oxygen content of the gas mixture.

Using these pumps and feedback control circuits, it has been possible to control the p_{O_2} in a flowing gas stream under laboratory conditions.[78,118] However, so far these attempts are restricted to small flows and laboratory conditions only. Micheli[195] has reported the use of such a pump for hot pressing of ferrites and other oxide ceramics under controlled oxygen partial pressures and obtained better material, as well as control of oxygen partial pressure during industrial production of ferrites.[195a]

Sandler[80] has employed an oxygen pump to measure the extent of adsorption of oxygen by various powders under Brunauer–Emmett–Teller (BET) conditions. The current through the pump was proportional to p_{O_2}. By integrating current–time data, the total amount of oxygen adsorbed was determined. Reasonable agreement was obtained with conventional BET data on a variety of materials. Spacil and Tedmon[103] have fabricated a high-temperature solid electrolyte cell in order to electrolytically dissociate water vapor to manufacture pure hydrogen. They considered thermodynamic and kinetic criteria for design of such a cell, and came to the conclusion that series connection of individual cells could reduce the electrical energy required for dissociation substantially, but the current density could not be high due to ohmic resistances. Use of discontinuous electrolyte segments and depolarized anodes gave the best overall performance.

3.3. Coulometric Control of Oxygen in Liquid Metals

It should be noted that although several high-temperature electro-chemical investigations of the transport of oxygen out of or into quiescent liquid-metal–oxygen alloys have been made, these studies have been mainly aimed at determining the diffusivities and solubilities of oxygen. Sufficient emphasis on the applied aspect of the subject still seems to be lacking.

A recent study by Oberg et al.[81] deals directly with electrochemical deoxidation of induction-stirred copper melts. The electrochemical cell used may be described as

$$\begin{array}{c|c|c} \begin{array}{c} \text{Cu(l)} \\ + \\ \text{O (dissolved)} \end{array} & \begin{array}{c} \text{ZrO}_2\text{(CaO)} \\ \text{solid electrolyte} \end{array} & \text{Pt–air } (p'_{O_2} = 0.21 \text{ atm}) \end{array} \qquad [6.5]$$

For deoxidation of copper, current is drawn from the copper electrode (i.e., electrons flow from the right to the left in the external circuit). The cell arrangement is schematically shown in Fig. 6.14.

The transport of oxygen from the metal–oxygen alloy to the reference gas during the deoxidation involves a series of kinetic steps.

It was found that at high oxygen concentrations in metal, transport of oxygen in the electrolyte essentially limited the oxygen removal, whereas in the lower oxygen concentration range, transport of oxygen from the metal

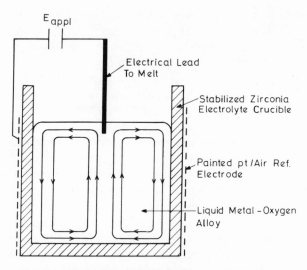

FIG. 6.14. Electrochemical deoxidation cell (Reference 81).

bulk to the metal/electrolyte interface was the rate-controlling step. The latter was so because the concentration gradient across the boundary layer is quite shallow.

It should be noted that the applied voltage for deoxidation cannot be exceeded beyond certain upper limits. With increasing voltage, and therefore current, the interfacial oxygen concentration approaches zero. With further increase in voltage, the electrolyte would start to become oxygen deficient and may dissociate ultimately. The maximum possible deoxidation rate, corresponding to control by limiting diffusion in the electrolyte, can be calculated for any melt–oxygen–crucible system using equations given by Oberg et al.[81] In engineering practice, it would be advisable to start with a high applied voltage and then progressively reduce it so as to always maintain the interfacial oxygen concentration safely above zero..

3.4. Other Applications

(a) *Oxygen Removal from Fused Salts.* It is known that the corrosiveness of fused salts such as alkali halides is sensitive to the oxygen content of the melt.[82–84] It may thus be advisable to deoxidize fused salts using an electrochemical deoxidation procedure, especially where corrosion of metals and alloys is relevant. No report along these lines has yet appeared in the literature.

(b) *Oxidation of Impurities in Metals and Alloys.* Many metals may be purified by bubbling oxygen or chlorine through the liquid metal. The impurities react preferentially and are removed either as a volatile product or a separate slag phase. For selective oxidation, mild oxidizing agents such as

NaNO$_3$ may be preferred to oxygen or air. Similarly, instead of chlorine gas, a chloride salt, especially those of the parent metal itself, has sometimes been used.

4. Energy Conversion

4.1. Introduction

Other forms of energy can be converted to electrical energy either directly or indirectly. A large number of books have appeared which underline the principles of energy conversion in general and direct conversion in particular.[85-88]

The devices employed in direct energy conversion can be classified under electrochemical, thermoelectric, thermoionic, photovoltaic, and magneto-hydrodynamic methods. Amongst the spectrum of these devices solid oxide electrolytes have found use in electrochemical, thermoelectric, and magneto-hydrodynamic generators.[66] The principles embodied in the fuel cells incorporating solid oxide electrolytes and the present state of development are briefly presented. With regard to use of solid electrolytes for other devices, the applicability appears to be very limited,[66] and so they will not be discussed here. However, a recent effort to construct a thermoelectric generator based on β-alumina solid electrolyte is worth mentioning.[66a] In this laboratory device, liquid sodium was circulated from one electrode to the other by heating and consequent thermal convection. The difference in sodium pressure between the two electrodes was the source of power. The device was operated successfully and it seems to have some promise.

4.2. Fuel Cells

The fuel cell is a primary electrochemical device where a fuel is fed to one electrode (anode) and oxidant to the other electrode (cathode). Generally the oxidant is oxygen or air. The chemical energy derived from the oxidation of fuel is continuously converted to electrical energy.

The fuel and the oxidant are separated by an electrolyte, which is predominantly oxygen ion conducting in the case of a solid oxide electrolyte fuel cell. The difference in the chemical potentials between the two electrodes manifests itself in the form of an electromotive force. When the electrodes are connected through a load, a current passes in the external circuit and the fuel is oxidized.

4.2.1. Principles Governing a Solid Oxide Fuel Cell

Figure 6.15 shows a schematic representation of the electrodes and the electrolyte of a fuel cell. The function of the electrolyte is to conduct oxygen

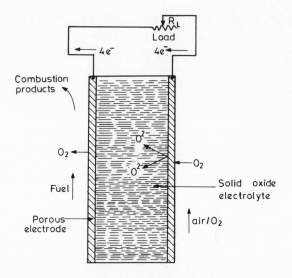

FIG. 6.15. Schematic diagram of the oxide–ion fuel cell.

ions from cathode to anode. The fuel is oxidized at the anode and oxygen or air is fed at the cathode.

The following reaction takes place in a fuel cell:

$$\text{fuel} + \text{oxygen} \rightarrow \text{combustion products} \qquad (6.29)$$

Since the operation is isothermal, the free-energy change accompanying this reaction is equal to the maximum work obtainable, W_e:

$$W_e = \Delta G = -nFE = \Delta H - T\Delta S \qquad (6.30)$$

where the symbols have their usual meaning.

The maximum thermal efficiency of a fuel cell, E_T, can be defined as

$$E_T = \Delta G/\Delta H = -nFE/\Delta H \qquad (6.31)$$

Sometimes it is more useful to define another efficiency term E_T' with reference to fuel cells working at high temperatures. This is defined as

$$E_T' = \Delta G/\Delta H_{298} = -nFE/\Delta H_{298} \qquad (6.32)$$

This definition takes into account the sensible heat of the combustion products.

There are, as mentioned earlier, several other methods of direct energy conversion. These are thermoelectric, thermionic, photovoltaic, and magneto-hydrodynamic methods. Of these, all the methods other than photovoltaic energy converters are limited by the Carnot efficiency. That is to say, the theoretical maximum efficiency achievable is $1 - (T_2/T_1)$, where T_1 and T_2 are, respectively, the temperature at which the heat is taken in (source) and the temperature at which excess heat is rejected (sink). On the other hand, since a fuel cell works isothermally, its efficiency bears no relationship to the Carnot efficiency. Bockris and Srinivasan[89] have compiled the data on fuel cells available until 1969. Recently, Takahashi[108] has reviewed literature pertinent to solid-electrolyte fuel cells.

4.2.2. Materials

(a) *Fuels.* Fuels for the high-temperature solid-electrolyte cells are gases like hydrogen, carbon monoxide, gas mixtures containing CH_4, CO, CO_2, H_2, H_2O, and N_2, as well as solid carbon. An additional reformer unit can also be incorporated which can produce producer gas, water gas, and cracked hydrocarbons for supply to the fuel cell.

Gas mixtures containing CO, CO_2, etc., often give rise to a carbon deposition problem. Carbon may tend to choke the fuel passage line and also deposit in the anode so that the electrochemical reaction occurring at the anode tends to be retarded. Thermodynamic calculations corresponding to equilibrium reactions between gases composed of C, H, and O can predict the gas composition corresponding to the carbon deposition range.[91,92]

(b) *Electrolytes.* The principal components of the fuel cell are the electrolyte and the two electronic leads for the anode and the cathode. Since this type of fuel cell should be so designed as to operate at high temperature under strongly reducing and oxidizing conditions for long durations of time, each of these components will have to satisfy some minimum requirements. Further, the cell assembly should be such as to incorporate features which would ensure trouble-free invariant source of power to be produced under stringent economic restrictions.

The important requirements that the solid oxide electrolyte should fulfil before it can be accepted as a component of the fuel cell are the following: (a) It should have high anionic conductivity at the operating temperature when operated under highly reducing and oxidizing atmospheres. (b) It should be impervious to gases. Permeability to gases would defeat the main purpose, as oxidation resulting from diffusing gases will not contribute to the generation of electricity. It would simply result in waste heat with low efficiency of fuel utilization. (c) When subjected to long durations of functioning the electrolytic properties should remain unaltered, for only then is the derived electrical energy amenable to better control. Solid electrolytes that exhibit electronic conduction, generally speaking, are not to be preferred, because as a result of electronic conductivity there is a loss in energy efficiency. On the other hand, electrolytes having higher total conductivity but also possessing a small electronic component compared to a purely ionic conductor may be acceptable because of the attendant advantages,[93] which are (i) a reduction in the operating temperature and (ii) an improved energy efficiency due to higher conductivity.[108] Zirconia-based solid electrolytes are considered to be more suitable primarily because they exhibit pure anionic conductivity over a wide range of oxygen partial pressure ($1-10^{-20}$ atm), thus satisfying the criterion of invariance. Other electrolytes studied include ThO_2-CeO, $ThO_2-Y_2O_3$, $ThO_2-La_2O_3$, $CeO_2-La_2O_3$, $CeO_2-ThO_2-La_2O_3$, and substituted perovskites.[66,108] Ternary oxides based on zirconia have also been studied and patented.[94] Ceria-based electrolytes have higher total conductivities than zirconia-based electrolytes. According to Takahashi,[108] $(CeO_2)_{0.6}(LaO_{1.5})_{0.4}$ has better characteristics in terms of power output than zirconia-based electrolytes. Partial reduction of ceria occurs under operating conditions of the fuel cell.[66] Partial reduction at the anode–electrolyte interface results in the loss of invariance of the electrolyte. It appears, therefore, that these electrolytes do not offer any additional advantage over zirconia-based electrolytes.

Recently Takahashi and Inahara[109] have reported that in some of the substituted perovskites the electrochemical properties are comparable to those of zirconia-based electrolytes. The transport number of oxygen in the case of $CaTi_{0.7}Al_{0.3}O_{3-x}$ was found to be greater than 0.9 and oxygen ion conductivity was greater than that of the zirconia-based electrolyte. Incorporation of $(CeO_2)_{0.6}(LaO_{1.5})_{0.4}$ as anode improved the anode characteristics during the discharge of the fuel cell.[109] In the experimental fuel cells that have been fabricated so far, the electrolytes are $CaO-ZrO_2$ or $Y_2O_3-ZrO_2$. Since the electrolyte to be used should not deteriorate with time easily, it is essential that dopant concentration should be such as to lessen the ordering during prolonged usage. $(ZrO_2)_{0.85}(CaO)_{0.15}$ also undergoes ordering at temperatures around 1000°C and suffers some loss in conductivity.[66]

It is evident that the smaller the thickness of the electrolyte, the lower would be the iR loss in the cell voltage. Thin wafers of the electrolytes were obtained by plasma arc spraying,[95] flame spraying,[96] and reactive sintering[97] (for further details, see Chapter 7). Reactive sintering appears to be a very potential method because of the low temperature needed for firing and better control over composition that is possible. The lower bound to the thickness of the electrolyte is, however, determined by the minimum mechanical strength that it should have during assembling and operation.

(c) *Electrodes.* Electrodes to be used in fuel cells must satisfy four basic needs: chemical, electrochemical, mechanical, and economical.

In the cathode, the basic electrochemical reaction that occurs is $\frac{1}{2}O_2(g) + 2e' \to O^{2-}$. It is obvious that the cathode material should be stable under strongly oxidizing conditions and must have phase stability in the temperature range of operation. Further, its vapor pressure should be low so as to retain an invariant electrode–electrolyte configuration. The electrochemical requirement is that it should be electronically conducting and should enable the reaction

$$O(ads) + 2e' \to O^{2-} \tag{6.33}$$

Table 6.2. Performance Data on Fuel Cell Cathodes Made from Electronically Conducting Oxides[98]

Material	Remarks
Pt	Properties very good, but because of highly uneconomical conditions and loss in vapor (large surface to volume ratio), useful only for short term purposes.
Liquid or solid Ag	High loss in vapor; otherwise good because of high diffusion coefficient for oxygen.
SnO_2 (doped)	Satisfactory if applied in thin coating.
In_2O_3 (doped)	Satisfactory if applied in thin coating. Difficulty arises due to cracking.
$ZnO(0.95)–ZrO_2(0.05)$ $ZnO(0.97)–Al_2O_3(0.03)$	Reported satisfactory at relatively low operating temperature. At higher temperatures, reacted with the electrolyte (800°C).
Ce and Pr oxides mixed with ZrO_2	Use of current-carrying grids (Pt) prohibits long-term use.
$PrCoO_3$	Satisfactory performance; thermal expansion mismatch.
$Li_{0.1}Ni_{0.9}O$	Li loss leading to loss in conductivity.
$Sr_{0.1}La_{0.9}CoO_3$	Reacts with the electrolyte; spalling and cracking during cooling cycles.
$LaCoO_3$	Chemical reaction with ZrO_2 prevents long-term application.
$PrO_{1.83}$	Thermal expansion mismatch.

to proceed at the maximum possible rate (almost reversible). The mechanical requirement is that the electrode must maintain the desired porosity for oxygen permeability. Further, physical and electrical contact with the electrolyte and lead must be maintained at all times during the cell operation. Finally, it should be attractive economically. Tedmon *et al.*,[98] who reviewed the literature in this context, presented a summary (Table 6.2). Recently Ag–Pd alloy with 80% Ag and 20% Pd was tried as cathode material.[110,111] Use of this alloy resulted in the reduction of Ag losses. A fully satisfactory cathode material is yet to be found. Judged by the present standards, SnO_2 and $PrCoO_3$ seem to have an edge over other materials.[98]

The requirements of an ideal anode material differ from that of the cathode material with respect to the expected chemical and electrochemical characteristics. These materials should withstand the severe reducing conditions ($p_{O_2} \simeq 10^{-20}$ atm) at high temperatures and should not have tendencies to form carbides or nitrides. Further, they should be able to promote the electrochemical reaction, viz.

$$\text{fuel} + O^{2-} \rightarrow \text{products of oxidation} + 2e' \qquad (6.34)$$

With these objectives, a number of materials were tested. These are C, Ni, TiO_{2-x}, mixed oxides such as $U(Ce, Pr)O_x$ mixed with ZrO_2, CeO_2–ThO_2, $U(Zr, Y)O_x$, etc. None of these materials was found satisfactory.[66]

First of all, lack of compatibility in thermal expansion characteristics was one stumbling block, added to the complexity of the electrochemical nature of the fuel reaction. Fuels, as was mentioned earlier, can range from pure hydrogen to CO. Therefore a single electrode material cannot be expected to ideally suit all types of fuels used. For instance, it is clear from the literature that even though no significant overpotential was observed when the reaction

$$H_2 + O^{2-} \rightarrow H_2O + 2e' \qquad (6.35)$$

was taking place at the platinum electrode, significant overpotential was present with CO oxidation to CO_2. Thus it is evident that in the case of CO oxidation, resort should be taken to a better electrocatalyst. The fuel should be appropriately treated too.[99] Therefore an alternative to platinum should basically satisfy the condition that the activation overpotential should be lower for it than for platinum. Further, the electrode–electrolyte contact should remain invariant during the operation of the fuel cell.

The use of oxides having mixed conduction as electrodes serves two purposes: firstly, as alternatives to platinum, and secondly, they tend to lower the polarization effect arising at the electrode–electrolyte interface. For a typical anode, Takahashi[108] presents a model as shown in Fig. 6.16 and cites experimental results for ceria-based oxide electrodes. From Fig. 6.16 it is clear that the effective electrode–electrolyte area of contact is much more compared to a purely electronic conductor.

4.2.3. Design of Fuel Cells

A schematic sketch of the fuel cell employed by Archer et al.[99] is shown in Fig. 6.17. Salient features of some experimental fuel cells developed so far have been reported in the literature.[90,99–102]

These cells have been largely successful with 60–70% efficiency. In all these cells, however, platinum functioned as the electrode component, an element whose supply cannot meet the demands of projected fuel cell consumption. As outlined already, attempts have been made to replace platinum. Fuel cells have been constructed with the minimal use of platinum. Of these, those developed by General Electric Company, Brown Boveri Company, and Westinghouse Electric Corporation[111–114] are particularly noteworthy. Detailed performance and design data of these cells are not available. The working temperature range of the cell developed by General Electric Company is 1000–1100°C. The assembly of cells gives 40-W output and 4-A current when short circuited. The open-circuit voltage was 40 V. The cell developed

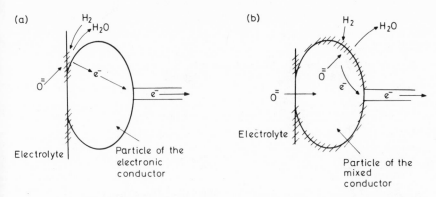

FIG. 6.16. Model of the electrode reaction zone: (a) in the case of the electronic conductor, (b) in the case of the mixed conductor (the hatched parts indicate the reaction zone).

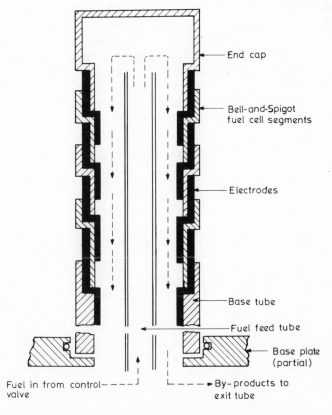

FIG. 6.17. Schematic diagram of the three-cell assembly used by Archer *et al.* (Reference 99).

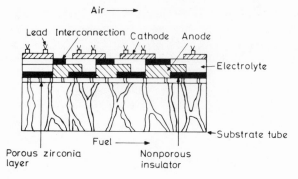

FIG. 6.18. Westinghouse thin-film battery (Reference 108).

by Brown Boveri Company[112] consists of $ZrO_2-Y_2O_3$ electrolyte, with 0.5-mm thickness. The cell yielded a power density of 250 kW/m³ at 900°C. The noteworthy feature of the cell developed by Westinghouse Electric Corporation is that the fabricated electrolytes are in the form of thin films (0.03–0.04 mm thick) of calcia-stabilized zirconia (Fig. 6.18). The thin films were fabricated by reactive sintering[97] and the interconnection of the tubes is made by oxides such as $LaCrO_3$.

Cells working with $CO-CO_2$ mixtures work better when a small amount of H_2O is injected into the stream.[99] H_2O injected thus reacts with CO as follows:

$$CO + H_2O \rightarrow CO_2 + H_2 \tag{6.36}$$

H_2, being a better fuel, reduces the activation overpotential.

4.2.4. Conclusions

The literature review shows clearly that the solid-electrolyte fuel cell is a reality, though at a high cost,[89] as of 1965, due to (a) the fabrication of the materials employed in the fuel cell construction and (b) the use of platinum as the electrode material.

Since then, major developments have taken place in the processing of thin-film electrolytes and incorporation of oxides and Cermets as electrodes. Estimates of Brown Boveri Company predict a possible efficiency of 40% with cheap hydrocarbon fuels and economically favorable power output.[113] Westinghouse Electric Corp. has been working on the development of a 100-kW system based on low Btu gasification of coal (to CO and H_2) and a high-temperature zirconia electrolyte fuel cell,[113] with a rated efficiency of 58%. It may also be noted that if hydrogen becomes available economically from nuclear or solar energy sources, the hydrogen–oxygen fuel cell will have to be reevaluated. At present, the areas of application of solid-electrolyte fuel cells seem to be on-site power generation for space, military, communication, or villages, or central power generation. With the advent of an electrolyte that can be operated at 600–800°C fuel cells may be employed for traction power also.

The possibility of obtaining high efficiency, the minimization of atmospheric pollution, the ruggedness and compactness in design that is achievable, and the absence of moving parts hold a good future for solid-electrolyte fuel cells. However, a lot more development is needed to achieve commercialization.

5. Batteries with β-Alumina Electrolytes

5.1. Sodium–Sulfur Battery

5.1.1. Introduction

The prevalent energy crisis and the rapidly multiplying ecological problems have caused a grave concern throughout the world and a search for cleaner energy sources based on renewable, nonfossil fuels. Even otherwise, there is a need for high-energy batteries, capable of acting as energy banks and as mobile power sources. Widespread resurgence of interest in rechargeable cells has resulted in the development of a number of high-energy battery systems. The sodium–sulfur battery which utilizes β-alumina ceramic as the electrolyte belongs to this class and is described here.

This cell, originally proposed by Kummer and Weber of Ford Motor Company in 1967,[118a–120] makes use of a revolutionary approach to the battery problem, that of using liquid electrodes and a solid electrolyte. Figure 6.19 shows comparison of this cell with that of the common lead–acid battery. The cell

$$\text{sodium (liquid)}/\beta\text{-alumina}/\text{sodium-polysulfides–graphite} \qquad [6.6]$$

operates at temperatures above 300°C and has an open-circuit potential of nearly 2 V. The reactions taking place in the cell are

$$2\text{Na} \rightleftharpoons 2\text{Na}^+ + 2e' \qquad \text{at the anode} \qquad (6.37)$$

$$2\text{Na}^+ + 2e' + x\text{S} \rightleftharpoons \text{Na}_2\text{S}_x \qquad \text{at the cathode} \qquad (6.38)$$

The sulfur-rich polysulfides formed at first react with sodium to give polysulfides richer in sodium:

$$2\text{Na} + (x - 1)\text{Na}_2\text{S}_x \rightleftharpoons x\text{Na}_2\text{S}_{x-1}, \qquad x = 3, 4, 5 \qquad (6.39)$$

The phase diagram of the Na_2S–S system, shown in Fig. 6.20, is proposed by Gupta and Tischer,[121] revising the diagram of Pearson and Robinson.[122]

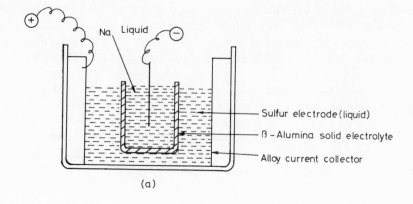

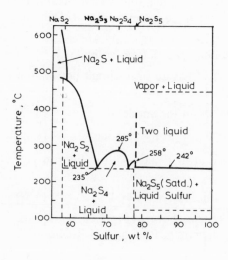

FIG. 6.19. Schematic diagram of a sodium–sulfur cell (a) in comparison with that of a lead–acid cell (b).

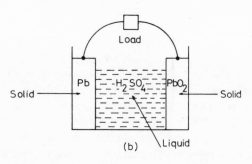

FIG. 6.20. Phase diagram of the Na_2S–S system (Reference 121).

The temperature at which the sulfur–polysulfide melt can be retained in a liquid phase sets up a lower limit on the working temperature. At a temperature of 300°C, a solid begins to separate out at a composition near that of Na_2S_3. At a temperature of 350°C, the composition $Na_2S_{2.7}$ would be the limit. However, in practice, the composition, Na_2S_3, is taken as the 100% discharged state. The melting point of Na_2S_3 (275°C) and the boiling point of sulfur (444°C) set the lower and upper limits to the operating temperature of a sodium–sulfur cell.

The free energy of formation of Na_2S_5 is about 96 kcal/mol[121] at 300°C and the corresponding emf of the couple is 2.08 V. At 100% discharge, i.e., corresponding to the melt composition of Na_2S_3, the emf of the cell falls to 1.76 V.

5.1.2. Inherent Advantages of the System

The heart of the Na/S battery is the β-alumina solid electrolyte. Since β-alumina can be made to conduct a range of species[123–125] (see Chapter 1), there is a choice in the selection of the anode material. Sodium is selected because its melting point is low, it is easily available, and the fabrication of sodium β-alumina assembly is easy. Sulfur is selected because it is inexpensive and plentiful, is in liquid state at 115°C, and the formation energy of polysulfides is quite high, which results in a high theoretical specific energy (energy released per unit mass of the reactants) of the system. Under proper conditions, the reaction between sodium and sulfur is electrochemically reversible. Also, the polysulfides can be retained in a liquid state at a relatively low temperature compared to other salts. These are semiconducting, and thus their accumulation does not affect the cathode reaction. Common materials can be used as containers, because, around 300°C, the reactants and products are not too corrosive. The highest pressure encountered in the system would be the vapor pressure of sulfur, which is about 60 mm of Hg at 300°C. The reactants and products being liquid, the cell reaction is fast, thus allowing high current densities. The reactants can be regenerated in the same physical form during charging. The cell reaction is simple and involves no gaseous products. Side reactions cannot take place in the system, thus ruling out the possibility of self-discharge.

5.1.3. Development of a Practical Battery[126,127]

The first laboratory Na/S cell was constructed using Kovar sealing glass for the assembly.[118a–120] The ceramic used for the electrolyte was MgO-stabilized β''-alumina with a resistivity of 5 ohm cm at 300°C. The ceramic was in the form of a thin-walled tube, to which a sodium reservoir was sealed at the top. Surrounding the tube was the sulfur electrode, consisting of a

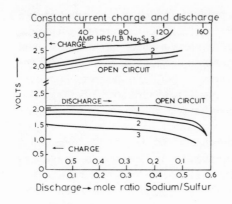

FIG. 6.21. Typical charge–discharge curves for a 0.8-mm-thick walled cell. Curve 1, 170 mA/cm² for 120 min; curve 2, 340 mA/cm² for 60 min; curve 3, 680 mA/cm² for 30 min.

3-mm-thick layer of carbon felt, holding the liquid sulfur by capillary forces. Since sulfur is not an electronic conductor, it was necessary to provide the graphite to act as the current collector. A metallic electrode was used as the lead at the cathode, while a wire dipped into the liquid sodium acted as the anode. Typical cell charge–discharge curves for a cell with a 0.8-mm-thick electrolyte are shown in Fig. 6.21. The voltage–current density relations drawn are essentially straight lines, and show the absence of overvoltages due to slow reaction or slow mass transfer. The results obtained from the first experiments showed the system to be very promising. This promise has not, however, fully materialized so far into commercially available batteries, though the development is in an advanced stage. Work on the systems is continuing in the U.S.A.,[128–130] Great Britain,[131–134] Japan,[135] the U.S.S.R.,[136–137] and France.[138–140]

Both flat-plate as well as tubular batteries have been tested by the British Railways.[133] They have standardized a working temperature of 350°C. The tubular cell design was adopted for initial experiments because the sealing area could be kept to a minimum. A number of single-tube and multiple-tube cells were constructed, and a specific energy of 400 W-hr/kg. was achieved. The flat-plate batteries were designed for high discharge rate. More recent cells have been tested for about 9000 hr covering about 2000 cycles.[140a] The Electricity Council Research Centre of Great Britain has achieved a lifetime of over 5000 hr in continuous charge/discharge cycling tests (4 cycles/day) where the maximum discharge current was restricted to 200 mA/cm² and maximum charge current to 100 mA/cm².[141] Chloride Silent Power Company of U.K. has operated 30 A-hr batteries for 25,000 hr (over 3 years) covering 2000 cycles.[140a]

Stainless steel cells have been fabricated in France.[138,139] The breakdown of the electrolyte has been prevented by controlling the rate of recharge, and by employing a new cell design.[140] Magnesia addition and presence of contaminants appear to lead to reduction of the lifetime of the β-alumina electrolyte.[142,142a] The role of sodium dendrites causing a crack growth in the electrolyte during charging is studied.[142a]

The alternative approach to increase the operating life of the system, namely, lowering the working temperature, has also been tried in the U.S.A.[143] By the addition of aluminum polysulfides to the sodium polysulfide melt, an electrode material with a melting point of 200°C has been achieved. This permits use of Teflon and Araldite for battery construction.

The Ford Motor Company seems to have found a solution to the cycle life problem and has achieved a life of over 1300 hr for over 2400 cycles. In a systematic study of the problems associated with the system, degradation of the electrolyte has been studied in detail and methods to avoid cracks in the ceramic by optimization of the composition and fabrication have been developed.[128]

The development of the Na/S battery has sparked off a search for better ceramic fabrication techniques, as well as for standardizing procedure for screening the same for battery use.[144-148] Investigations into the properties of the polysulfides and the nature of the sulfur electrode are being undertaken.[149,150] The kinetics of the electrode reactions at the graphite–polysulfide electrode are also being investigated.[151]

5.1.4. Comparison with Other Systems[152-157a]

The performance characteristics of some of the tested battery systems are shown in Table 6.3. The sodium/sulfur battery uses a moderately high temperature but gives a very high specific energy and power. The theoretical specific energy of the Na/S system is 762 W-hr/kg (based on Na_2S_3 as 100% discharged state), of which, it is believed, a value of about 330 W-hr/kg can be achieved for practical systems compared to a value of 15 W-hr/kg for the common lead–acid battery. The system gives energy densities comparable to the high energy density systems such as that of Li/Cl_2, which require a higher operating temperature, and is associated with more severe materials problems. The Na/S cells are capable of yielding very high current densities (~ 400 mA/cm^2). The batteries can be recharged at a high efficiency even at very high charging rates.

5.1.5. Limitations

The limitations of the Na/S battery are summarized below.

(i) A high operating temperature is required. In case a battery cools below the operating temperature, it has to be heated above the operating temperature in order to start it again. Hence thermal shield as well as heating coils are needed for a practical battery.

Table 6.3. Estimated Properties of Electrical Energy Storage Cells

Cells	Typical temperature (°C)	Resistivity of electrolyte (ohm cm)	Cell voltage at discharge (V)	Current density (A/cm²)	Specific power (W/kg)	Specific energy (W-hr/kg)	Operating life (cycles)
Lead–acid	40–50	1.53	2.1–1.46	0.010	7–30	4–33	10–400
Nickel–iron	0–40	1.96	1.3–0.75	—	7–40	30–35	100–3000
Nickel–cadmium	40–60	1.96	1.3–0.75	0.01	7–44	35–40	100–2000
Silver–zinc	0–40	1.96	1.55–1.1	0.45	24–150	80–100	100–300
Silver–cadmium	40–60	1.96	1.3–0.8	1–3	20–66	50–60	500–1100
Lithium–chlorine	650	0.17	3.40	1–3	80–150	330	—
Sodium–sulfur	300	5	2.08–1.76	0.7	330	330	2000
Sodium–air	130	3.58	2.3	0.07	90	350	—
Zinc–air	25	—	1.4	—	60	200–300	300–400
Lithium–tellurium	450–480	0.26	1.79–1.67	2–5	280	180	—
Sodium–bismuth	540–580	0.44	0.8–0.44	0.5–1	80	40	500

(ii) The resistance of the presently manufactured electrolyte is high, which limits the current density.

(iii) Ceramic degradation limits the life of the electrolyte in terms of the number of cycles of charge and discharge.

(iv) The durability of the seals and cathodic current collectors under the working conditions are not good.

(v) Ceramics are sensitive to mechanical and thermal shocks as well as to thermal cycling.

(vi) Better materials of construction which are cheap and corrosion resistant must be found.

(vii) The battery is not at present reliable and safe. The breakdown of the electrolyte can result in the mixing of the reactants, causing an explosion.

5.1.6. Applications

Large Na/S battery units are under serious consideration and intense development for two major applications: load leveling and electric vehicles.

For load leveling and peaking purposes,[157a,b,c] the following characteristics are sought in a Na/S battery to compete with the cost of delivered energy from pumped hydroelectric storage: First cost $10–20/kW-hr, life > 150 cycles, efficiency > 85%, operational and maintenance expense < 0.5 mill/kW-hr. Based on present experience, no outstanding fundamental scientific or technical problems remain and the above characteristics appear achievable. Megawatt bulk energy storage units are currently under development. Some of the practical problems being faced are a decline in cell capacity with cycling with a consequent increase in cell cost, requirement of increasing cell life by a factor of about 10, and an increase in resistivity with time. Overcharge damages the ceramic electrolyte, while overdischarge leads to irreversible formation of insoluble products. These aspects are being examined with a view to demonstrate load-leveling Na/S batteries in the early 1980's.

Electric vehicles have assumed importance due to the increasing prices of petroleum fuels, which are of the nonreplaceable type, and due to objectionable conventional vehicle emissions. Electric vehicles are particularly advantageous for driving modes involving many start/stop operations. These include urban mass transit, delivery vans or forklift trucks, and taxicabs. The characteristics of batteries for this purpose are presented by Gay et al.[157d] Somewhat more distant applications are in automobiles and railroad cars.

5.2. Other Systems Using β-Alumina

The use of β-alumina in primary cells using sodium–mercury amalgam as the anode and water, air, or halogens as the cathode material has been tried.[158,159] The cells operate at temperatures around the ambient and have been proved to give high cell voltage and energy densities. Long operating lifetimes (up to 10 years) are expected for the Na/Br_2 cell, making it ideally suitable for heart pacer applications. Iron-doped sodium β-alumina was also tried as cathodes in galvanic cells based on β-alumina.[159a] β-alumina has a potential for direct energy conversion in a hybrid system coupled to a sodium-cooled nuclear reactor.

6. Solid State Ionics

6.1. General

The term "solid state ionics" has been in use for a long time, though the practical applications were limited to the laboratories until recent years. However, the study of solid state ionics has opened up new vistas in solid state chemistry, energy conversion devices, and the development of a host of "ionic" devices.

Compared to the low ionic conductivity $[< 10^{-10}$ (ohm cm)$^{-1}]$ at room temperature of most ionic solids, the so-called "superionic conductors" have a high value of conductivity $[0.1$ (ohm cm)$^{-1}]$ at room temperature (comparable to aqueous solutions of ionic compounds) and these can be used as electrolytes for galvanic cells as well as for such new devices which could not be conceived ten years ago.

Several review articles have appeared recently on solid electrolyte batteries[152,160-164d] and solid state ionics in general.[165-167] The solid-electrolyte Na/S battery has already been discussed. Other ionic devices which are complete solid state systems will be discussed now.

6.2. Energy Conversion Devices

6.2.1. Introduction

Work done before 1950 was of theoretical interest only and consisted in measurement of single-cell potentials. During the 1950's pilot plant production techniques were developed, and the advantages of the solid electrolyte batteries were confirmed. Because of the high internal resistance and high polarization losses during discharge, the current density was limited to about 10^{-6} A/cm^2 at room temperature. With the discovery of the superionic conductors, the interest in the field was revived. Takahashi and Yamamoto[168] reported cells of the type $Ag/Ag_3SI/I_2$, which yielded currents of 1 mA/cm^2 at 25°C. Later, cells of the type $Ag/RbAg_4I_5/RbI_3$ are described.[169,170] Iodine and reaction products of iodine and tetrabutyl ammonium iodide[171] were used as cathodes in cells to give currents in the milliampere region. Recently, organic semiconductors, normally known as charge transfer complexes (CTC),[172,173,173a] possessing a number of intrinsic properties which make them very promising as halogen electrodes in solid state cells, have been reported and used in solid state cells. A family of cyanide–iodide solids whose conductance is essentially ionic and differs but little from that of the aqueous battery electrolytes have been employed along with a charge transfer complex perylene iodine cathode in the cell[174]

$$Ag/KAg_4I_4CN/2 \text{ perylene} \cdot 3I_2 \qquad [6.7]$$

Cells using substituted ammonium iodides combined with silver iodide and silver-iodide–silver-oxyacid salts have also been reported.[175,176]

6.2.2. Desired Characteristics of the Electrolyte and the Electrode Materials

An electrolyte should have a high conductivity, so that the internal loss of voltage and power is low. The electronic transference number should be as low as possible because electron conduction within the electrolyte promotes reaction within the electrolyte, which does not contribute to the current in the outer circuit. Finally, a solid electrolyte should have good mechanical properties and should be chemically stable.

The selection of electrode materials for a galvanic cell based on a silver-ion-conducting electrolyte is restricted to silver at the anode. At the cathode the choice is wider, but two criteria need to be fulfilled, viz., (i) the cathode reversible potential must be less than the decomposition voltage of the electrolyte and (ii) the cathode half-cell reaction must be compatible with the silver ion. These requirements are satisfied by the use of molecular iodine, formed into a state of reduced activity in an appropriate polyiodide crystalline lattice or a complex organic compound. Special attention must be paid to the possible reactions between the electrolyte and the electrode materials, which may not be obvious.

6.2.3. Solid State Primary Batteries

The reported solid state battery systems, having both electrodes and electrolytes in solid form, are listed in Table 6.4. A typical cell of the $Ag/Rb\text{-}Ag_4I_5/RbI_3$ type suitable for battery fabrication is shown in Fig. 6.22a. The cathode, the anode, and the electrolyte are all in the form of pellets. The negative current collector can be copper, silver, or titanium. A titanium foil is placed at the cathode to serve as the electronic contact and intercell connector. The completed cell is encapsulated by a plastic ring and batteries are fabricated by stacking an appropriate number of cells into a can and then sealing them. The cell potential is 0.66 V at 25°C and an achievable energy density of about 30 W-hr/kg has been predicted. A cutaway view of a five-cell stack battery is shown in Fig. 6.22b.

Table 6.4. Solid State Cells Using Silver Anode

Number	Cathode	Electrolyte	Electrolyte conductivity (ohm cm)$^{-1}$	Cell emf (V)	Theoretical specific energy (W-hr/kg)	Reference
1	I_2 (Pt)	AgI	1 at 150°C	0.68	74	183
2	Br_2	—	—	0.994	140	184
3	CuBr	AgBr	5×10^{-4} at 200°C	0.74	—	185
4	I_2(C)	Ag_3SI	10^{-2}	0.68	74	168
5	I_2	$RbAg_4I_5$	0.26	0.66	78, 18^a	170, 171, 186, 187
	tetraalkyl ammonium iodide					187
6	Te	$RbAg_4I_5$		0.217	34	188
	Se			0.265	58	188
7	RbI_3(C)	$RbAg_4I_5$		0.635	50, 33^a	162, 169, 170
	CsI_3	$RbAg_4I_5$		0.314	45	189, 190
	$I_2 + Rb_2AgI_3$	$RbAg_4I_5$		0.68	74	
8	$AuBr_3$	Ag_3SBr	2×10^{-3} at 25°C	0.91	91	179
	$AuBr_x$ ($x = 2.4 \pm 0.1$)			0.84	—	
	K–$AuBr_3$			0.82	74	
9	2perylene·$3I_2$	KCN–4AgI	1.4×10^{-1}	0.64	—	174
10	I_2, Hex·I_2(C)	Hex·$Ag_{12}I_{14}$		0.635	—	178
11	I_2	$Ag_6I_4WO_4$	0.047	0.687	78	191

^{a} Indicates achieved values.

A solid state battery system using a powdered silver anode, either KCN–4AgI or RbCN–4AgI as the solid electrolyte, and the charge transfer complex (CTC) 2perylene·$3I_2$ as cathode, has been developed.[174] The CTC, which provides iodine for electrochemical reduction at a low vapor pressure, thus preserving high-temperature shelf life, is well suited for use as a cathode due to its high electronic conductivity.

The cell equations of discharge are

$$Ag \rightarrow Ag^+ + e' \quad \text{at the anode} \tag{6.40}$$

$$\tfrac{1}{2}I_2 \text{ (perylene complex)} + e' \rightarrow I^- + \text{perylene} \quad \text{at the cathode} \tag{6.41}$$

giving the overall cell reaction as

$$Ag + \tfrac{1}{2}I_2 \text{ (perylene complex)} \rightarrow AgI + \text{perylene} \tag{6.42}$$

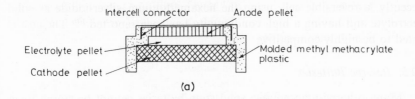

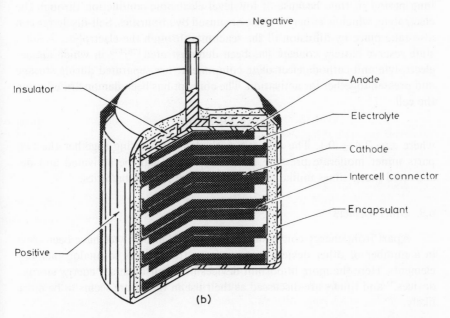

FIG. 6.22. (a) Ag/RbAg$_4$–I$_5$/RbI$_3$-type solid state cell (Reference 169). (b) Section view of a 5-stack cell (Reference 170).

Cell construction and molding techniques have been developed that permit use of electrolyte thickness as low as 0.1 mm without sacrificing shelf life. Satisfactory cell discharge has been demonstrated over the temperature range of −40 to 120°C.

6.2.4. Chargeable Cells

Using the CTC electrodes with RbAg$_4$I$_5$ electrolyte, chargeable cells have been fabricated by Scrosati.[173] The silver anode has been replaced by other materials. Zn/RbAg$_4$I$_5$/AgI,P- and C/RbAg$_4$I$_5$/AgI,P-type cells reduce cost of chargeable cells. Here P represents perylene. The number of cycles achieved, however, is still small. Electrolytic cells of the type Ag/RbAg$_4$I$_5$/Au were prepared with a vacuum-deposited electrolyte film and cycled over 1000 times before degradation started due to the coalescence of the RbAg$_4$I$_5$ films.[207]

Recently, a reversible cell, using the hexamethonium–silver-iodide as solid electrolyte and having a high conductivity has been reported.[178] The cost is stated to be highly competitive.

6.2.5. Reserve Batteries

Many otherwise acceptable solid state batteries cannot be stored for a long period of time because of low-level electronic conduction through the electrolyte, which is either intrinsic or caused by impurities. Self-discharge can also take place by diffusion of the reactants through the electrolyte. A solid state reserve battery concept has been demonstrated[173,179] in which anode/electrolyte and cathode/electrolyte components are separated during storage and pressed together for activation. The concept has been demonstrated using the cell

$$Ag/Ag_3SBr/AuBr_x \qquad [6.8]$$

where $x = 2.4 \pm 0.1$. The cells are activated by pressing together the two parts under moderate pressure of a few psi and can be activated and de-activated many times unlike most conventional reserve batteries.

6.3. Other Devices

Apart from energy conversion devices, solid electrolytes have been tried in a number of other devices ranging from transducers to analog memory elements. Here the more important devices like transducers, "energy storage devices," and timers are discussed as their use in the future seems to be more likely.

6.3.1. Capacitors, Timers, and Coulometers

These devices exploit the interface properties of superionic and electronic conductors. When an electronic conductor is placed in contact with an ionic conductor, a blocking contact is formed, the blocking nature arising from the fact that ions cannot penetrate the electronic conductor and electrons cannot enter the ionic conductor. When a voltage is applied across such an interface, charges accumulate as in the charging of a capacitor whose plates are separated by distances of only atomic dimensions. As a result of this closeness, remarkably high capacitances can be achieved in small volumes. When the surface area is maximized by the use of powders, a capacitance density of $10 \ F/cm^3$ can be achieved. Because of this high capacitance, a new term "energy storage device" has been coined for this. An electrochemical cell which utilizes a β-alumina electrolyte and a simultaneously ion- and electron-conducting electrode [$Na_2 \ 0.5(Fe_{0.95}Ti_{0.05})AlO_3$] has been tested[180] and found to have a high value of capacitance ($\sim 80 \ F/cm^3$ of electrode volume). The leakage

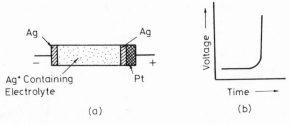

FIG. 6.23. (a) Schematic diagram of a timing device. (b) Variation of voltage across a timing device as constant current is passed through it.

current of these devices is very low, but the breakdown voltage is also quite low (~ 0.6 V for $RbAg_4I_5$-based devices).

An electrode which permits a flow of ions, e.g., silver in the case of Ag^+-ion-conducting electrolytes, together with a blocking electrode can be used to make a timing device. Figure 6.23a shows the schematic diagram of such a timing device. A layer of silver is deposited on a blocking electrode (Pt) and a constant current passed so as to erode the silver metal at a constant rate. After a certain time depending upon the amount of the deposited silver, the nature of the electrode changes to a blocking one. The voltage across the device increases very sharply, as shown in Fig. 6.23b, when the blocking electrode is completely stripped of silver. Exploiting this effect, timers with time settings from minutes to months can be made.

Historically, Tubandt[181] used weight measurements to verify Faraday's law for AgI and $PbCl_2$ solid electrolytes. These measurements can be used basically in a coulometer, where the material accumulating at an indicator electrode is weighed and related to the amount of charge. Most practical coulometers, like timers, make use of electrical readout, which is accomplished by monitoring the voltage across the cell. AgBr, Ag_3SBr, Ag_3SI, and $RbAg_4I_5$ solid electrolytes have been used in coulometers.[182] Solid state timing devices, giving very large time settings, can be used for military applications, where resistance to shock and vibration, and operation under extreme temperatures and after extended storage are desired.

6.3.2. Transducers

Solid electrolytes have a vast potential for use in sensing devices. These can be extremely small, passive devices which are activated by a perturbation, the power generated being proportional to the magnitude of the perturbation. A silver-ion-conducting electrolyte can be used in temperature- and pressure-sensing elements by using two silver electrodes across a thin pellet of the electrolyte. The magnitude and the sense of the generated voltage depend upon the temperature conditions to which the electrodes are subjected. Strain gages utilizing solid electrolytes do not need any auxiliary power supply and can be used for long-term monitoring of deformable elements in inaccessible places.

6.3.3. Miscellaneous

Calcia-stabilized zirconia and other such ionic conductors also possess much higher electrical conductivity as compared to other ceramic materials. Calcia-stabilized zirconia is also a stable refractory up to very high temperatures. These facts have led to its use as heating elements[196-199] and as electrodes in magnetohydrodynamic generators.[200-206] Since this review is concerned with technological applications where the electrolytic conduction has been made use of, no further discussion will be attempted.

6.4. Advantages and Limitations

The advantages offered by the solid state cells and battery systems can be summarized as follows:

(i) Absence of liquids extends the temperature range of operation and storage.

(ii) The systems are leak proof.

(iii) There cannot be any anode corrosion and loss of solvent due to evaporation, hence a long shelf life.

(iv) Miniaturization is possible.

(v) Fabrication and quality control are easy.

(vi) These have exceptional resistance to mechanical shock and vibration.

(vii) The devices are amenable to low cost as well as automated manufacture.

The cost of the silver-ion-conducting electrolytes is high, the energy densities obtained in practice are low, and the shelf lives have not yet been fully ascertained in most of the systems.

MAg_4I_5 group materials are somewhat unstable to moisture, handling, and especially to an iodine atmosphere. The melting points of these are not very high ($232°C$ for $RbAg_4I_5$), and hence the cells cannot be used above certain limiting temperatures.

References

1. L. Meites, ed., *Handbook of Analytical Chemistry*, McGraw-Hill, New York (1963).
2. J. P. Hoare, *The Electrochemistry of Oxygen*, Interscience, New York (1968), p. 146.
3. G. J. Janz, F. Colom, and F. Saegusa, *J. Electrochem. Soc.* **107**, 581 (1960).

4. V. I. Mineuko, S. M. Petrov, and H. S. Ivanova, *Zh. Fiz. Khim.* **35**, 1534 (1961).

5. R. Didschenko and E. G. Rochow, *J. Am. Chem. Soc.* **76**, 3291 (1954).

6. A. Ghosh and T. B. King, *Trans. Metall. Soc. AIME* **245**, 145 (1969).

6a. J. Fouletier, P. Fabry, and M. Kleitz, *J. Electrochem. Soc.* **123**, 204 (1976).

7. T. H. Etsell and S. N. Flengas, *J. Electrochem. Soc.* **118**, 1890 (1971).

8. J. B. Clegg, *J. Chromatogr.* **52**, 367 (1970).

9. T. H. Etsell and S. N. Flengas, *Met. Trans.* **3**, 27 (1972).

10. C. B. Alcock, private communication (December 1973).

11. H. S. Spacil, *Met. Progr.* **96**(5), 106, 108, 111, 112, 114, 116, 118 (1969).

12. M. Sato, in *Research Techniques for High Pressure and High Temperature*, ed. G. C. Ulmer, Springer-Verlag, Berlin (1971).

12a. A. Ghosh, private communication (September 1976).

13. B. C. H. Steele and C. B. Alcock, *Trans. Metall. Soc. AIME* **233**, 1359 (1965).

14. K. P. Jagannathan, Ph.D. thesis, Indian Institute of Technology, Kanpur, 1972.

15. D. A. Rudd, *Iron Steel* **44**(5), 331 (1971).

15a. L. M. Handman, C. E. Spangter, and R. F. Thompson, U.S. Pat. No. 3,864,232 (1975).

16. J. D. Tretyakov and A. Muan, *J. Electrochem. Soc.* **116**, 331 (1969).

17. J. Weissbart and R. Ruka, *Rev. Sci. Instrum.* **32**, 593 (1961).

18. T. Katsura and M. Hasegawa, *Bull. Chem. Soc. Jpn.* **40**, 561 (1967).

19. M. Sato, *Am. Mineral.* **55**, 1424 (1970).

20. W. M. Hickam and J. F. Zamoria, *Instrum. Control Syst.* **40**, 87 (1967).

21. C. R. Edwards, A. Lambert, A. Massey, and J. Laskey, *J.I.S.I.* **205**, 142 (1967).

22. V. A. Leitske and D. L. Schroeder, *Iron Steel Eng.* **XLIV**, 121 (August 1967).

23. Brahma Deo and V. B. Tare, *J. Sci. Ind. Res. (India)* **30**, 465 (1971).

24. K. Kiukkola and C. Wagner, *J. Electrochem. Soc.* **104**, 379 (1957).

25. J. F. Elliott, M. Gleiser, and V. Ramakrishna, *Thermochemistry of Steelmaking*, Addison-Wesley, Reading, Massachusetts (1960).

26. R. Hultgren, R. L. Orr, P. D. Anderson, and K. K. Kelley, *Selected Values of Thermodynamic Properties of Metals and Alloys*, John Wiley, New York (1963).

27. T. C. Wilder, *Trans. Metall. Soc. AIME* **236**, 1035 (1966).

28. C. F. Diaz and F. D. Richardson, in *Electromotive Force Measurements in High Temperature Systems*, ed. C. B. Alcock, Elsevier, New York (1968).

29. J. Osterwald, *Z. Phys. Chem. N.F.* **49**, 138 (1966).

30. W. Puschkell and H. J. Engell, *Z. Metallkd.* **56**, 450 (1965).

31. W. A. Fischer and W. Ackermann, *Arch. Eisenhuttenwes.* **37**, 43 (1966).

32. H. Rickert and H. Wagner, *Electrochim. Acta* **11**, 83 (1966).

33. J. Dompas and J. Van Melle, *J. Inst. Met.* **98**, 304 (1970).

34. A. Achari, private communication (1973).

35. R. J. Fruehan and F. D. Richardson, *Trans. Met. Soc. AIME* **245**, 1721 (1969).

36. T. N. Belford and C. B. Alcock, *Trans. Faraday Soc.* **61**, 443 (1965).

37. T. A. Ramanarayanan and R. A. Rapp, *Met. Trans.* **3**, 3239 (1972).

38. C. R. Nanda and G. M. Geiger, *Met. Trans.* **1**, 1235 (1970).

39. K. T. Jacob, S. K. Seshadri, and F. D. Richardson, *Trans. Inst. Min. Met.* **79**, C27 (1970).

40. K. T. Jacob and J. H. E. Jeffes, *Trans. Inst. Min. Met.* **80**, C32 (1971).

41. C. B. Alcock and T. N. Belford, *Trans. Faraday Soc.* **60**, 822 (1964).

42. G. K. Bandyopadhyay and H. S. Ray, *Met. Trans.* **2**, 3055 (1971).

43. K. T. Jacob and J. H. E. Jeffes, *J. Chem. Thermodyn.* **3**, 433 (1971).

44. W. A. Fischer and D. Janke, *Arch. Eisenhuttenwes.* **41**, 361 (1970).

45. R. Baker and J. M. West, *J.I.S.I.* **204**, 212 (1966).
46. U. V. Choudary and A. Ghosh, *J. Electrochem. Soc.* **117**, 1027 (1970).
47. P. J. Roychowdhury and A. Ghosh, *Met. Trans.* **2**, 2171 (1971).
48. S. K. Das and A. Ghosh, *Met. Trans.* **3**, 803 (1972).
49. W. A. Fischer and D. Janke, *Arch. Eisenhuttenwes.* **39**, 89 (1968).
50. G. R. Fitterer, *J. Met.* **18**, 961 (1966).
51. G. R. Fitterer, *J. Met.* **19**(9), 92 (1967).
52. G. R. Fitterer, C. D. Cassley, and V. L. Vierbiesky, *J. Met.* **21**(8), 46 (1969).
53. K. Schwerdtfeger, *Trans. Metall. Soc. AIME* **239**, 1276 (1967).
54. R. J. Fruehan, L. J. Martonic, and E. T. Turkdogan, *Trans. Metall. Soc. AIME* **245**, 1501 (1969).
55. C. K. Russell, R. J. Fruehan, and J. Rittiger, *J. Met.* **23**(11), 44 (1971).
56. D. A. Dukelow, J. M. Stelzer, and G. F. Koons, *J. Met.* **23**(12), 22 (1971).
57. E. T. Turkdogan, *J.I.S.I.* **210**, 21 (1972).
58. Y. Matsushita and K. Goto, *Trans. Iron Steel Inst. Jpn.* **6**, 131 (1966).
59. M. Ihida and Y. Kawai, *Trans. Iron Steel Inst. Jpn.* **12**, 269 (1972).
60. D. A. J. Swinkels, S. R. Richards, and J. B. Henderson, "Oxygen Probe Applications in Steelmaking," (1972), unpublished.
61. S. R. Richards, D. A. J. Swinkels, and J. B. Henderson, *Proc. ICSTIS Section 2, Suppl. Trans. Iron Steel Inst. Jpn.* **11**, 371 (1971).
62. V. I. Yavoiskii, V. P. Luzgin, A. G. Svyazhim, N. S. Grigorev, E. V. Merker, I. M. Konovalov, E. A. Nechaev, and N. N. Zakharov, *Steel USSR* **1**, 513 (1971).
63. V. I. Yavoiskii, L. P. Luzgin, E. E. Merker, A. G. Svyazhin, and N. S. Grigorev, *Steel USSR* **2**, 20 (1972).
64. G. N. Oiks and T. M. Asadov, *Steel USSR* **2**, 192 (1972).
65. R. Littlewood, *Can. Met. Quart.* **5**(1), 1 (1966).
66. T. H. Etsell and S. N. Flengas, *Chem. Rev.* **70**, 339 (1970).
66a. N. Weber, *Energy Convers.* **14**, 1 (1974).
67. D. A. J. Swinkels, *J. Electrochem. Soc.* **117**, 1267 (1970).
68. H. Schmalzried, *Z. Phys. Chem. N.F.* **38**, 87 (1963).
68a. P. H. Scaife, D. A. J. Swinkels, and S. R. Richards, *High Temp. Sci.* **8**, 31 (1976).
68b. H. W. den Hartog and B. Slangen, *Ironmaking Steelmaking* **3**, 64 (1976).
68c. C. E. A. Shanahan, *Proc. Chem. Conf.* **25**, 48 (1972).
69. H. Rickert, in *Electromotive Force Measurements in High Temperature Systems*, ed. C. B. Alcock, Elsevier, New York (1968).
70. H. Rickert and A. El Miligy, in *Reactivity of Solids*, eds. J. Mitchell, R. DeVries, R. Roberts, and P. Cannon, Wiley Interscience, New York (1969), p. 17.
71. K. Goto and Y. Matsushita, *J. Electrochem. Soc. Jpn.* **35**, 10 (1967).
72. R. Rapp and D. Shores, in *Techniques in Metals Research*, Vol. 4, Part 2, ed. R. A. Rapp, Interscience, New York (1970), p. 123.
73. V. B. Tare and S. H. Ghude, *J. Sci. Ind. Res. (India)* **29**, 190 (1970).
74. H. Rickert and Z. Steiner, *Z. Phys. Chem.* **49**, 127 (1966).
75. R. A. Rapp, in *Proceedings of the Conference on Thermodynamics of Nuclear Materials*, I.A.E.A., Vienna (1967–1968).
76. L. Heyne, in *Mass Transport in Oxides*, eds. J. B. Wachtman, Jr. and A. D. Franklin, *Natl. Bur. Stand. (U.S.) Spec. Publ.* **296**, 149 (1968).
77. D. Yuan and F. A. Kröger, *J. Electrochem. Soc.* **116**, 594 (1969).
78. Y. K. Agarwal, D. W. Short, R. Gruenke, and R. A. Rapp, *J. Electrochem. Soc.* **121**, 354 (1974).
79. C. B. Choudhary, H. S. Maiti, and E. C. Subbarao, private communication (1973).
80. Y. L. Sandler, *J. Electrochem. Soc.* **121**, 764 (1974).

81. K. E. Oberg, L. M. Friedman, W. M. Boorstein, and R. A. Rapp, *Met. Trans.* **4**, 75 (1973).
82. R. Littlewood, *Trans. Metall. Soc. AIME* **233**, 772 (1965).
83. R. Littlewood and E. J. Argent, *Electrochim. Acta* **4**, 114, 155 (1961).
84. R. Littlewood, *J. Electrochem. Soc.* **109**, 525 (1962).
85. C. R. Russell, *Elements of Energy Conversion*, Pergamon Press, New York (1967).
86. E. M. Walsh, *Energy Conversion*, Ronald Press, New York (1967).
87. G. W. Sutton, ed., *Direct Energy Conversion*, McGraw-Hill, New York (1966).
88. S. L. Soo, *Direct Energy Conversion*, Prentice-Hall, Englewood Cliffs, New Jersey (1968).
89. J. O'M. Bockris and S. Srinivasan, *Fuel Cells — Their Electrochemistry*, McGraw-Hill, New York (1969).
90. H. Binder, A. Koehling, H. Krupp, K. Richter, and G. Sandstede, *Electrochim. Acta* **8**, 781 (1963).
91. H. A. Liebhafsky and E. J. Cairns, *Fuel Cells and Fuel Batteries — A Guide to Their Research and Development*, John Wiley, New York (1968).
92. E. J. Cairns and A. D. Tevebaugh, *J. Chem. Eng. Data* **9**, 453 (1964).
93. S. V. Karpachev and S. F. Palguev, in *Fuel Cells — Their Electrochemical Kinetics*, eds. V. L. Bagotskii and Yu. B. Vasilev, Consultants Bureau, New York (1966), p. 97.
94. N. W. Panney, *Fuel Cells — Recent Developments*, Noyes Development Corp., Park Ridge, New Jersey (1969).
95. D. W. White, Electrochemical Society Extended Abstracts, Fall Meeting, Montreal, October 6–11, 1968, Abs. 354, p. 187.
96. J. L. Bliton, H. S. Rechter, and Y. Harada, *Am. Ceram. Soc. Bull.* **42**, 6 (1963).
97. N. J. Maskalick and C. C. Sun, *J. Electrochem. Soc.* **118**, 1386 (1971).
98. C. S. Tedmon, Jr., H. S. Spacil, and S. P. Mitoff, *J. Electrochem. Soc.* **116**, 1170 (1969).
99. D. H. Archer, L. Elikan, and R. L. Zahradnik, in *Hydrocarbon Fuel Cell Technology*, ed. B. S. Baker, Academic Press, New York (1965), p. 51.
100. J. Weissbart and R. Ruka, *J. Electrochem. Soc.* **109**, 723 (1962).
101. R. E. Carter, W. A. Rocco, H. S. Spacil, and W. E. Tragert, *Chem. Eng. News*, Jan. 14 (1963), p. 47.
102. D. H. Archer, J. J. Alles, W. A. English, L. Elikan, E. F. Sverdrup, and R. L. Zahradnik, *Adv. Chem. Ser.* **47**, 332 (1965).
103. H. S. Spacil and C. S. Tedmon, Jr., *J. Electrochem. Soc.* **116**, 1618, 1627 (1969).
104. P. A. Chreskasov and W. A. Fischer, *Arch. Eisenhuttenwes.* **42**, 873 (1971).
105. H. W. den Hartog, P. J. Kreyger, and H. Shrink, *Iron Steel Int.* **46**, 332 (1973).
106. H. Rickert, in *Physics of Electrolytes*, Vol. 2, ed., J. Hladik, Academic Press, New York (1972), p. 519.
107. J. H. Kennedy, in *Physics of Electrolytes*, ed. J. Hladik, Academic Press, New York (1972), p. 931.
108. T. Takahashi, in *Physics of Electrolytes*, ed. J. Hladik, Academic Press, New York (1972), p. 989.
109. T. Takahashi and H. Inahara, *Energy Convers.* **11**, 105 (1971).
110. F. Mortier and J. Fally, Fr. Demande 2,109,106; 2,109,015.
111. W. M. Heap, Brit. Pat. No. 1,261,317 (1972).
112. W. Fisher, H. Kleinsmager, F. J. Rohr, R. Steiner, and H. H. Eigsel, *Chem. Ing. Tech.* **44**, 726 (1972).
113. University of Oklahoma, *Energy Alternatives: A Comprehensive Analysis*, U.S. Government Printing Office, Washington, D.C. (1975).

114. E. F. Sverdrup et al., Fuel Cell Research and Development, Report No. 57, U.S. Government Printing Office, Washington, D.C. (1970).
115. H. S. Isaacs, J. Electrochem. Soc. 119, 455 (1972).
116. B. Marincek and G. Heinke, Ger. Offen. 2,041,836.
117. M. Kleitz, C. Deportes, and P. Fabry, Rev. Gen. Therm. 97, 19 (1970).
118. N. M. Beekmans and L. Heyne, Philips Tech. Rev. 31, 112 (1970).
118a. J. T. Kummer and N. Weber, paper presented at the Automotive Engineering Congress, Detroit, January 1967.
119. J. T. Kummer and N. Weber, in Energy Conversion Engineering, The American Society of Mechanical Engineers, New York (1967), p. 913.
120. J. T. Kummer and N. Weber, in Proceedings of the 21st Annual Power Sources Conference, Atlantic City, New Jersey, PSC Publications Committee (1967), p. 37.
121. N. K. Gupta and R. P. Tischer, J. Electrochem. Soc. 119, 1033 (1972).
122. T. G. Pearson and P. L. Robinson, J. Chem. Soc. 132, 1473 (1940).
123. J. T. Kummer, in Progress in Solid State Chemistry, vol. 7, eds. H. Reiss and J. O. McCaldin, Pergamon Press, New York, (1972), p. 141.
124. M. S. Whittingham and R. A. Huggins, in Solid State Chemistry, eds. R. S. Roth and S. J. Schneider Jr., Natl. Bur. Stand. U.S. Spec. Publ. 364 (1971).
125. Y. Y. Yao and J. T. Kummer, J. Inorg. Nucl. Chem. 29, 2453 (1967).
126. J. D. Busi, in Proceedings of the Symposium on Batteries, Traction and Propulsion, 1972, ed. R. L. Kerr, Columbus Section, Electrochemical Society, Battelle Memorial Institute, Columbus, Ohio (1972), p. 195.
127. G. H. Gelb and N. A. Richardson, in Proceedings of the Symposium on Batteries, Traction and Propulsion, 1972, ed. R. L. Kerr, Columbus Section, Electrochemical Society, Battelle Memorial Institute, Columbus, Ohio (1972), p. 178.
128. S. Gratch, J. V. Petrocelli, R. P. Tischer, R. W. Minck, and T. J. Whalen, Proceedings of the 7th Intersociety Energy Conversion Engineering Conference, 1972, ACS, Washington, D.C. (1972), p. 38.
129. R. W. Minck, Proceedings of the 7th Intersociety Energy Conversion Engineering Conference, 1972, ACS, Washington, D.C. (1972), p. 42.
130. S. P. Mitoff, U.S. Pat. No. 3,740,268 (1973).
130a. W. Dale Compton, in Energy, Environment and Productivity, Proceedings of the First Symposium on Research Applied to National Needs, National Science Foundation, Washington, D.C., NSF 74-19 (1974).
131. J. L. Sudworth, Sulphur Inst. J. 8, 12 (1972).
132. J. L. Sudworth and M. D. Hames, in Power Sources, Vol. 3, ed. D. H. Collins, Oriel Press, Newcastle-upon-Tyne (1971), p. 227.
133. J. L. Sudworth, M. D. Hames, M. A. Storey, M. F. Azim, and A. R. Tilley, in Proceedings of the Eighth International Power Sources Symposium, Brighton, Sussex, England, 1972, International Power Sources Committee, Croydon, Surrey, England (1972), p. 1.
133a. J. L. Sudworth, in Fast Ion Transport in Solids, ed. W. van Gool, North-Holland, Amsterdam (1973), p. 581.
134. L. J. Miles and I. Wynn Jones, in Proceedings of the Eighth International Power Sources Symposium, Brighton, Sussex, England, 1972, International Power Sources Committee, Croydon, Surrey, England (1972), p. 245.
135. T. Nakabayashi, Ger. Offen. 2,240,278 (1973).
136. N. G. Bukun, E. A. Ukshe, and V. V. Evtushenko, Elektrokhimiya 9, 406 (1973).
137. N. G. Bukun, E. A. Ukshe, N. S. Lidorenko, and A. A. Lanin, U.S.S.R. Pat. 366,517 (1973).
138. J. Fally, J. Richez, Y. Lazzennec, and C. Lasne, Fr. Demande, 2,129,864 (1972).

139. J. Fally and J. Richez, Fr. Demande, 2,140,318 (1973).
140. J. Fally, C. Lasne, and V. Lezennev, Fr. Demande 2,142,695 (1973).
140a. R. S. Gordon, presented at Workshop on Ceramics for Energy Applications, Columbus, Ohio (1975).
141. I. Wynn Jones, in *Fast Ion Transport in Solids*, ed. W. van Gool, North-Holland, Amsterdam (1973), p. 559.
142. J. Fally, C. Lasne, Y. Lazennec, and P. Margotin, *J. Electrochem. Soc.* **120**, 1296 (1973).
142a. J. Lazennec, C. Lasne, P. Margotin, and J. Fally, *J. Electrochem. Soc.* **122**, 734 (1975).
143. J. J. Werth, U.S. Pat. No. 3,718,505 (1973).
144. J. M. Thomas and A. J. White, *J. Mater. Sci.* **7**, 838 (1972).
145. N. S. Choudhary, NASA TN D7322, 16 pp. (1973).
146. H. E. Kautz, J. Singer, W. Fielder, and J. S. Fordyc, NASA TN D7146, 10 pp. (1973).
147. N. Kimura, S. Suzuki, and S. Toshina, *Denki Kagaku* (*Japan*) **41**, 22 (1973).
148. R. D. Armstrong, T. Dickinson, and J. Turner, *J. Electroanal. Chem. Interfacial Electrochem.* **44**, 157 (1973).
149. B. Cleaver and A. J. Davies, *Electrochim. Acta* **18**, 727, 733, 741 (1973).
150. G. J. Janz and R. P. T. Tomkins, Current Awareness Bulletin: Polysulfides, July 5, 1973, Molten Salts Data Chemistry Center, Polytechnic Institute, Troy, New York.
151. K. D. South, J. L. Sudworth, and J. G. Gibson, *J. Electrochem. Soc.* **119**, 554 (1972).
152. M. N. Hull, *Energy Convers.* **10**, 215 (1970).
153. E. J. Cairns and H. Shimotake, *Science* **64**, 1347 (1969).
154. M. Barak, *Proc. IEEE* **117**, 1561 (1970).
155. M. Barak, *Electron. Power* **18**, 290 (1972).
156. S. Gross, in *Proceedings of the Symposium on Batteries, Traction and Propulsion, 1972*, ed. R. L. Kerr, Columbus Section, Electrochemical Society, Battelle Memorial Institute, Columbus, Ohio (1972), p. 9.
157. E. H. Hietbrink, J. McBreen, S. M. Selis, S. B. Tricklebank, and R. R. Witherspoon, in *Electrochemistry of Cleaner Environments*, ed. J. O'M. Bockris, Plenum Press, New York (1972), p. 47.
157a. E. J. Cairns, in *Critical Materials Problems in Energy Production*, ed. C. Stein, Academic Press, New York (1976), p. 684.
157b. J. R. Birk, in *Superionic Conductors*, eds. G. D. Mahan and W. L. Roth, Plenum Press, New York, (1976), p. 1.
157c. S. P. Mitoff and J. B. Bush, Jr., *J. Electrochem Soc.* **122**, 457 (1975).
157d. E. C. Gay, D. R. Visser, F. J. Martino, and K. E. Anderson, *J. Electrochem. Soc.* **123**, 1591 (1976).
158. F. G. Will and S. P. Mitoff, Abstract No. 39, Boston, Massachusetts Meeting of the Electrochemical Society, October 7–11, 1973.
159. F. G. Will, R. R. Dubin, and J. J. Hess, Abstract No. 40, Boston, Massachusetts Meeting of the Electrochemical Society, October 7–11, 1973.
159a. J. H. Kennedy and A. F. Sammells, *J. Electrochem. Soc.* **121**, 1 (1974).
160. R. T. Foley, *J. Electrochem. Soc.* **116**, 13c (1969).
161. J. N. Mrgudich, in *Encyclopedia of Electrochemistry*, Reinhold, New York (1964).
162. B. B. Owens, in *Advances in Electrochemistry and Electrochemical Engineering*, Vol. 8, ed. C. W. Tobias, Wiley Interscience, New York, (1971), p. 1.
163. R. T. Foley, in *Physics of Electrolytes*, Vol. 2, ed. J. Hladik, Academic Press, London (1972), p. 960.

164. K. O. Hever, in *Physics of Electrolytes*, Vol. 2, ed. J. Hladik, Academic Press, London (1972), p. 809.
164a. C. C. Liang, in *Fast Ion Transport in Solids*, ed. W. Van Gool, North-Holland, Amsterdam (1973), p. 19.
164b. R. M. Dell, in *Electrode Processes in Solid State Ionics*, eds. M. Kleitz and J. Dupuy, Reidel, Dordrecht, (1976), p. 387.
164c. T. Tahahashi, in *Super Ionic Conductors*, eds. G. D. Mahan and W. L. Roth, Plenum Press, New York (1976), p. 379.
164d. B. B. Scholtens and W. van Gool, in *Solid State Electrolytes*, eds. P. Hagenmuller and W. van Gool, Academic Press, New York (1978), p. 451.
165. L. Heyne, *Electrochim. Acta* **15**, 1251 (1970).
166. P. D. Greene, *Electron. Power* **18**, 395 (1972).
167. T. Takahashi, *J. Appl. Electrochem.* **3**, 79 (1973).
168. T. Takahashi and O. Yamamoto, *Electrochim. Acta* **11**, 779 (1966).
169. B. B. Owens, G. R. Argue, and I. J. Groce, in *Power Sources 2*, ed. D. H. Collins, Pergamon Press, London (1970).
170. J. E. Oxley and B. B. Owens, Gould Ionics Inc., Report 70-ER-067A, (April 1970).
171. M. De Rossi, G. Pistoia, and B. Scrosati, *J. Electrochem. Soc.* **116**, 1642 (1969).
172. B. Scrosati, M. Torroni, and A. D. Butherus, in *Proceedings of the Eighth International Power Sources Symposium, Brighton, Sussex, England, 1972*, International Power Sources Committee, Croydon, Surrey, England (1972).
173. B. Scrosati, *J. Electrochem. Soc.* **120**, 78 (1973).
173a. M. Pampallona, A. Ricci, B. Scrosati, and C. A. Vincent, *J. Appl. Electrochem.* **6**, 269 (1976).
174. D. V. Louzos, W. G. Darland, and G. W. Mellors, *J. Electrochem. Soc.* **120**, 1151 (1973).
175. B. B. Owens, J. H. Christie, and G. T. Tiedeman, *J. Electrochem. Soc.* **118**, 1144 (1971).
176. T. Takahashi, S. Ikeda, and O. Yamamoto, *J. Electrochem. Soc.* **119**, 477 (1972).
177. K. Kitazawa and R. L. Coble, *J. Am. Ceram. Soc.* **57**, 360 (1974).
178. M. De Rossi, M. L. Berardelli, and G. Fonseca, *J. Electrochem. Soc.* **120**, 149 (1973).
179. J. H. Kennedy, F. Chen, and R. C. Miles, *J. Electrochem. Soc.* **120**, 171 (1973).
180. K. O. Hever, *J. Electrochem. Soc.* **115**, 830 (1968).
181. C. Tubandt, *Handb. Exp. Phys.* **12**(1), 383 (1932).
182. J. H. Kennedy, in *Physics of Electrolytes*, Vol. 2, ed. J. Hladik, Academic Press, London (1972).
183. J. L. Weininger, U.S. Pat. No. 2,923,546 (1960).
184. J. L. Weininger and H. Liebhafsky, U.S. Pat. No. 2,987,568 (1961).
185. B. Reuter and K. Hardel, *Naturwissenschaften* **48**, 161 (1966).
186. B. B. Owens, J. S. Sprouse, and D. L. Warburton, *Proc. of the 25th Power Sources Symposium*, Atlantic City, New Jersey, May 1972, p. 8.
187. B. B. Owens, G. R. Argue, I. J. Groce, and L. D. Hermo, *J. Electrochem. Soc.* **116**, 312 (1969).
188. T. Takahashi and O. Yamamoto, *J. Electrochem. Soc.* **117**, 1 (1970).
189. G. R. Argue, B. B. Owens, and I. J. Groce, *Proc. Ann. Power Sources Conf.* **22**, 103 (1968).
190. G. R. Argue and B. B. Owens, U.S. Pat. No. 3,443,997 (1969).
191. T. Takahashi, S. Ikeda, and O. Yamamoto, *J. Electrochem. Soc.* **120**, 647 (1973).
192. J. K. Pargeter, *J. Met.* **20**, 27 (1968).
193. W. A. Fischer, *Arch. Eisenhuttenwes.* **38**, 442 (1967).

194. C. B. Alcock and S. Zador, *J. Appl. Electrochem.* **2**, 289 (1972).
195. L. Micheli, Res. Publ. GMR-1352 of General Motors Research Lab., Warren, Michigan, 1973.
195a. E. C. Subbarao, in *Proceedings of the International Symposium on Industrial Electrochemistry*, Madras, India (1976).
196. E. K. Keler and E. N. Nikitin, *J. Appl. Chem. USSR* **32**, 2033 (1959).
197. W. H. Davenport, S. S. Kistler, W. M. Wheildon, and O. J. Whittmore, *J. Am. Ceram. Soc.* **30**, 333 (1959).
197a. S. M. Lang and R. F. Geller, *J. Am. Ceram. Soc.* **34**, 193 (1951).
197b. E. Rothwall, *J. Sci. Instrum.* **38**, 191 (1961).
197c. A. M. Anthony, in *Solid Electrolytes*, eds. P. Hagenmuller and W. van Gool, Academic Press, New York (1978), p. 519.
198. M. Leipold and J. Taylor, in *Temperature: Its Measurement and Control in Science and Industry*, Vol. 3, Reinhold, New York (1962), p. 1150.
199. T. H. Neilson and M. Leipold, *J. Am. Ceram. Soc.* **46**, 381 (1963).
200. P. Ya. Gokhstein and R. A. Khakin, *Teplofiz. Vys. Temp.* **7**, 1031 (1968).
201. V. D. German, Yu. P. Kukota, and G. A. Lyubimov, *Teplotekh. Probl. Pyramogo Preobrazov. Energy* (1969), p. 79.
202. M. Guillou, *Rev. Gen. Therm.* **8**, 751 (1959).
203. L. N. Popova, L. M. Demidenko, D. N. Poluboyarinov, R. Ya. Popilskii, and V. S. Bakernov, *Tr. Mosk. Khim. Tekhnol. Inst.* **63**, 132 (1969).
204. P. Ya. Gokhstein and A. A. Safonov, *Teplofiz. Vys. Temp.* **8**, 398 (1970).
205. S. K. Adams and R. E. W. Casselton, *J. Am. Ceram. Soc.* **53**, 117 (1970).
206. R. E. W. Casselton, *Phys. Stat. Solidi A* **2**, 571 (1970).
207. J. H. Kennedy, F. Chen, and J. Hunter, *J. Electrochem. Soc.* **124**, 454 (1973).

7

Fabrication

A. K. Ray and E. C. Subbarao

1. Introduction

Before a solid electrolyte is utilized for research purposes or practical devices, it has to be fabricated into the desired shape and size out of a single crystal or a polycrystalline material. The common shapes are disks, tubes, and crucibles. The electrolyte sometimes has to be joined to metallic, ceramic, or glass components. In all applications, the electrolyte has to be contacted with an appropriate metallic or nonmetallic electrode in solid or liquid state. These aspects are discussed in the present chapter.

2. Single Crystals

Single crystals are needed primarily for basic research on solid electrolytes, e.g., their structure, properties, anisotropy, etc. No serious application based on single crystals of solid electrolytes has emerged so far. Therefore only a brief discussion of the growth of single crystals of stabilized ZrO_2, ThO_2, CaF_2, β-alumina, and silver ion conductors is included here.

2.1. Stabilized Zirconia

Single crystals of stabilized zirconia containing 10 mol % CaO have been grown by Reed[1] using a Verneuil-type geometry. Melting of the material at 2600–2700°C may be accomplished by an induction plasma torch in an

A. K. Ray and E. C. Subbarao • Department of Metallurgical Engineering, Indian Institute of Technology, Kanpur 208016, India. Dr. Ray's present address: IBM Thomas J. Watson Research Center, Yorktown Heights, New York 10598.

atmosphere of 50% oxygen–50% argon or in a carbon arc image furnace.[2] The resulting boule, 10 mm in diameter and 45 mm long, had some cracks but contained large single-crystal regions.

2.2. Thoria

Laszlo et al.[3] vaporized a small area of polycrystalline ThO_2 pellet in a solar furnace with a heat flux of 500 cal/cm² sec, corresponding to a sample temperature of about 3800°C, and condensed the vapors around the molten crater in the form of crystals. The growth rate was about 0.5 mm/min.

Fused thoria can be made by melting thoria powder covering an electric arc struck between two tubular carbon electrodes through which oxygen gas is passed to prevent the formation of a carbide.[4]

Due to the high melting point of ThO_2 (about 3300°C), flux methods are, however, favored for the growth of its single crystals. A variety of fluxes ($PbO–PbF_2$, $Bi_2O_3–PbF_2$,[5] PbF_2,[5,6] $Na_2O–B_2O_3$, $PbO–PbF_2–B_2O_3$, $NaF–PbF_2–B_2O_3$,[7] $NaF–B_2O_3$,[7] and $Li_2O·2WO_3$[8]) were used to accomplish melting at 1250–1300°C. Cooling rates of the order of 1°C/min were employed up to 1000°C. The crystals are recovered by digesting the mass with 50% acetic acid. The crystals were 2–5 mm in size[5] with larger sizes obtained by employing localized cooling.[6] Crystals grown from lead fluxes suffered damage from UV radiation and hence lead-free fluxes were used to obtain good optical quality crystals.[7] Rare-earth-doped crystals of ThO_2 are also grown.[8] The liquid inclusions in flux-grown crystals were eliminated by a temperature gradient method.[9]

2.3. Calcium Fluoride

Single crystals of CaF_2 are grown by a variety of techniques including Bridgeman, Czochralski, sublimation, flux ($CaCl_2/NaF$), and gel methods. The relative merits of the different methods have recently been reviewed.[10,11]

2.4. β-Alumina

Single crystals of sodium β-alumina are obtained by fusing an appropriate mixture of Na_2CO_3 and Al_2O_3 in an arc furnace or by carbon resistance or high-frequency induction heating.[12,13] The fusion-cast material is broken and transparent single crystals of β-alumina are obtained by screening. These crystals were up to 1 cm in diameter and 0.03 cm in thickness in the c direction.[14] Any soda-rich second phase present may be removed by treating with concentrated H_2SO_4. The crystals generally contained soda in excess of the stoichiometric composition. Flux methods (e.g., Bi_2O_3) are also used.[15]

Cocks and Stormont[16] have applied a recently developed technique called the edge-defined, film-fed growth (EFG) to grow sodium β-alumina crystals having constant cross-sectional area. An inert gas at a pressure of 300 psi fills the crystal growth chamber to minimize the vaporization of sodium. The crucible and the crystal-pulling die are made of iridium. Heating is accomplished by water-cooled radiofrequency (RF) coils and pulling rates of 0.01 to 10 in./hr were used. With the help of a suitable die, single-crystal tubes of β-alumina, 0.5 cm in outer diameter, were grown with the high-conductivity (cleavage) planes normal to the tube axis. A Czochralski pulling method under a slight argon pressure was also utilized to grow large crystals of β-alumina.[17]

Yao and Kummer[14] have found that substituted β-alumina single crystals can be obtained by placing sodium β-alumina crystals in respective molten nitrates and allowing ion exchange to take place. Single crystals of potassium,[14a] silver,[12,14a] ammonium, rubidium, and thallium β-alumina are produced in this way. In the case of monovalent ions which differ substantially from Na^+ ion in size (e.g., Li^+, Rb^+, Cs^+), silver β-alumina was preferred as the starting material for ion exchange. For indium β-alumina, In–Hg amalgam was used instead of a molten indium salt.

Since β''-Al_2O_3 decomposes above 1500°C, it is not possible to grow single-phase crystals of β''-Al_2O_3 by fusion. However, Fally *et al.*[18] have obtained a mixture of β- and β''-alumina crystals by rapid cooling of an induction-melted mass of compositions within a narrow range of Na_2O to Al_2O_3 ratios (1:5.33 to 1:11) in the Na_2O–Al_2O_3 system. On the other hand, crystals of stabilized β''-alumina (ideal formula $Na_2O \cdot MgO \cdot 5Al_2O_3$) were obtained[19] by first prereacting a mixture of 10 wt % Na_2O, 4.5 wt % MgO, and 85.5 wt % Al_2O_3 at 1200°C in air and then melting it at about 2000°C in a nitrogen atmosphere. Rapid cooling of the melt yielded an ingot containing single crystals of β- and β''-alumina. The β''-alumina crystals were formed in the region where solidification occurred first. Thin platelets of β''-Al_2O_3 of composition $0.84Na_2O \cdot 0.84MgO \cdot 5Al_2O_3$ were grown by heating a mixture of 35 wt % Na_2CO_3, 3.2 wt % MgO, and 62 wt % Al_2O_3 in a sealed platinum crucible at 1660°C for 24 hr and then cooling to 25°C over 1 hr.[19a] This can be converted to hydronium β''-Al_2O_3 crystals of average composition $0.84H_2O \cdot 0.84MgO \cdot 5Al_2O_3 \cdot 2.8H_2O$ by treating in conc. H_2SO_4 at 240°C for several days.[19a] β''-alumina crystals stabilized by magnesia were also grown by the EFG method.[20] Transmission electron microscopic studies of these crystals have revealed the presence of dislocations and antiphase boundaries.[21–23]

The related β- and β''-sodium-gallate crystals have been grown by the slow cooling of a molten bath[24] or by using a flux (e.g., Na_2O or NaF–NaCl),[25] temperature gradient,[25a] or Verneuil method.[25b]

2.5. AgI-Type Compounds

AgI crystals are grown by a variety of techniques, only the more recent ones of which are mentioned here. They may be grown from a solution of

AgI in concentrated aqueous KI by convective circulation of the solution between two chambers held at different temperatures[26] as well as by the gel method.[27] Small crystals of AgI are also grown by the slow dilution of the above solution with water from the vapor phase.[28]

MAg_4I_5 (where M = K, NH_4, Rb) crystals are grown using HI as a solvent.[29] To a hot (50°C) solution of 0.805 g AgI in 1 g of 57% HI, 0.166 g of RbI (or 0.124 g of KI or 0.108 g of NH_4I) is added and the solution cooled to 30°C (or 39 or 35°C in the case of K and NH_4 salts). Orthorhombic crystals of MAg_4I_5 appear, which were filtered out. Further cooling of the solution to 27°C (or 36 or 35°C for K and NH_4 salts) resulted in crystals which are octahedral in shape and $\frac{3}{8}$ to $\frac{1}{2}$ in. in diameter. Scrosati[30] has grown crystals of $RbAg_4I_5$ from an acetone solution by using a similar method.

3. Polycrystalline Materials

For most applications, solid electrolytes in polycrystalline form are employed.[31] These are prepared by ceramic or powder metallurgy techniques consisting of the following steps: (1) powder preparation, (2) mixing, (3) calcination, (4) fabrication, and (5) sintering. Sometimes, some of these steps may be combined, as in hot pressing when fabrication and sintering are carried out together. Disks are made by pressing, while tubes are made by slip casting, extrusion, isostatic pressing, or electrophoretic deposition. Crucibles may be fabricated by pressing as well as by slip casting. A good summary of the preparation techniques is presented by Mitoff.[32] The finished product should be dense, impervious to the gaseous or liquid species on either side of the electrolyte, mechanically strong, and thermally and chemically stable under conditions of use. These are the nonelectrical characteristics sought in a polycrystalline solid electrolyte.

3.1. Zirconia-Base Materials

The zirconia-base solid electrolytes of importance are stabilized by lime, yttria, or magnesia in the defect cubic fluorite-type solid solution (Chapter 1). The relevant portions of the binary phase diagrams of ZrO_2 with CaO, $YO_{1.5}$ and MgO are presented in Chapter 1 (Figs. 1.22 and 1.24).

3.1.1. Powder Preparation, Mixing, and Calcination

The most important aspect of the fabrication of polycrystalline materials is the preparation of fine powders in an active state. Measurement of particle size and surface area of the dispersed powder is used to characterize the powders and their sinterability. However, an actual sintering experiment is the best test for sinterability. Decrease of surface area leads to sintering and hence the need for small particle size of the starting powder. Further, a uniformly small grain size also aids in sintering. However, it must be noted that a higher green density results from a material with a distribution of particle sizes than from powders of uniform size.

The common methods of preparation of zirconia powders are: (a) thermal decomposition of oxalates, acetates, carbonates,[33] or hydroxides,[34] (b) hydrolysis of chlorides or organic oxychlorides, (c) vapor phase reaction of halides with steam,[35] (d) plasma [36] or electric arc processes,[37] (e) hydrolysis of alkoxides,[38] and (f) coprecipitation.[39]

For example, $Zr(OH)_4$ reacts with $Ca(OH)_2$ at 250–320°C in a co-precipitation reaction to form a cubic solid solution.[39] Under the same conditions, ZrO_2 does not react with $Ca(OH)_2$. Also, a mechanical mixture $CaCO_3$ and $Zr(OH)_4$ requires 1000–1300°C for the formation of solid solution.[40] Mixed oxide precipitates of ZrO_2 and $YO_{1.5}(Y_2O_3)$ as ultrahigh purity submicron powder may be prepared by the alkoxide route. Alkoxides are compounds of the type $M(OR)_n$, where M is the metal, R is the alkyl group such as *n*-propyl, isopropyl, *n*-butyl, etc., and the subscript *n* is an integer. Alkoxides are synthesized according to the reaction

$$MCl_4 + 4ROH + 4NH_3 \xrightarrow[C_6H_6]{5°C} M(OR)_4 + 4NH_4Cl \qquad (7.1)$$

where M is Zr, Th, Hf, etc., and R is the isopropyl radical. When the isopropoxide is a solid at room temperature as in the case of Zr, a more complex radical is employed such as zirconium tetratertiary butoxide $[Zr(OC_4H_9)_4]$, which is a liquid at room temperature and pressure.[41,42] Mixed oxides are prepared by starting with a solution of the high-purity isopropoxides of Zr and Y in a mutual solvent such as *n*-hexane, C_6H_{14}, or benzene, C_6H_6.[43] While the solution is vigorously stirred, drops of triply distilled or demineralized water are added. The hydrolysis of the alkoxides leads to the precipitation of mixed oxides of high purity and submicron size:

$$Zr(OC_3H_7)_4 + 2H_2O \rightarrow ZrO_2 + 4C_3H_7OH \qquad (7.2)$$

$$2Y(OC_3H_7)_3 + 3H_2O \rightarrow Y_2O_3 + 6C_3H_7OH \qquad (7.3)$$

The mixed-oxide powders, prepared by Mazdiyasni *et al.*,[43] who pioneered this alkoxide method, had an average particle size of 50 Å on drying at 100–110°C for 24 hr. At this stage, the powder is in a highly reactive state due to the large surface area. On calcination in air at 850°C for about 30 min, the particle size increases to about 400 Å. The x-ray diffraction pattern of the calcined powder agreed with that of sintered stabilized zirconia, except that the x-ray lines were somewhat broad due to the small particle size of the calcined material. The advantages of this process are (1) the smaller (6%) Y_2O_3 needed for stabilization than in the conventional methods (7–15% Y_2O_3; see also Chapter 1, Section 4.1.1), due to the intimate mixing on a fine scale and (2) the lower sintering temperature (1450°C for a fully dense, translucent product) than the usual 2000°C, because of the active state of the fine-particle material.

Markin[44] has pointed out that with the sol–gel process, which is another novel method, solid solutions of Y_2O_3-stabilized ZrO_2 can be prepared by heating to only 800°C.

The conventional method, extensively employed, consists of mixing ZrO_2 powder with an appropriate amount of the dopant as oxide, carbonate, or sulfate. Usually wet mixing is employed, followed by a drying operation. When the dopant is in the form of carbonate or sulfate, the mixture is calcined at about 1000°C to decompose the salts to the oxide. Sometimes, in order to ensure the homogeneity of the calcined powder, the calcination step is repeated with an intermediate step of grinding and pelletizing.

3.1.2. Fabrication

(a) *Pressing.* Disks, crucibles, and boats are prepared by cold pressing[45] and sometimes by isostatic pressing. The material for pressing contains 3%–5% water, some binder, such as zirconium oxychloride, and a lubricant, e.g., carbowax. In an isostatic press, the pressure is applied more uniformly from all directions than in a unidirectional press and therefore the density of isostatically pressed samples is more uniform and higher.

(b) *Extrusion.* Tubes and rods are extruded from a paste made by mixing the powder with organic binders like gum tragacanth or methyl cellulose in water, or rubber solution in xylene. Extrusion under vacuum conditions, called deairing, decreases the porosity and leads to higher green density.

(c) *Slip Casting.*[46] Slip casting of crucibles, tubes, and boats is carried out in plaster of Paris molds, using a slip consisting of stabilized zirconia powder and a binder (e.g., fresh zirconia powder or 1% polyvinyl alcohol solution). The pH of the slip is maintained between 3 and 4 by the addition of acid.

(d) *Thin Films.* Stabilized zirconia is used as an electrolyte in fuel cells (see Chapter 6, Section 4.2). Since the internal resistance of the electrolyte limits the current drawn, as thin a ceramic as is consistent with mechanical strength and impermeability to gases is utilized for this purpose. Thicknesses of less than 0.04 cm cannot be achieved by machining or plasma spraying. Electrolytes of 5-μm thickness can be prepared by sputtering but the low sputtering rate makes the economic viability of the process on an industrial scale questionable. Chemical vapor deposition techniques have been employed to make 30-μm-thick films of yttria-stabilized zirconia, but there are limitations of thickness control, high-temperature masking, chemical attack of cell materials, etc.

To overcome these difficulties, Maskalick and Sun[47] developed a reactive sintering process in which monoclinic ZrO_2 and calcium zirconate react quickly and completely at 1000°C to form dense, stabilized cubic zirconia according to the reaction

$$x ZrO_2 + y CaZrO_3 \rightarrow (ZrO_2)_{x+y}(CaO)_y \qquad (7.4)$$

Submicron ZrO_2 (76 wt %) and -325 mesh $CaZrO_3$ (24 wt %) together with butyl acetate as a grinding medium was ball milled for about 20 hr. The slurry, corresponding to a composition of $Ca_{0.15}Zr_{0.85}O_{1.85}$, was used to fabricate the electrolyte. The fuel electrode, consisting of a two-phase mixture of cobalt and stabilized zirconia, was coated on the outer surface of a porous zirconia tube and was sintered in an atmosphere of nitrogen with 5% H_2 saturated with water at 26°C. On the sintered electrode, a thin layer of the electrolyte slurry was applied either by slip casting or spraying. The assembly was sintered at 1400°C in a reducing (H_2/H_2O mixture) or inert atmosphere (to prevent the oxidation of the metal in the electrode). The grain size was 5–10 μm after sintering for 1 hr and 15 μm after 10–15 hr. The apparent porosity of the electrolyte film was less than 5%. The leak rate of a typical 60-μm-thick electrolyte was 6×10^{-7} mol/min over 40 cm² area when the initial vacuum on the other side of the electrolyte film was 10^{-2} Torr.

Of these various methods, plasma spraying and reactive sintering appear promising for preparing thin electrolyte films at the present time.[48] Y_2O_3-doped zirconia blankets are reported to be marketed in the U.S.A.[49]

3.1.3. Sintering

Sintering of polycrystalline-stabilized zirconia is carried out at such temperature and for such time as to give dense, low-porosity, and impermeable ceramic. However, sintering of cubic zirconia is difficult, as shown by the data on $Ca_{0.16}Zr_{0.84}O_{1.84}$[50] (Table 7.1). The poor sintering rate even at these elevated temperatures is attributed to the low cation diffusion rates, which become the rate-determining step. It may be recalled that the cation diffusion coefficients are at least five orders of magnitude lower than the anion diffusion coefficient (Fig. 1.7). The kinetics of isothermal grain growth in $Ca_{0.16}Zr_{0.84}O_{1.84}$, which seems to be limited by the cation diffusion rate, is given by[50]

$$D = (Kt)^{0.4} \qquad (7.5)$$

where D is the grain diameter, K is a constant, and t is time. The activation

Table 7.1. Sintering Data on
$Ca_{0.16}Zr_{0.84}O_{1.84}$

Temperature (°C)	Time (hr)	Density (%, theoretical)
1700	2	90
1800	2	92
1900	8	94

energy for grain boundary migration was 80 kcal/mol, which is close to that for cation diffusion.

Hague[51] has found that the maximum densification in the ZrO_2–CaO system was obtained for the lowest CaO content needed for stabilization and that the density decreased linearly with increasing CaO content in the stabilized zirconia. This is analogous to the variation of ionic conductivity with CaO content. Similarly, Kainarskii et al.[52] have found that the shrinkage and compaction sharply increased with the amount of stabilizing agent up to 6–7 mol % CaO, 3–4% Y_2O_3, and 12%–18% CeO_2, when sintered at 1500°C for 6 hr. Since these compositions do not represent the lowest amount of additives needed for stabilization, it appears that maximum densification is achieved by the formation of 60–70% solid solution. The optimum compositions for maximum densification shifted to lower concentration of additives when the sintering temperature is raised. When the optimum addition of stabilizers is exceeded, shrinkage and densification of the specimen are reduced.

Densification is more intense when stabilization takes place during sintering at temperatures up to 1500°C than with prestabilized material.[53,54] However, addition of unfired monoclinic zirconia to prestabilized zirconia up to 30% increases the thermal stability and densification of the sintered product.

The source of the constituent materials also has an influence on the sinterability. For example, oxides obtained by thermal decomposition of compounds sinter more easily than precalcined oxides. Thus ZrO_2 obtained by thermal decomposition of the nitrate at 200–400°C is in a more imperfect and active state than that calcined at 600°C when the stable monoclinic ZrO_2 is formed.[52]

Young and Cutler[55] found that the particle size distribution affects the sintering kinetics of yttria-stabilized zirconia. Addition of fine particles to uniformly coarse material greatly affects sinterability, whereas addition of coarse particles to a fine matrix has negligible effect. Sintering in this case is attributed to a grain boundary diffusion mechanism with an activation energy of 90 kcal/mol. Spacil and Tedmon[56] produced impervious ZrO_2-base electrolyte ceramics at temperatures as low as 1225°C by adding 1% Fe_3O_4 or Co_3O_4. The fabrication consists of plasma-spraying aggregates of 50-μm-size particles, obtained by spray-drying a mixture of ZrO_2, Y_2O_3, and Fe_3O_4. The stabilization of the zirconia occurs during the melting and freezing of the plasma-sprayed powder. This material exhibited essentially anionic conduction down to an oxygen partial pressure of 10^{-20} atm.

The microstructure of stabilized zirconia may be observed by etching the surface with potassium hydrogen sulfate (bisulfate) at about 450°C and using a reflected light microscope.[45]

3.1.4. Partially Stabilized Zirconia (PSZ)

Partially stabilized zirconia is made from compositions in the two-phase region of monoclinic and cubic ZrO_2 phases (Fig. 1.22). The thermal expansion of PSZ is intermediate between that of monoclinic ZrO_2 and fully stabilized cubic phases[57,58] (Fig. 7.1). Low thermal expansion coefficient together with low Young's modulus and high thermal conductivity favors better thermal shock resistance. These properties for some oxide solid electrolytes and other relevant materials are listed in Table 7.2. PSZ is therefore more attractive than fully stabilized cubic zirconia from this standpoint.[59–62] It has already been pointed out (Chapter 6) that the most severe limitation in the use of stabilized zirconia is its poor thermal shock resistance. Karaulov *et al.*[63] and Alekseenko *et al.*[64] have found that the spalling resistance is maximum when the PSZ contains about 30% monoclinic ZrO_2. A number of explanations

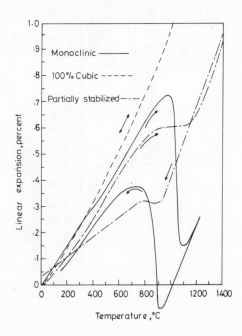

FIG. 7.1. Thermal expansion behavior of monoclinic, cubic, and partially stabilized zirconia (adapted from References 57 and 58).

Table 7.2. Properties of Oxide Solid Electrolytes and Related Materials

Property[a]	CaO-stabilized zirconia		Thoria	Alumina	18-8 stainless steel	Borosilicate glass
Coefficient of thermal expansion ($\times 10^6$ per °C, R.T. $-$ 1000°C)	8.8–11.8		8.2–9.9	8.4	9	2.7
Thermal conductivity (cal sec^{-1} cm^{-1} °C^{-1})						
R.T.	—		28×10^{-3}	70×10^{-3}	35×10^{-3}	2.5×10^{-3}
400°C	4.7×10^{-3} (100°C)		14	31.4	—	3.45
1000°C	5.5		7	14.7	—	—
Young's modulus (kg/cm^2)	10% CaO	15.5% CaO				
R.T.	1.47×10^6	0.74×10^6	2.43×10^6	3.08×10^6	2×10^6	0.69×10^6
600°C	1.13	0.59	2.22	2.85		
1000°C	1.1		2.05	2.69		

[a] R.T. = room temperature.

have been advanced to account for the improved thermal shock resistance of PSZ. According to King and Yavorsky,[65] in a single grain of PSZ, alternate monoclinic and cubic domains exist with regular structural relation between them. The microstructure study of Garvie and Nicholson[60] confirmed the presence of monoclinic domains of 1000 Å size embedded in a cubic ZrO_2 matrix. During the tetragonal–monoclinic transformation, high stress is developed due to the volume change. For the material to be thermally stable, this stress has to be relieved. Two mechanisms are sugested for this:[60,65] (a) The stress may be relieved through a deformation mechanism via slip. The reported superplasticity of monoclinic ZrO_2 at the monoclinic–tetragonal phase transformation enables the easy deformation of the monoclinic domains in the two-phase composite and thereby relieves the stresses.[66] On the other hand, the cubic phase is brittle and develops cracks during thermal cycling. (b) Microcracks develop in monoclinic ZrO_2 domains during thermal shock treatment through the transformation temperature. Because of their large numbers, these cracks propagate only quasistatically and the ceramic maintains a large fraction of its strength after thermal cycling. The relationship of microstructure to fracture in this material has been studied.[67] Recently, more detailed studies of specimens subject to different heat treatments were carried out using electron microscopy, fractography, selected-area electron diffraction, etc., which show that the extent of precipitation of monoclinic ZrO_2 can cause brittle or ductile behavior.[68,69] A sample given an optimum aging treatment contains particles of tetragonal symmetry which are coherent with the matrix and has toughness which is about twice that of quenched overaged samples. The toughness is attributed to particle–crack interactions which result in a transformation zone around the crack tip.[70–75] The role of liquid phase formed due to the presence of SiO_2 as an impurity was also discussed.[76]

PSZ ceramics were found to be satisfactory anionic conductors for oxygen partial pressures down to 10^{-15} atm at 1000°C, but electronic conductivity sets in when lower pressures (e.g., 10^{-22} at 1000°C) are employed.[77]

Garvie[78] has described a method of fabricating sintered PSZ ceramics. Zirconia of submicron particle size was mixed with 3–4 wt % CaO (in the form of Ca acetate monohydrate). Using carbowax as a binder, forms were cold pressed at 3000 psi, followed by isostatic pressing at 20,000 psi. They were heated at 600°C for 4 hr and then slowly to 1800–1900°C and held for 2 hr. The specimens were cooled to 1200–1300°C in 6 hr and annealed at that temperature for 1–2 days to precipitate fine monoclinic zirconia in the cubic grains. The specimens were then cooled to room temperature in 13 hr. The modulus of rupture of as-machined specimens as well as after repeated thermal shocking between 816 and 1038°C was found to improve threefold due to the annealing treatment. Marked increase in the modulus of elasticity was also noted.

3.2. Thoria-Base Materials

3.2.1. Powder Preparation, Mixing, and Calcination

Thoria powders can be prepared by the thermal decomposition of compounds like thorium tannate, carbonate, chloride, oxalate,[79,80] benzoate, and hydroxide[81] obtained by precipitation from thorium nitrate solution. Harada et al.[82] found that the surface area and volume shrinkage decreased with increasing calcination temperature. Thoria obtained from oxycarbonate had the highest surface area among the various materials calcined in the temperature range of 400–800°C and gave the highest fired bulk density.

Moorthy and co-workers[83-85] have extensively studied the sintering behavior of thoria powders from different origins with results generally in agreement with those of Harada et al.[82] The crystallite size of thoria powders derived by calcination at 600°C decreased in the order tannate, carbonate, oxalate, benzoate. Increase of weight due to hydration was also in the same order. A smaller crystallite size and a greater hydration in powder resulted in high densification. Thoria powders of different origins differ in their crystallinity. All the powders are amorphous at room temperature and on calcination at 400°C. Calcining at 1000°C makes thoria from oxalate crystalline though that from tannate still remains amorphous.

Ferguson et al.[86] have developed a sol–gel process for producing microspheres of ThO_2. In this process, an aqueous thorium nitrate solution is decomposed by steam. A low temperature evaporation converts the activated ThO_2 sol dispersed in dilute HNO_3 solution to a gel. The gel has a surface area of 90 m^2/g and contains particles of 100-Å size.[87,88] On heating, the gel loses additional water. On sintering between 1000 and 1150°C, it obtains a density of 9.9 g/cm^3 with a grain size of 0.5 μm. The ThO_2 microspheres may also be obtained by the removal of water by organic liquids.[89,90]

Solid-solution compositions can be prepared by coprecipitation from the constituent component solutions. For example, Subbarao et al.[91] coprecipitated thorium and yttrium as hydroxide and oxalate from the nitrate solutions mixed in proper proportion. Disks made from powder obtained by the hydroxide method achieved, on sintering at 1800°C, a density of 7.9 g/cm^3 compared to 9.7 g/cm^3 for disks made by the oxalate method. Segregation could be a problem in this method. Thompson et al.[92] observed differences in particle characteristics of ThO_2 + 12 mol % Y_2O_3 prepared in CO_2 and argon atmospheres. Particle size, green packing density, and sintered density were influenced by these differences.

The general method for preparing specimens of thoria consists of mixing the components in proper proportion, cold pressing, and then sintering. Thus Bransky and Tallan[93] prepared specimens from pure thoria (99.999%) by pressing and then sintering them, covered by thoria powder in a thoria crucible, in air at 1800°C for 4 hr to obtain 96% of theoretical density. Wimmer *et al.*[94] wet-mixed ThO_2 with 7 mol % Y_2O_3 in alcohol, dried and cold pressed into disks at 30,000 psi. The specimens were vacuum sintered at 2000°C in a thoria crucible. There was about 7–8% loss of Y_2O_3 on firing.

Mehrotra *et al.*[95] and Maiti and Subbarao[96] wet-mixed ThO_2 and $CaCO_3$, dried and calcined at 1100°C for 3–4 hr. Pellets were cold pressed at 75,000 psi and presintered at 1400°C. The presintering was repeated with intermediate grinding and pelletizing, until a homogeneous composition was obtained. Final sintering was carried out at 2000°C in a gas-fired furnace to achieve densities greater than 90% of the theoretical.

Transparent polycrystalline thoria-base ceramic has recently been made with its density approaching that of a single crystal.[97] Thoria of particle size 0.05–2.0 μm is mixed with 0.5–0.8 mol % CaO having particle size <0.1 μm and pressed at 10,000–100,000 psi. The body is fired at temperatures of about 2000°C in a hydrogen–water-vapor atmosphere until complete densification is achieved.

3.2.2. Fabrication

All the usual methods of ceramic fabrication are used for making thoria-base material. For example, cold pressing is carried at pressures from 10,000 to 100,000 psi. Isostatic pressing, sometimes of partially sintered compact, is also employed. The improved densification on resintering is attributed to the dislocations generated in isostatic pressing.[98]

Due to the high specific weight of ThO_2, particles less than 0.5 μm in size are employed to prepare a slip of specific gravity between 4 and 5. Its pH is adjusted to 3 by adding HCl. (1% polyvinyl alcohol to increase the stability of the slip.) A small amount of octyl alcohol as an antifroth agent[45] and 1% polyvinyl alcohol to increase the stability of the slip are added. The slip must be evacuated to remove entrapped air and agitated to prevent larger thoria particles from settling down. As usual, plaster of Paris molds are used for casting.

For extrusion, temporary organic plasticizers like rubber solution in xylene can be used.[45]

3.2.3. Sintering

Thoria is generally sintered at temperatures up to 2000°C in air or oxidizing atmosphere using thoria crucibles as containers. Densities in the range 9.5–9.8 (95%–98% of theoretical) are achieved.[99] Densities of 99% or higher of theoretical value are obtained by incorporating 0.5–1.0% CaO. The sintering temperature may then be lowered to 1800°C. Hydrogen, hydrogen–water-vapor, air, and vacuum[100] have been employed as the sintering atmosphere. The presence of CaO introduces oxygen ion vacancies and aids sintering.[101,102]

Halbfinger and Kolodney[103] studied the effect of small amounts of NiO, ZnO, and CuO on the sintering behavior of ThO_2 and ThO_2–Y_2O_3 solid solutions. Shrinkage was found to increase and sintering temperature to decrease with increasing addition of NiO. For example, high-density bodies of $Th_{0.85}Y_{0.15}O_{1.925}$ could be obtained by sintering at 1500°C by the addition of 0.8 wt % NiO.

Factors affecting sintering of ThO_2 between 1700 and 1900°C were determined by Oel.[104] Morgan and co-workers[105–108] found that matter transport during sintering of ThO_2 takes place by a dislocation motion. In the final stages of sintering, ThO_2 was found to exhibit discontinuous grain growth, whereas ThO_2 with 2% CaO had a uniform grain structure, presumably due to precipitation of excess CaO inhibiting grain boundary migration.[109]

Isothermal grain growth and intermediate stage sintering in ThO_2 and ThO_2 doped with 2.31, 4.25, and 8.8 mol % CaO was studied by Laha and Das[110] in the temperature range 1300–1600°C. The activation energy for diffusion calculated from the sintering data for the four compositions is 93, 56, 47, and 45 kcal/mol. These values may be compared with the activation energy for Th^{4+} diffusion obtained from a creep study (112 kcal/mol)[111] and from a tracer study (58 kcal/mol).[112,113] It is interesting to note the decrease in activation energy with CaO content.

Moorthy and co-workers[114,115] carried out hot-pressing experiments on thoria powders obtained from different sources and found that the hot-pressing behavior was in general similar to their sintering behavior following cold pressing. No significant shrinkage was found below the temperature at which the powder was calcined and shrinkage ceased beyond a certain temperature. Effect of pressure on densification was significant in powders calcined at higher temperature. Densification of powders calcined at low temperature occurred in early stages of heating due to pore entrapment.

Powders obtained from tannate could be hot pressed to 95% of theoretical density at 1200°C and 3000 psi, whereas under the same conditions, the density of pellets from powders of oxalate origin gave densities of only about 70% of theoretical.

3.3. β-Alumina

There are two high-conductivity, nonstoichiometric phases in the Na_2O–Al_2O_3 system (Fig. 1.31). These are β-alumina, with its formula varying from $Na_2O \cdot 11Al_2O_3$ to $Na_2O \cdot 9.5Al_2O_3$, and β''-Al_2O_3, richer in soda than β-Al_2O_3 and having compositions between $Na_2O \cdot 7Al_2O_3$ and $Na_2O \cdot 5Al_2O_3$. β''-Al_2O_3

decomposes into β-Al$_2$O$_3$ and NaAlO$_2$ above the eutectic temperature, 1595°C. Incorporation of a small amount of MgO or Li$_2$O stabilizes the β''-Al$_2$O$_3$ phase, so that it can be sintered at higher temperatures without decomposition. At the same time, addition of MgO to β- and β''-Al$_2$O$_3$ increases their electrical conductivity (see Chapter 1, Section 4.2).

3.3.1. Preparation

Ground fused-cast bricks (from, e.g., Harbison Carborundum Company or Electro Refractaire) or calcined product from Aluminum Company of America are extensively used as the source of β-Al$_2$O$_3$ powder. Though β- and β''-Al$_2$O$_3$ can in principle be prepared by the calcination of appropriate mixtures of Na$_2$CO$_3$ and α-Al$_2$O$_3$, there are many conflicting and confused reports in the early literature, which was carefully reviewed by DeVries and Roth,[116] and, more recently, by Kummer.[117] Ray and Subbarao[118] have pointed out the difficulties of segregation and nonequilibrium phases inherent in such synthesis. Due to the great disparity in the melting points of the constituents, segregation takes place on heating to temperatures above 850°C (melting point of Na$_2$CO$_3$), such that molten sodium carbonate with α-Al$_2$O$_3$ embedded in it settles to the bottom of the crucible, while loose powder of α-Al$_2$O$_3$ more or less free of soda is present at the top. Any amount of thorough mixing of the initial constituents does not seem to eliminate the problem of inhomogeneity completely, though repeated calcination with intermediate grinding minimizes the compositional nonuniformity. Starting with NaAlO$_2$ and α-Al$_2$O$_3$ is an improvement, since calcination at 1300°C for 10 hr then gives rise to β- or β''-Al$_2$O$_3$ depending upon the initial composition and the soda loss.[119] Mitoff[32] has emphasized the use of low-melting salts or hydrated salts, such as hydrated ammonium aluminum sulfate, hydrated sodium aluminum sulfate, and magnesium sulfate, which melt together at about 100°C. This material foams as it dehydrates at about 200°C and can be calcined at a high temperature without sintering, so that grinding into a fine powder is easy. The formation of β-alumina by the sulfate route was found to be a slow process. For example, calcination at 1000°C leaves the material in an amorphous state. The β-Al$_2$O$_3$ structure is gradually formed at higher temperatures, and finally at 1825°C completely crystallized β-Al$_2$O$_3$ results. Decomposition of nitrates, alkoxides, and oxalates, and coprecipitation are the other methods which may be employed. The effect of powder preparation techniques on the final products is discussed by Powers and Mitoff.[125]

Grinding β- and β''-Al$_2$O$_3$ with alumina balls, which is usually required, introduces α-Al$_2$O$_3$ as a contaminant.[32] Ethylene glycol, among many other media, has been investigated as a grinding aid.[120] Generally, particle sizes in the range of 0.1–1 μm are sought. For example, the calcined β-Al$_2$O$_3$ prepared by Francis *et al.*[120] had a surface area of 1.2 m^2/g and a particle size distribution of $+100$ mesh 13%, $+200$ mesh 91%, $+325$ mesh 99%.

Stabilized β''-Al_2O_3 of composition $Na_2O \cdot MgO \cdot 5Al_2O_3$ was prepared[121] by streaming a solution of $MgCl_2$ and $AlCl_3$ having proper cation ratio into a stirred solution of NH_3 at 80°C, pH being kept between 9.5 and 10. The precipitate was dried and then calcined at 1300°C for $\frac{1}{2}$ hr. About 10% Na_2O in the form of Na_2CO_3 was added to it, thoroughly mixed, and calcined for 3 hr at 1300°C. Excess soda was washed away by dilute HCl.

Addition of fluorine (as NaF) to a 1:5 mixture of γ-Al_2O_3 and Na_2CO_3 gave β- or β''-Al_2O_3 on heating at 1250°C. With 0.1% F only β''-Al_2O_3 was obtained and at 1% F only β-Al_2O_3, with mixture of β- and β''-Al_2O_3 at intermediate values of F.[122] Further, β''-Al_2O_3 synthesized from γ-Al_2O_3 and Na_2CO_3 had defect structure (stacking faults) which was absent when α-Al_2O_3 was used.[122] The effects of other impurities (e.g., Cd, Pb, etc.) on the stability of β- and β''-Al_2O_3 has also been studied[123] as well as the stoichiometry.[124]

3.3.2. Fabrication

Many of the usual and some uncommon fabrication methods are employed with β-aluminas. Uniaxial pressing can be used for certain shapes, though shape distortions on firing are not uncommon due to friction of the wall dies.[32] On the other hand, such distortions can be avoided and a uniform microstructure achieved by using isostatic pressing. In this method, the space between a metal mandrel and a stiff rubber mold is filled with powder and subjected to hydrostatic pressure. After pressing, the specimen slides off the mandrel and sintered in the usual way. While no binder is necessary for small sizes, 2% wax is added for larger shapes. The organic binder is burnt out by slowly heating to 300°C.

Electrophoresis was employed by Fally *et al.*,[18] Powers,[126] and Kennedy and Foissy[127,128] to produce β-Al_2O_3 tubes. In this method the fused β-alumina was ball milled and metallic impurities and any sodium aluminate present removed by treatment with HCl. The powder was suspended in an organic solvent (amyl alcohol or dichloromethane) and electrically charged by adsorption of positive charges from dissociation of a dissolved organic acid. Trichloracetic acid may be added to the suspension to facilitate particle charging. Homogeneous closed-end tubes of thickness ranging from 200 to 2000 μm, 1 cm in diameter, and up to 10 cm in length may be obtained.

Other fabrication methods considered include extrusion for tubes and rods, plasma spraying on metal mandrels, injection molding, fusion casting, slip casting, and hot pressing. The fabrication technique employed can give rise to preferred orientation of grain structure, e.g., in hot pressing. This can have an important effect on the electrical properties due to the anisotropy of electrical resistivity.[125]

3.3.3. Sintering

Sintering temperatures in the range 1600–1900°C are necessary to obtain dense, impervious bodies of β-Al_2O_3 and stabilized β''-Al_2O_3. Loss of soda at these temperatures causes compositional changes and appearance of α-Al_2O_3 as a second phase surrounded by pores. Francis *et al.*[120] have found that burying the specimens under β-alumina powder during sintering minimizes the loss of soda. Sintering for 1 hr at 1600, 1650, 1700, and 1780°C in a gas-fired furnace showed a decrease in the total and apparent porosity with increasing sintering temperature up to 1700°C, and then porosity increases at 1780°C. Bodies sintered from ground fused bricks appear to be brittle, porous, and permeable to gases.[119] Long, thin electrolyte tubes fabricated by iso-static pressing were sintered at 1600–1900°C in a unique zone-sintering furnace to achieve zero open porosity and $< 5\%$ closed porosity.[129] Imai and Harata[130,131] investigated the sintering behavior and the phases formed when ground fusion-cast β-Al_2O_3 bricks are mixed with varying amounts of MgO, NiO, and ZnO. The sintering of the disks was carried out at 1700°C for 4 hr in closed platinum crucibles to prevent evaporation of sodium oxide. For example, in the compositions $(1.16 + y)Na_2O \cdot xMgO \cdot 11Al_2O_3$, β-Al_2O_3, and $NaAlO_2$ are present for $x = 0$ and $y \geqslant 0.79$. Some unknown phase was present in compositions with large x ($x \geqslant 1.25$) and small y. When both MgO and Na_2O are present sufficiently (i.e., $2x = y$ and $x \geqslant 1.25$), β''-Al_2O_3 was observed. The sodium content in β-Al_2O_3 may be increased by incorporating Mg^{2+} ions in Al^{3+} sites in the spinel lattice. Kennedy and Sammells[132,133] reported that the sintering temperature increases from 1720°C for β-Al_2O_3 to 1750°C for 7 hr for β-Al_2O_3 with 1%–2% MgO and to 1830°C for 6 hr for β-Al_2O_3 with 4% MgO.

Hot pressing at 1200–1300°C and 50–100 kg/cm² with a final, quick treatment at 1600–1650°C and 150–200 kg/cm² was successfully employed to obtain impervious β-alumina-type bodies.[119] More recently, Virkar *et al.*[134] used hot pressing, followed by annealing at 1300–1400°C to increase the electrical conductivity by converting β-Al_2O_3 into β''-Al_2O_3. Incomplete conversion to β''-Al_2O_3 after hot pressing may also be due to the relative stability of β- and β''-Al_2O_3 as a function of pressure.[135] Above 5 kbar and 560°C, Na^+ β-Al_2O_3 decomposes to α-Al_2O_3 and $NaAlO_2$. At slightly higher temperature, magnesia-stabilized β''-Al_2O_3 decomposes to α-Al_2O_3, $MgAl_2O_4$, and $NaAlO_2$. Gordon and co-workers[136,137] have developed a process in which lithium is introduced as $Li_2 0.5Al_2O_3$ (ζ-lithium aluminate) into the reaction mixture for better distribution of the minor constituent. Using this method and rapid sintering (2–10 min) at 1600°C and then annealing at 1475–1550°C for 1 hr, high density ($> 98\%$), low resistivity (~ 4 ohm cm), high strength ($\geqslant 30$ kpsi), β''-Al_2O_3 ceramic is obtained.

The effect of sintering time on the electrical conductivity was studied by

Whalen *et al.*[139] Atmosphere control is found to be important in sintering. Pure oxygen is better than air since oxygen diffuses easily through β-Al_2O_3, but not nitrogen which gets trapped in pores. Similarly, water vapor is undesirable, as α-Al_2O_3 and NaOH appear in the ceramic sintered in moist atmosphere.[125] Since the electrical conductivity of these materials is strongly anisotropic, the microstructure of the polycrystalline material has an important bearing on the conductivity of the ceramics. Uniaxed grains are preferred to duplex structure (large, elongated crystals embedded in a matrix of fine particles) and uniform particle size to uneven grain size Sintering procedures and starting materials have to be optimized to achieve this.[125]

Though conduction in β-Al_2O_3 is strongly anisotropic, grain size has only a small effect on the conductivity. For example, a 100-fold increase in grain size increases the conductivity only by a factor of 2.[138]

The variable composition of β- and β''-Al_2O_3 usually requires the determination of Na_2O and MgO content, if any, by chemical analysis of the sintered product using atomic absorption techniques. Among the physical properties of a β-Al_2O_3 ceramic electrolyte, the most important is its impermeability to species on both sides of the electrolyte. The progress of sintering in this regard can best be followed through microstructural study[32] or measurement of permeability with a suitable apparatus.[119] Permeability, particularly through cracks, in the presence of corrosive media employed in batteries or other galvanic cells, may be evaluated through mechanical properties measured during exposure to the corrosive environment. In this connection, electrolysis of simple sodium salts, followed by fracture–stress measurements have been proposed as screening tests for β-alumina electrolytes.[139]

3.4. AgI-Type Electrolytes

The discussion here is confined to superionic conductors based on AgI. The thermodynamic properties and phase diagrams of this class of materials, recently summarized by Owens,[140] provides a useful background for their preparation.

(*a*) *Ag_3SI.* Polycrystalline Ag_3SI can be prepared[141] by grinding together equal mole fractions of AgI and Ag_2S and heating the mixture in a sealed evacuated glass tube at 165°C until the reaction is complete (in about 17 hr). Ag_2S decomposes in vacuum to metallic silver and sulfur and this metallic silver imparts electronic conductivity to the electrolyte Ag_3SI. To prevent the decomposition of Ag_2S, sulfur is placed at a suitable position in the glass tube to maintain the vapor pressure of sulfur at 1 atm. Powdered Ag_3SI is cold pressed into pellets at 4000 kg/cm².

(*b*) *Ag_2HgI_4.* This can be prepared[142] by coprecipitation from a solution of HgI_2 in KI and $AgNO_3$. The powder is dried by leaving it in vacuum at room temperature for 24 hr. Pellets are usually pressed at 10,000 kg/cm^2.

(c) MAg_4I_5. In these superionic conductors, M may be K^+, Rb^+, or NH_4^+. $RbAg_4I_5$ has higher conductivity and is more stable than the other two. $RbAg_4I_5$ can be prepared in two ways: Owens and Argue[143] prepared it by melting stoichiometric mixture of RbI and AgI and then quenching the melt. The quenched product was ground, made into pellets, and then annealed at 165°C for 16 hr. The process of pulverizing, compaction, and annealing was repeated to ensure formation of a single phase. KAg_4I_5 was found to be hygroscopic and hence a dry box was used. On the other hand, Scrosati[30] prepared the compound by crystallization from a solution of AgI and RbI (in the ratio of 2:1) in acetone. Excess RbI, which precipitates spontaneously, is filtered out and the solution is slowly evaporated:

$$4AgI + 2RbI \rightarrow 4AgI \cdot 2RbI \rightarrow RbAg_4I_5 + RbI \downarrow \qquad (7.6)$$

(d) $Ag_7I_4PO_4$. This can be prepared[144] by mixing stoichiometric amounts of AgI and Ag_3PO_4 and then heating the mixture in a sealed Pyrex tube under vacuum at 400°C (above the liquidus temperature) for about 18 hr. It is then slowly cooled to room temperature at a cooling rate of about 1.5°C/min. It must, however, be noted that on cooling the melt below the eutectic temperature of 235°C, solidification starts with the formation of AgI and the nonconductive $Ag_5I_2PO_4$. It is interesting to note that these phases recombine at the low temperature of 79°C to form conducting single-phase $Ag_7I_4PO_4$, as confirmed by x-ray studies.

(e) $Ag_{19}I_{15}P_2O_7$. Heating an intimate mixture of stoichiometric amounts of AgI and $Ag_4P_2O_7$ in a sealed glass tube at 250°C for about 20 hr and then slowly cooling to room temperature yields this compound.[144]

(f) $(Me_2Et_2N)Ag_7I_8$. Recently Owens[145,146] has developed a new class of high-conductivity solid electrolytes formed by the combination of tetra-alkyl-ammonium iodide with AgI. The double salts are prepared by the combination reaction

$$QI + nAgI = QAg_nI_{n+1} \qquad (7.7)$$

where QI is the quarternary tetra-alkyl-ammonium iodide salt. Highest conductivity was found with a double salt in the Me_2Et_2NI–AgI system for 87.5% AgI. The corresponding formula may be $(Me_2Et_2N)Ag_7I_8$. This can be made by intimately mixing the reactants in stoichiometric amount, then compacting and heating in a sealed tube in argon atmosphere at a temperature of 125–165°C. Multiple compaction and annealing steps are used to achieve an equilibrium product.

Vacuum-deposited thin films of $RbAg_4I_5$ have been prepared.[147,148] These films exhibit a conductivity of 0.25 $(ohm\ cm)^{-1}$ at 25°C and an activation energy of 2.3 kcal/mol.[148]

4. Electrodes

The chief characteristics which a satisfactory electrode for solid electro-lyte systems should possess are electronic conductivity, ionic conductivity for the conducting ionic species, and stability in the environment in which it is used. These requirements make a particular electrode specific to a given electrolyte. The reversible electrodes fit the above description, but the non-reversible ones do not permit the diffusion of the conducting ions and there-fore serve as blocking electrodes. The pile-up of charged species at such electrodes leads to polarization. This can be overcome only by employing sufficiently high frequency electric fields.[149]

4.1. Oxygen Ion Conductors

Platinum, prepared and applied in different ways, is the most common electrode used with thoria- and zirconia-type electrolytes. In order to evaluate various kinds of platinum electrodes applied to one side of $Ca_{0.15}Zr_{0.85}O_{1.85}$ disk as the cathode, a thin sputtered platinum anode was applied to the other side and their I–V characteristics studied by Kröger and co-workers.[150,151] The cathodes studied were the following.

(a) *Nonporous.* (i) Platinum foil, 30 μm thick, attached under 600-g load by heating at 1350°C for 1 hr; (ii) 60% Pt–40% Rh alloy foil applied the same way as in (i) [not much different in behavior from (i)]; (iii) thick sputtered electrodes; and (iv) paste containing 10 wt % lead borosilicate glass as flux.

(b) *Porous.* Platinum paste, containing submicron platinum powder suspended in an organic vehicle, painted and fired to different temperatures to vary porosity and grain size. This study led to the following conclusions: (1) For measurement of bulk ionic conductivity of zirconia electrolytes, thin sputtered[152] or unfluxed pastes[153] are suitable. (2) For polarization measure-ments, thick foils or fluxed paste electrodes are needed.

Karapachov *et al.*[154] used evaporated Pt layers, in which they made scratches to create triple-phase boundaries (electrode–electrolyte–gas). The reactions at the Pt electrodes are studied by Kröger and co-workers,[150] among others, and are discussed in Chapter 5, Section 2.1.

When stabilized zirconia is used as an electrolyte in a fuel cell operating at about 1000°C, the electrodes have to satisfy stringent chemical, electro-chemical, mechanical, and economic conditions, which are summarized in Chapter 6, Section 4.2.2. Tedmon *et al.*[155] have evaluated metals (Pt, Pd, Au, Ag), oxides (porous stabilized zirconia) with current collector grids made of Pt or Pd wires, and electronically conducting oxides as potential cathodes in fuel cells. Low melting points, high vaporization rates, and economics essentially rule out the noble metals with or without the oxide matrix. The interest therefore centers around oxide conductors such as Li-doped NiO, Sr-doped perovskites (e.g., $LaCoO_3$, $LaMnO_3$), Sn-doped In_2O_3, ZnO doped with 3% Al_2O_3, or ZrO_2-doped SnO_2, $PrCoO_3$, etc. The conduction mechanism in these oxides is not always well understood. The temperature variation of conductivity shows the cobaltates and SnO_2 as promising cathode materials (Fig. 7.2). The rapid loss of lithium from Li-doped NiO and the reaction of doped ZnO with the electrolyte at temperatures above 1000°C are limitations in these cases. Doped In_2O_3 or SnO_2 was applied by vapor deposition (which may not be a practical proposition in the case of real fuel cell systems). When the thickness of the air electrode based on these materials exceeds about 3×10^{-2} mm, it develops cracks during cooling due to thermal expansion mismatch. Perovskite-type cobaltates of La and Pr offer interesting possibilities as air electrodes. These materials are usually applied as a slurry and heated, sometimes *in situ*, to develop a good bond with the electrolyte. The thermal expansion coefficient of $LaCoO_3$ is 28×10^{-6} °C^{-1} (that of $PrCoO_3$ is nearly the same) compared to that of stabilized zirconia of about 10×10^{-6} °C^{-1}. This mismatch leads to cracking and spalling of the electrode during cooling and thermal cycling. At temperatures above 1100°C these oxides react with the electrolyte causing local destabilization of the zirconia. Sometimes, a more complex air electrode has been used in high-temperature fuel cell systems, for example, a three-layer electrode consisting of stabilized ZrO_2 porous skeleton, plasma-sprayed In_2O_3 doped with 2 mol % SnO_2, and a thin layer of $PrCoO_3$.[47] Electrical conduction in metal oxides, including the cobaltates, has recently been reviewed.[156]

Cermets based on a mixture of stabilized zirconia and metal (e.g., Co) powders are coated on zirconia electrolyte and used as fuel electrodes in fuel

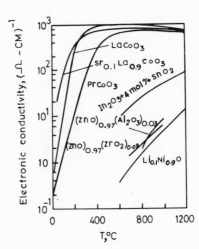

FIG. 7.2. Temperature dependence of electrical conductivity of promising cathode materials in high-temperature fuel cells (Reference 155).

cells.[47] Ni–yttria-stabilized zirconia Cermet electrodes are used for electrochemical dissociation of water vapor in zirconia electrolyte cells.[50,157] These are prepared by spray-drying a solution obtained by mixing fine, reagent-grade NiO and yttria-stabilized zirconia with water, alcohol, and a suspending agent. The powder of the mix is then screened to agglomerate to 50-μm size, calcined in air at 1325°C, and plasma sprayed. The NiO in the electrode was subsequently reduced in H_2 at about 400°C to produce a Ni–yttria-stabilized zirconia electrode. Co–yttria-stabilized zirconia electrode may be prepared in a similar way. The 20 wt % yttria-stabilized zirconia present in these Cermets serves to minimize sintering and growth of the metal powders during use. Other anode materials that hold promise are nickel, manganese, titanium oxide, and ceria doped with lanthana or yttria.[158]

For emf measurements, one sometimes uses metal–metal-oxide electrodes, which are prepared by intimately mixing suitable proportions of the metal and metal oxide powders and cold pressing into pellet form. This type of electrode cannot be used for conductivity measurement as the area of contact of the metal particles does not conform to the specimen area. These composites exhibit an oxygen environment (indicated by p_{O_2}) which varies with temperature. The data for the more common metal–metal-oxide electrodes are given in Table 7.3 at 800 and 1000°C.[159] The thermodynamic equations for these and many other electrodes are given in Table 4.1.

4.2. β-Alumina

A variety of solid and liquid electrodes have been used in conjunction with single-crystal and polycrystalline β- and β″-alumina, as summarized in

Table 7.3. Metal–Metal-Oxide Electrodes[159]

Electrode	$-\log p_{O_2}$	
	1000°C	800°C
Cu, CuO	0.9	2.9
Cu, Cu_2O	6.3	8.8
Ni, NiO	10.3	13.9
Co, CoO	12.0	15.1
Fe, FeO	14.9	18.9
Cr, Cr_2O_3	21.8	27.5
Nb, NbO	25.1	31.5
V, VO	26.4	32.9
Ti, TiO	34.8	42.8

Table 7.4. Electrodes for Conductivity Measurements on Sodium β-Aluminas

Material	Electrode	Frequency (Hz)[a]	$\sigma \times 10^3$ (ohm cm)$^{-1}$	Temp. (°C)	Activation energy (kJ/mol)	Reference
Single-crystal β-Al_2O_3	In	500 K	30 ± 3	25	11.1	14
	In	500 K	285	300		
	aqueous 5 M NaNO₃	10 K–1 M	23	25	14.0	160
	molten NaNO₃/NaNO₂	—	355	300		
	Na	—	76.5	100	—	164
	Na$_x$WO$_3$	—	14	25	15.8	165
Sintered disk β-Al_2O_3	Aqueous 5 M NaNO₃	1 M	1.84	25	14.0	160
	NaNO₃/NaNO₂	10 K–1 M	28.5	300	—	160
	Na	dc	4.5	25	—	14
	Na	dc	55	300	—	14
MgO-doped (1%)	Evaporated Ag	0.75–10 K	0.66	25		132
MgO-doped (2%)	Evaporated Ag	0.75–10 K	3.5	25	14.6	132
MgO-doped (4%)	Evaporated Ag	0.75–10 K	8.0	25	14.6	132
MgO-doped (2%)	Evaporated Ag	0.75–10 K	4.4	25	—	132
MgO-doped	Na	1592	36	300	13.7	160
	Ag	1000	125	350	16.0	18
MgO- and ZrO₂-doped	Ag	1000	200	350	16.0	18
Pure β-alumina	Na	20 K	62.5	350	—	163
Single-crystal β''-Al_2O_3	Pt	1 M	1453	300	—	117
Sintered disk β''-Al_2O_3	Na	dc	247	300		117
MgO-doped (2%)	In	dc	117	300	16.8	166
Li₂O-doped (0.7%)	In	dc	79	300	28.8	166

[a] K stands for kilohertz, M stands for megahertz.

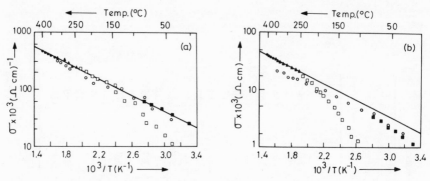

FIG. 7.3. Conductivity data of β-alumina measured with different electrode systems (Reference 160). (a) Single-crystal electrolyte, (b) sintered electrolyte. $\square$, Na/Hg; $\blacksquare$, aqueous $NaNO_3$; $\bigcirc$, Au; $\blacktriangle$, $NaNO_3/NaNO_2$; $\bullet$, $NaNO_3$.

Table 7.4. The results of Armstrong et al.[160] are presented in Fig. 7.3 where the temperature dependence of conductivity of single-crystal and poly-crystalline β-alumina, measured with various electrodes, are plotted. The data on single crystals show less scatter than those on sintered electrolytes. Molten salt contacts, aqueous $NaNO_3$ in contact with a single crystal, Au and Na/Hg contacts (under an argon atmosphere) at high temperature are satisfactory at measuring frequencies of the order of 1 MHz. The Na/Hg deviates from the expected behavior in the case of sinters at low temperatures. Aqueous 5 M $NaNO_3$ (30–80°C) and Au give Arrhenius-type variation of conductivity with temperature, though exhibiting somewhat lower conduc-tivity values. Because of the blocking or nonreversibility of Pt electrodes, conductivity measurements on single-crystal and polycrystalline specimens of β-Al_2O_3 are strongly frequency-dependent in the temperature range 30–518°C.[161] Sodium amalgam has been successfully used as electrodes in concentration cell to determine the electronic transference number of β-alumina.[162]

Recently, Tiku[167] carried out a detailed study of the electrical conduc-tivity of β- and β''-Al_2O_3 in pure and MgO-doped form as a function of fre-quency and temperature and using a variety of electrodes. The conductivity values obtained with molten $NaNO_3$ electrodes were 5–10 times higher than those with silver paste electrodes, while liquid sodium electrodes are inter-mediate between these two. $NaNO_3$ and Na electrodes being liquids are re-versible to a large extent, whereas the painted silver electrode is essentially a blocking electrode. Painted silver electrodes exhibit polarization effects up to a frequency of at least 1 MHz. On the other hand, sodium electrodes have the least frequency dependence and therefore can be used at relatively low fre-quency. However, handling of molten sodium poses problems. Liquid $NaNO_3$ therefore appears to be the most convenient electrode for routine testing of

Na$^+$ β-alumina specimens, though it can be used only over a limited temperature range (the melting point of NaNO$_3$ is 310°C). These aspects are discussed in greater detail by Powers.[125,168]

Kennedy and Sammells[132] cleaned the surface of polycrystalline sodium β-alumina by glow discharge and then vapor-deposited Ag or Au to carry out ac conductivity measurements at a frequency of about 1 MHz to avoid polarization at the electrodes. The measured resistance varies as $\omega^{-1/2}$, where ω is the frequency.[133] Extrapolation of this straight line to infinite frequency gives the true resistance of the sample. The pellet resistance was found to increase with time, apparently due to deterioration of the contact. Silver electrodes obtained by baking at 500°C a coating of silver resinate covered by a silver foil was employed in the case of silver β-alumina single crystals.[169]

For dc measurements, sodium tungsten bronzes[170] or liquid sodium or molten sodium salts[160,161] such as nitrate or nitrite may be used. Tungsten bronzes can be prepared[171] by reaction of Na$_2$WO$_4$, W, and WO$_3$ mixed in proper proportion according to the reaction

$$\frac{x}{2} \, Na_2WO_4 + \frac{3-2x}{3} \, WO_3 + \frac{x}{6} \, W \xrightarrow[10\,hr]{850°C} Na_xWO_3 \qquad (7.8)$$

The heating is done in an alumina boat in a purified argon atmosphere. The bronze powder is crushed, pressed into pellet at about 20 tsi, and sintered at 850°C under purified argon. These pellets can be used as electrodes at 10 KHz even at low temperatures without polarization effects, provided an anhydrous helium atmosphere is maintained in the cell. For the cathode in galvanic cells based on β-alumina, iron-doped β-alumina of the composition Na$_2$O$\cdot$11 (Fe$_x$Al$_{2-x}$O$_3$) was pelletized and sintered at about 1100°C. The electronic conductivity may be further increased by the addition of TiO$_2$.[133,172] The role of Fe$_2$O$_3$ or other transition metal oxides in the cathode of these cells is not clearly understood.

In Na–S batteries based on sodium β-alumina, molten sodium metal, and graphite felt or porous graphite impregnated with sulfur are used.[163,173] The graphite felt, or carbon cloth, electrode is inexpensive and has the same electrochemical properties as other carbon electrodes.[174] The electrochemical area is a few times the geometrical area. The electrode can stand up to high current density (e.g., 100 mA/cm^2). However, being a woven fabric shredding takes place and the mechanical strength is poor. The conductivity of sulfur may be increased by a suspension of carbon black in sulfur.[175] The penetration of sodium and sulfur into the electrolyte during the operation of the battery leads to the formation of dendrites, depending upon the amount of the current passed through the cell.[138] Impurities and solid reaction products of discharge experiments set up stresses and cause failure to the electrolyte.[176] Recently Armand[177] discussed the preparation and characterization of a new family of interstitial compounds derived from graphite. Their general formula is

$M_x - (C_\alpha - M'X_n)$, where M is an alkali metal, M' is a transition metal in a high valence state (Cr, Fe, Mn), X is an electronegative nonmetal (O, F, Cl) or a grouping, with a ratio n to the M' atoms, α is the ratio of carbon (graphite) to transition metal atoms, and x is the alkali metal content, which can vary widely. These materials exhibit mixed conductivity due to the movement of alkali ions and electrons. A solid state battery $Na/\beta\text{-}Al_2O_3/C_8CrO_3$ exhibited an open-circuit voltage of 3.9 V at 25°C with an energy storage capacity of 1.1 kW-hr/kg of electrodes. The electronic conductivity of these compounds is higher than that of graphite and the ionic conductivity is about 10^{-2} $(ohm\ cm)^{-1}$ at 25°C. The fact that these materials are solids, are safe and cheap, and represent a variety of compounds with different ions and variable stoichiometry makes them promising electrodes for β-alumina-type electrolytes. More recently, Na_xTiS_2-type conductors are being studied for electrode applications.[178,179]

4.3. Fluorides

In the case of fluoride electrolytes, molybdenum or tantalum electrodes may be used since they do not react with the fluorides.[180]

4.4. Silver Ion Conductors

Silver foil and evaporated silver electrodes usually have a high contact resistance.[143] Silver powder pressed over the electrolyte gives better results.[181] Polarization was found to be minimum with a liquid interface as well as with silver amalgam.[149,182] But the conductivity of $RbAg_4I_5$ electrolyte was found to decrease with time when Ag–Hg electrodes were used. The best performance was obtained using a mixture of 20% $RbAg_4I_5$ and Ag powder pressed as the electrode.[183] The mixtures were prepared by the reduction of Ag_2O *in situ* in a carbon–$RbAg_4I_5$ matrix. Micropolarization studies showed the silver electrodes to be reversible but not the mixed silver–electrolyte electrodes when used with $RbAg_4I_5$ solid electrolyte.[184] Dendritic growth of silver along the preferential deposition points accounts for this behavior. The cathode discharge reaction in the cell $Ag\ |\ RbAg_4I_5\ |\ RbI_3$ formed $RbAg_4I_5$ and Rb_2AgI_3 when the cell operated above 27°C and Rb_2AgI_3 and AgI at temperatures below 27°C.[185,186] Therefore, while the Ag–$RbAg_4I_5$ pellet was used as the anode, the cathode was a pellet of a mixture (by weight) of Te and Ag_2Te (1:2), Se, Ag_2Se (1:2), Te, Ag_2Te, graphite, and $RbAg_4I_5$ (0.3:0.1:0.1:0.7) or Se, Ag_2Se, graphite, and $RbAg_4I_5$ (0.3:0.1:0.1:0.7).[187]

Takahashi and Yamamoto[141] used an iodine cathode with Ag_3SI electrolyte but Owens *et al.*[188] have shown that, under these conditions, Ag_3SI decomposes into AgI and solid sulfur, the AgI serving as a blocking electrode for electrons.

5. Joining of Materials

Whether it is an experimental setup or a commercial device, a solid electrolyte has often to be joined to other materials. A successful seal has to meet some general requirements for electrical, chemical, thermal, and stress behavior, whereas other requirements are[189]: a matching of the thermal expansion coefficients of the materials to be joined, high electronic resistivity ($> 10^6$ ohm cm), chemical compatibility with the environment, impermeability to liquids and gases present, resistance to thermal cycling, and mechanical strength to withstand normal mechanical shocks and stresses. Graded seals may sometimes be employed to overcome thermal expansion mismatch.

Calcia-stabilized zirconia tubes are joined to a graded seal (Corning No. 6460) which in turn was sealed to a standard ultrahigh vacuum flange through Kovar to make a selective oxygen leak device.[190] Pizzini *et al.*[191] have described a method of joining stabilized zirconia tube to a tube of nickel–iron alloy which has the same thermal expansion coefficient as the zirconia in a butt joint making use of a soft glass. The composition of the glass, which has a softening point of 570°C, is 70% SiO_2, 4% Al_2O_3, 1% B_2O_3, the rest being Na_2O, CaO and traces of K_2O. The glass is ground to an average grain size of·10 μm. The powder is uniformly distributed on the mechanically polished surface of the metal and heated to obtain a thin layer of glass (about 0.5 mm). The metal and ceramic are then brought into contact and heated with a glass blower's flame, avoiding thermal shocking of the seal or the ceramic. The seal is cooled carefully over 10 min. Stabilized zirconia has been sealed to itself using Pyrex[192] or plasma-sprayed Ni or Co.[56] Zirconia stabilized by 15% CaO may be joined to an alumina tube using a glass of composition 12 wt % Na_2O, 20% Al_2O_3, and 68% SiO_2.[193]

The sealing of zirconia electrolyte pellet to a fused silica tube in order to fabricate a probe to measure oxygen content in molten steel has been described in Chapter 6 (Section 2.7.3). Briefly, it consists of heating a slightly oversized silica tube (7-mm diameter) in an oxygen hydrogen flame until it softens and collapses onto the pellet (5-mm diameter). The assembly is cooled to cherry red heat and finally quenched in water. Oxygen probes have also been made by fusing zirconia pellets to an Al_2O_3 tube. A stabilized zirconia tube has also been sealed to solt glass and then to Pyrex through graded glass seals.[194] Cobalt chromite may be used to interconnect cells of zirconia electrolyte to build a fuel cell.[195] It is an electronic conductor stable in both air and mixtures of H_2–H_2O and CO–CO_2 having oxygen partial pressures greater than 10^{-15} atm at 1000°C. The resistivities of manganese-doped cobalt chromite are sufficiently low to make it useful for the fuel cell battery application. $LaCrO_3$ and $LaMnO_3$ are also used for interconnection.[196]

Sintered β-alumina ceramic tube can be sealed to Pyrex with "Autostic" (Carlton Brown Ltd., Elbord, England), which is an alumina-based cement of negligible conductivity and apparently inert to molten salts and sodium amalgam.[160] It is stable at least up to 1100°C. Graphite packing ("Grafoil," Union Carbide Corporation) is also utilized for this purpose.[173] In the case of Na/S battery systems based on β-aluminas, the sealant between the electrolyte and an insulating ceramic such as α-alumina must resist attack by sodium as well as by the sulfur/sodium polysulfide system. A borosilicate glass frit has been employed for this purpose, though resistance to thermal cycling and to high temperatures has yet to be established.[197]

6. Conclusions

The poor thermal shock resistance and the limited sources of supply of calcia-stabilized zirconia ceramics need attention as industrial uses of this electrolyte expand. Thoria-base electrolytes are more stable but have poorer thermal shock resistance. Development of fabrication techniques for β- and stabilized β''-alumina shapes of high density, particularly with some degree of preferred orientation, are warranted. In spite of their high cost, the recently discovered Ag^+ ion conductors are likely to be attractive for small-size battery applications due to their excellent conductivity around room temperature. Hence methods of obtaining high-density compacts of these need to be improved.

Platinum electrodes, which have been used extensively so far, need to be replaced by less expensive electronic conductors such as oxides and Cermets. Joining techniques have received minimal attention so far. As device interest in these electrolytes increases, better joining methods have to be developed.

ACKNOWLEDGMENTS

The authors are grateful to the U.S. National Bureau of Standards and the U.S. Aerospace Laboratory, who have supported their work in this field, in part, over the years.

References

1. T. B. Reed, *J. Appl. Phys.* **32**, 2534 (1961).
2. G. J. Goldsmith, M. Hopkins, and M. Kestigian, *J. Electrochem. Soc.* **111**, 260 (1964).
3. T. S. Laszlo, P. J. Sheehan, and R. E. Gannon, *J. Phys. Chem. Solids* **28**, 313 (1967).
4. W. Swanger and F. R. Caldwell, *J. Res. Natl. Bur. Stand.* **6**, 1131 (1931).
5. A. B. Chase and J. A. Osmer, *Am. Mineral.* **49**, 1469 (1964).
6. A. B. Chase and J. A. Osmer, *J. Am. Ceram. Soc.* **50**, 325 (1967).
7. R. C. Linares, *J. Phys. Chem. Solids* **28**, 1285 (1967).

8. C. B. Finch and G. W. Clark, *J. Appl. Phys.* **36**, 2143 (1965).
9. A. B. Chase and W. R. Wilcox, *J. Am. Ceram. Soc.* **49**, 460 (1966).
10. K. Recker and R. Leckabusch, *J. Cryst. Growth* **9**, 274 (1971).
11. R. Leckabusch and K. Recker, *J. Cryst. Growth* **13**, 276 (1972).
12. W. L. Roth, *J. Solid State Chem.* **4**, 60 (1972).
13. J. Antoine, D. Vivien, J. Livage, J. Thery, and R. Collongues, *Mater. Res. Bull.* **10**, 865 (1975).
14. Y. Y. F. Yao and J. T. Kummer, *J. Inorg. Nucl. Chem.* **29**, 2453 (1967).
14a. M. M. Breiter and G. C. Farrington, *Mater. Res. Bull.* **13**, 1213 (1978).
15. P. D. Dernier and J. P. Remeika, *J. Solid State Chem.* **17**, 245 (1976).
16. F. H. Cocks and R. W. Stormont, *J. Electrochem. Soc.* **121**, 596 (1974).
17. R. J. Baughman and R. A. Lefever, *Mater. Res. Bull.* **10**, 607 (1975).
18. J. Fally, C. Lasne, Y. Lazennec, Y. Le Cars, and P. Margotin, *J. Electrochem. Soc.* **120**, 1296 (1973).
19. M. Bettman and C. R. Peters, *J. Phys. Chem.* **73**, 1774 (1969).
19a. G. C. Farrington and J. L. Briant, *Mater. Res. Bull.* **13**, 763 (1978).
20. A. D. Morrison, R. W. Stormont, and F. H. Cocks, *J. Am. Ceram. Soc.* **58**, 41 (1975).
21. Y. Le Cars, J. Thery, and R. Collongues, *Rev. Int. Haute Temp. Refract.* **9**, 153 (1972).
22. R. Stevens, *J. Mater. Sci.* **9**, 801 (1974).
23. Y. Le Cars, D. Gratias, R. Portier, and J. Thery, *J. Solid State Chem.* **15**, 218 (1975).
24. J. P. Boilot, J. Thery, and R. Collongues, *Mater. Res. Bull.* **8**, 1143 (1973).
25. L. M. Foster and J. E. Scardefield, *J. Electrochem. Soc.* **123**, 141 (1976).
25a. L. M. Foster and J. E. Scardefield, *J. Electrochem. Soc.* **124**, 434 (1977).
25b. T. Harwig and J. Schoonman, *J. Solid State Chem.* **23**, 205 (1978).
26. D. R. Mills, C. M. Perrott, and N. H. Fletcher, *J. Cryst. Growth* **6**, 266 (1970).
27. J. L. Caslavsky and S. K. Suri, *J. Cryst. Growth* **6**, 213 (1970).
28. M. E. Hills, *J. Cryst. Growth* **7**, 257 (1970).
29. M. R. Manning, C. J. Venuto, and D. P. Boden, *J. Electrochem. Soc.* **118**, 2031 (1971).
30. B. Scrosati, *J. Electrochem. Soc.* **118**, 899 (1971).
31. W. Van Gool, ed., *Fast Ion Transport in Solids*, North-Holland, Amsterdam (1973).
32. S. P. Mitoff, in *Fast Ion Transport in Solids*, ed. W. Van Gool, North-Holland, Amsterdam (1973), p. 415.
33. M. L. Nielsen, P. M. Hamilton, and R. J. Walsh, in *Ultrafine Particles*, eds. W. E. Kuhn, H. Lamprey, and C. Sheer, John Wiley and Sons, New York (1963), p. 181.
34. R. C. Rau, *J. Am. Ceram. Soc.* **47**, 179 (1964).
35. L. J. White and G. J. Duffy, *Ind. Eng. Chem.* **51**, 232 (1959).
36. H. J. Hedger and A. R. Hall, *Powder Metall.*, No. 8, 65 (1961).
37. C. Sheer and S. Karman, in *Arcs in Inert Atmospheres and Vacuum*, ed. W. E. Kuhn, John Wiley and Sons, New York (1956), p. 169.
38. K. S. Mazdiyasni, C. T. Lynch, and J. S. Smith, *J. Am. Ceram. Soc.* **48**, 372 (1965).
39. B. G. Cherkhlantsev and Yu. M. Galkin, *Zh. Neorg. Khim.* **14**, 311 (1969).
40. D. S. Rutman, Yu. S. Toropov, Yu. M. Galkin, and G. S. Matvlichuk, *Dokl. Akad. Nauk SSSR* **199**, 1367 (1971).
41. K. S. Mazdiyasni, C. T. Lynch, and J. S. Smith, *J. Am. Ceram. Soc.* **49**, 286 (1966).
42. K. S. Mazdiyasni, C. T. Lynch, and J. S. Smith, *Inorg. Chem.* **5**, 342 (1966).
43. K. S. Mazdiyasni, C. T. Lynch, and J. S. Smith, *J. Am. Ceram. Soc.* **50**, 532 (1967).

44. T. L. Markin, in *Fast Ion Transport in Solids*, ed. W. Van Gool, North-Holland, Amsterdam (1973), p. 298.
45. E. Ryshkewitch, *Oxide Ceramics*, Academic Press, New York (1960).
46. P. D. S. St. Pierre, *Trans. Brit. Ceram. Soc.* **51**, 260 (1952).
47. N. J. Maskalick and C. C. Sun, *J. Electrochem. Soc.* **118**, 1386 (1971).
48. T. L. Markin, in *Fast Ion Transport in Solids*, ed, W. Van Gool, North-Holland, Amsterdam (1973), p. 453.
49. R. S. Roth, in *Fast Ion Transport in Solids*, ed. W. Van Gool, North-Holland, Amsterdam (1973), p. 453.
50. T. Y. Tien and E. C. Subbarao, *J. Am. Ceram. Soc.* **46**, 489 (1963).
51. J. R. Hague, cited by R. C. Garvie in *High Temperature Oxides*, Part II, ed. A. M. Alper, Academic Press, New York (1970), p. 117.
52. I. S. Kainarskii, L. S. Alekseenko, and E. V. Degtyareva, *Refractories (USSR)*, 434 (1967).
53. E. V. Degtyareva, I. S. Kainarskii and L. S. Alekseenko, *Refractories (USSR)*, 362 (1968).
54. B. Ya. Sukharevskii, B. G. Alapin, and A. M. Gavrish, *Izv. Akad. Nauk. SSSR Neorg. Mater.* **1**, 1537 (1965).
55. W. S. Young and I. B. Cutler, *J. Am. Ceram. Soc.* **53**, 859 (1970).
56. H. S. Spacil and C. S. Tedmon, Jr., *J. Electrochem. Soc.* **116**, 1627 (1969).
57. R. F. Geller and P. J. Yavorsky, *J. Res. Natl. Bur. Stand.* **35**, 87 (1945).
58. O. J. Whittemore and N. M. Ault, *J. Am. Ceram. Soc.* **39**, 443 (1956).
59. C. E. Curtis, *J. Am. Ceram. Soc.* **30**, 180 (1947).
60. R. C. Garvie and P. S. Nicholson, *J. Am. Ceram. Soc.* **55**, 152 (1972).
61. R. G. Cooke and D. E. Lloyd, in *Special Ceramics*, Vol. 5, ed. P. Popper, British Ceramics Research Association, London (1972), p. 125.
62. R. E. Jaeger and R. E. Nickell, in *Ceramics in Severe Environments, Materials Science Research*, Vol. 5, eds. W. W. Kriegel and H. Palmour, Plenum Press, New York (1971), p. 163.
63. A. G. Karaulov, A. A. Grebenyuk, and I. N. Rudyak, *Izv. Akad. Nauk SSSR Neorg. Mater.* **3**, 1101 (1967).
64. L. S. Alekseenko, I. S. Kainarskii, and E. V. Degtyareva, *Refractories USSR*, 721 (1966).
65. A. G. King and P. G. Yavorsky, *J. Am. Ceram. Soc.* **51**, 38 (1968).
66. J. L. Hart and A. C. D. Chaklader, *Mater. Res. Bull.* **2**, 521 (1967).
67. D J. Green P. S. Nicholson and J. D. Embury, *J. Am. Ceram. Soc.* **56**, 619 (1973).
68. A. M. El-Shiakh and P. S. Nicholson, *J. Am. Ceram. Soc.* **57**, 19 (1974).
69. G. K. Bansal and A. H. Heuer, *J. Am. Ceram. Soc.* **58**, 235 (1975).
70. D. L. Porter and A. H. Heuer, *J. Am. Ceram. Soc.* **60**, 183 (1977).
71. N. Claussen, *J. Am. Ceram. Soc.* **59**, 179 (1976).
72. D. L. Porter, G. K. Bansal, and A. H. Heuer, *J. Am. Ceram. Soc.* **59**, 179 (1976).
73. R. C. Garvie, R. H. Hannik, and R. T. Pascoe, *Nature* **258**, 703 (1975).
74. R. W. Rice, *J. Am. Ceram. Soc.* **60**, 280 (1977).
75. D. L. Porter and A. H. Heuer, *J. Am. Ceram. Soc.* **60**, 280 (1977).
76. D. J. Green, D. R. Maki, and P. S. Nicholson, *J. Am. Ceram. Soc.* **57**, 136 (1974).
77. A. Kontopoulos and P. S. Nicholson, *J. Am. Ceram. Soc.* **54**, 317 (1971).
78. R. C. Garvie, U.S. Pat. No. 3,620,781 (1971).
79. S. K. Kantan, R. V. Raghavan, and G. S. Tendolkar, *Proc. Intern. Conf. Peaceful Uses At. Energy, 2nd Geneva* **6**, 132 (1958).
80. R. W. M. d'Eye and P. G. Sellman, *J. Inorg. Nucl. Chem.* **1**, 143 (1955).

81. F. Hund and R. Mezger, *Z. Phys. Chem.* **201**, 268 (1952).
82. Y. Harada, Y. Baskin, and H. J. Handwerk, *J. Am. Ceram. Soc.* **45**, 253 (1962).
83. V. K. Moorthy and A. K. Kulkarni, *Trans. Ind. Ceram. Soc.* **22**, 116 (1963).
84. V. K. Moorthy, A. K. Kulkarni, and S. V. K. Rao, *Proc. Intern. Conf. Peaceful Uses At. Energy, 3rd Geneva* **11**, 386 (1964).
85. V. K. Moorthy, G. Eswaraprasad, and A. K. Kulkarni, *Trans. Ind. Ceram. Soc.* **24**, 103 (1965).
86. D. E. Ferguson, O. C. Dean, and D. A. Douglas, *Proc. Intern. Conf. Peaceful Uses At. Energy, 3rd 1964* **10**, 307 (1965).
87. M. J. Bannister, *J. Am. Ceram. Soc.* **50**, 619 (1967).
88. M. J. Bannister, *J. Am. Ceram. Soc.* **51**, 228 (1968).
89. P. A. Haas and S. D. Clinton, *Ind. Eng. Chem. Product Res. Dev.* **5**, 236 (1966).
90. A. G. Faochini, *Energ. Nucl. (Milan)* **17**, 225 (1970).
91. E. C. Subbarao, P. H. Sutter, and J. Hrizo, *J. Am. Ceram. Soc.* **48**, 443 (1965).
92. M. A. Thompson, D. R. Young, and E. R. McCartney *J. Am. Ceram. Soc.* **56**, 648 (1973).
93. I. Bransky and N. M. Tallan, *J. Am. Ceram. Soc.* **53**, 11 (1970).
94. J. M. Wimmer, L. R. Bidwell, and N. M. Tallan, *J. Am. Ceram. Soc.* **50**, 198 (1967).
95. A. K. Mehrotra, H. S. Maiti, and E. C. Subbarao, *Mater. Res. Bull.* **8**, 899 (1973).
96. H. S. Maiti and E. C. Subbarao, *J. Electrochem. Soc.* **123**, 1713 (1976).
97. P. J. Jorgensen, cited by R. C. Anderson in *High Temperature Oxides*, Part II, ed A. M. Alper, Academic Press, New York (1970), p. 1.
98. C. S. Morgan and L. L. Hall, *J. Am. Ceram. Soc.* **50**, 382 (1967).
99. S. M. Lang and F. P. Knudsen, *J. Am. Ceram. Soc.* **39**, 415 (1956).
100. C. A. Arenberg, *Am. Ceram. Soc. Bull.* **36**, 302 (1957).
101. J. R. Johnson and C. E. Curtis, *J. Am. Ceram. Soc.* **37**, 611 (1954).
102. C. E. Curtis and J. R. Johnson, *J. Am. Ceram. Soc.* **40**, 63 (1957).
103. G. P. Halbfinger and M. Kolodney, *J. Am. Ceram. Soc.* **55**, 519 (1972).
104. H. J. Oel, *Z. Metallkd.* **8**, 569 (1967).
105. C. S. Morgan and C. S. Yust, *J. Nucl. Mater.* **10**, 182 (1963).
106. C. S. Morgan, C. J. McHargue, and C. S. Yust, *Proc. Brit. Ceram. Soc.*, No. 3, 977 (1965).
107. C. S. Morgan, in *Modern Developments in Powder Metallurgy Processes*, Vol. 4, ed. H. Hausner, Plenum Press, New York (1971), p. 231.
108. C. S. Morgan, *J. Am. Ceram. Soc.* **56**, 393 (1973).
109. P. J. Jorgensen and W. G. Schmidt, *J. Am. Ceram. Soc.* **53**, 24 (1970).
110. S. N. Laha and A. R. Das, *J. Nucl. Mater.* **39**, 285 (1971).
111. I. E. Poteat and C. S. Yust, *J. Am. Ceram. Soc.* **49**, 410 (1966).
112. R. J. Hawkins and C. B. Alcock, *J. Nucl. Mater.* **26**, 112 (1968).
113. A. D. King, *J. Nucl. Mater.* **38**, 347 (1971).
114. A. K. Kulkarni and V. K. Moorthy, *Trans. Ind. Ceram. Soc.* **24**, 87 (1965).
115. V. K. Moorthy and J. K. Bahl, *Trans. Ind. Ceram. Soc.* **26**, 111 (1967).
116. R. C. DeVries and W. L. Roth, *J. Am. Ceram. Soc.* **52**, 364 (1969).
117. J. T. Kummer, in *Progress in Solid State Chemistry*, Vol. 7, eds. H. Reiss and J. O. McCaldin, Pergamon Press, New York (1972), p. 141.
118. A. K. Ray and E. C. Subbarao, *Mater. Res. Bull.* **10**, 583 (1975).
119. R. Galli, F. A. Tropeano, P. Bazzarin, and U. Mirarchi, in *Fast Ion Transport in Solids*, ed. W. Van Gool, North-Holland, Amsterdam (1973), p. 573.
120. T. L. Francis, F. E. Phelps, and G. Maczura, *Am. Ceram. Soc. Bull.* **50**, 615 (1971).
121. R. H. Radzilowski, *J. Am. Ceram. Soc.* **53**, 699 (1970).
122. C. N. Poulieff, R. Krachkov, and I. M. Balkanov, *Mater. Res. Bull.* **13**, 323 (1978).

123. J. P. Boilet, A. Khan, J. Thery, R. Collongues, J. Antoine, D. Vivien, C. Chevrette, and D. Gourier, *Electrochim. Acta* **22**, 741 (1977).
124. R. Collongues, *Mater. Res. Bull.* **13**, 128 (1978).
125. R. W. Powers and S. P. Mitoff, in *Solid Electrolytes*, eds. P. Hagenmuller and W. Van Gool, Academic Press, New York (1978), Chapter 9.
126. R. W. Powers, *J. Electrochem. Soc.* **122**, 490 (1975).
127. J. H. Kennedy and A. Foissy, *J. Electrochem. Soc.* **122**, 482 (1977).
128. J. H. Kennedy and A. Foissy, *J. Am. Ceram. Soc.* **60**, 33 (1977).
129. I. Wynn Jones and L. J. Miles, *Proc. Brit. Ceram. Soc.*, No. 19, 161 (1971).
130. M. Harata, *Mater. Res. Bull.* **6**, 461 (1971).
131. A. Imai and M. Harata, *Jpn. J. Appl. Phys.* **11**, 180 (1972).
132. J. H. Kennedy and A. F. Sammells, *J. Electrochem. Soc.* **119**, 1609 (1972).
133. J. H. Kennedy and A. F. Sammells, in *Fast Ion Transport in Solids*, ed. W. Van Gool, North-Holland, Amsterdam (1973), p. 563.
134. A. V. Virkar, G. J. Tennenhouse, and R. S. Gordon, *J. Am. Ceram. Soc.* **11**, 508 (1974).
135. W. L. Roth and R. C. DeVries, *J. Solid State Chem.* **20**, 111 (1977).
136. G. E. Youngblood, A. V. Virkar, W. R. Cannon, and R. S. Gordon, *Am. Ceram. Soc. Bull.* **56**, 206 (1977).
137. A. V. Virkar, G. R. Miller, and R. S. Gordon, *J. Am. Ceram. Soc.* **61**, 251 (1978).
138. I. Wynn Jones, in *Fast Ion Transport in Solids*, ed. W. Van Gool, North-Holland, Amsterdam (1973), p. 560.
139. T. J. Whalen, G. J. Tennenhouse, and C. Meyer, *J. Am. Ceram. Soc.* **57**, 497 (1974).
140. B. B. Owens, in *Fast Ion Transport in Solids*, ed. W. Van Gool, North-Holland, Amsterdam (1973), p. 593.
141. T. Takahashi and O. Yamamoto, *Electrochim. Acta* **11**, 779 (1966).
142. J. A. A. Ketelaar, *Z. Kristallogr.* **80**, 190 (1931).
143. B. B. Owens and G. R. Argue, *Science* **157**, 308 (1967).
144. T. Takahashi, S. Ikeda, and O. Yamamoto, *J. Electrochem. Soc.* **119**, 477 (1972).
145. B. B. Owens, *J. Electrochem. Soc.* **117**, 1536 (1970).
146. B. B. Owens, J. H. Christie, and G. T. Tiedeman, *J. Electrochem. Soc.* **118**, 1144 (1971).
147. T. Takahashi, O. Yamamoto, and S. Ikeda, *J. Electrochem. Soc. Jpn.* **44**, 796 (1971).
148. J. H. Kennedy, F. Chen, and J. Hunter, *J. Electrochem. Soc.* **120**, 454 (1973).
149. D. O. Raleigh, in *Progress in Solid State Chemistry*, Vol. 3, ed. H. Reiss, Pergamon Press, Oxford (1967), p. 83.
150. H. Yanagida, R. J. Brook, and F. A. Kröger, *J. Electrochem. Soc.* **117**, 593 (1970).
151. R. J. Brook, W. L. Pelzmann, and F. A. Kröger, *J. Electrochem. Soc.* **118**, 185 (1971).
152. L. Heyne, in *Mass Transport in Oxides*, eds. J. B. Wachtman and A. D. Franklin, *Natl. Bur. Stand. (U.S.) Spec. Publ.* **296**, 149 (1968).
153. D. Yuan and F. A. Kröger, *J. Electrochem. Soc.* **116**, 594 (1969).
154. S. V. Karapachev and A. T. Filyaev, *Elektrokhimiya* **2**, 1330 (1966).
155. C. S. Tedmon, Jr., H. S. Spacil, and S. P. Mitoff, *J. Electrochem. Soc.* **116**, 1170 (1969).
156. C. N. R. Rao and G. V. Subbarao, *Phys. Stat. Solidi (A)* **1**, 597 (1970).
157. H. S. Spacil and C. S. Tedmon, Jr., *J. Electrochem. Soc.* **116**, 1618 (1969).
158. T. Takahashi, in *Physics of Electrolytes*, Vol. 2, ed. J. Hladik, Academic Press, New York (1972), p. 989.
159. J. W. Patterson, E. C. Borgen, and R. A. Rapp, *J. Electrochem. Soc.* **114**, 752 (1967).
160. R. D. Armstrong, T. Dickinson, and J. Turner, *J. Electrochem. Soc.* **118**, 1135 (1971).

161. D. S. Demott and P. Hancock, *Proc. Brit. Ceram. Soc.* **19**, 193 (1969).
162. R. Galli, P. Lorghi, T. Mussini, and F. A. Tropeano, *Electrochim. Acta* **18**, 1013 (1973).
163. J. L. Sudworth, A. R. Tilley and K. D. South, in *Fast Ion Transport in Solids*, ed. W. Van Gool, North-Holland, Amsterdam (1973), p. 581.
164. A. Imai and M. Harata, 137th Meeting of the Electrochemical Society, Los Angeles, May 1970, Abs. No. 277, quoted in C. S. Morgan, *J. Am. Ceram. Soc.* **56**, 393 (1973).
165. M. S. Whittingham and R. A. Huggins, *J. Chem. Phys.* **54**, 414 (1971).
166. L. J. Miles and I. Wynn Jones, in *Power Sources*, Vol. 3, ed. D. H. Collins, Oriel Press, Newcastle-upon-Tyne (1971), p. 245.
167. S. K. Tiku, M.Tech. thesis, Indian Institute of Technology, Kanpur, 1974.
168. R. W. Powers, in *Superionic Conductors*, eds. G. D. Mahan and W. L. Roth, Plenum Press, New York (1976), p. 351.
169. M. S. Whittingham and R. A. Huggins, *J. Electrochem. Soc.* **118**, 1 (1971).
170. M. S. Whittingham and R. A. Huggins, in *Solid State Chemistry*, eds. R. S. Roth and S. J. Schneider, *Natl. Bur. Stand. (U.S.) Spec. Publ.* **364**, 51 (1972).
171. T. A. Ramanarayanan and W. L. Worrell, *J. Electrochem. Soc.* **121**, 1530 (1974).
172. M. O. Hever, *J. Electrochem. Soc.* **115**, 830 (1968).
173. J. C. Sudworth and M. D. Hames, in *Power Sources*, Vol. 3, ed. D. H. Collins, Oriel Press, Newcastle-upon-Tyne (1971), p. 227.
174. R. Hand, A. K. Carpenter, C. J. O'Brien, and R. F. Nelson, *J. Electrochem. Soc.* **119**, 74 (1972).
175. T. K. Wiewiorowski, *J. Electrochem. Soc.* **118**, 1711 (1971).
176. J. Fally, C. Lasne, Y. Lazennec, and P. Margotin, *J. Electrochem. Soc.* **120**, 1292 (1973).
177. M. B. Armand, in *Fast Ion Transport in Solids*, ed. W. Van Gool, North-Holland, Amsterdam (1973), p. 665.
178. B. C. H. Steele, in *Superionic Conductors*, eds. G. D. Mahan and W. L. Roth, Plenum Press, New York (1976), p. 47.
179. M. S. Whittingham, in *Solid Electrolytes*, eds. P. Hagenmuller and W. Van Gool, Academic Press, New York (1978), p. 367.
180. R. J. Heus and J. J. Egan, *Z. Phys. Chem.* **49**, 38 (1966).
181. B. Scrosati, G. Germano, and G. Pistoia, *J. Electrochem. Soc.* **118**, 86 (1971).
182. J. N. Mrgudich, *J. Electrochem. Soc.* **107**, 475 (1960).
183. B. B. Owens and G. R. Argue, *J. Electrochem. Soc.* **117**, 898 (1970).
184. B. Scrosati and A. Duane Butterns, *J. Electrochem. Soc.* **119**, 128 (1972).
185. L. Topol and B. B. Owens, *J. Phys. Chem.* **72**, 2106 (1968).
186. D. A. Nissen and R. W. Carlsten, in *Fast Ion Transport in Solids*, ed. W. Van Gool, North-Holland, Amsterdam (1973), p. 675.
187. T. Takahashi and O. Yamamoto, *J. Electrochem. Soc.* **177**, 1 (1970).
188. B. B. Owens, G. R. Argue, I. J. Groce, and L. D. Hermo, *J. Electrochem. Soc.* **116**, 312 (1969).
189. G. H. Gell, N. A. Richardson, E. T. Sao, and H. P. Silverman, in *Proceedings of the Symposium on Batteries, Traction, and Propulsion*, ed. R. L. Kerr, Columbus Section, Electrochemical Society, Battelle Memorial Institute, Columbus, Ohio (1972), p. 178.
190. C. J. Mogab, *Rev. Sci. Instrum.* **43**, 1605 (1972).
191. S. Pizzini, A. Bonomi, and P. Colombo, *J. Phys. E.* **3**, 832 (1970).
192. H. Schmalzried and H. G. Sockel, *Ber. Bunsen. Ges.* **72**, 745 (1968).
193. N. Sano, S. Honura and Y. Matsushita, *Metall. Trans.* **1**, 301 (1971).

194. Y. K. Agrawal, D. W. Short, R. Gruenke, and R. A. Rapp, *J. Electrochem. Soc.* **121**, 354 (1974).
195. C. C. Sun, E. W. Hawk, and E. F. Sverdrup, *J. Electrochem. Soc.* **119**, 1433 (1972).
196. F. J. Rohr, in *Solid Electrolytes*, eds. P. Hagenmuller and W. Van Gool, Academic Press, New York (1978), p. 431.
197. S. Gratch, J. V. Petroceeri, R. P. Tischer, R. W. Minck, and T. J. Whalen, in Proceedings of the 7th Intersociety Energy Conversion Engineering Conference, 1972, p. 38.

Index